IET ENERGY ENGINEERING 555

Bioelectromagnetics in Healthcare

Other volumes in this series:

Volume 1	**Geometrical theory of diffraction for electromagnetic waves, 3rd edition** G.L. James
Volume 10	**Aperture antennas and diffraction theory** E.V. Jull
Volume 11	**Adaptive array principles** J.E. Hudson
Volume 12	**Microstrip antenna theory and design** J.R. James, P.S. Hall and C. Wood
Volume 15	**The handbook of antenna design, volume 1** A.W. Rudge, K. Milne, A.D. Oliver and P. Knight (Editors)
Volume 16	**The handbook of antenna design, volume 2** A.W. Rudge, K. Milne, A.D. Oliver and P. Knight (Editors)
Volume 18	**Corrugated horns for microwave antennas** P.J.B. Clarricoats and A.D. Oliver
Volume 19	**Microwave antenna theory and design** S. Silver (Editor)
Volume 21	**Waveguide handbook** N. Marcuvitz
Volume 23	**Ferrites at microwave frequencies** A.J. Baden Fuller
Volume 24	**Propagation of short radio waves** D.E. Kerr (Editor)
Volume 25	**Principles of microwave circuits** C.G. Montgomery, R.H. Dicke and E.M. Purcell (Editors)
Volume 26	**Spherical near-field antenna measurements** J.E. Hansen (Editor)
Volume 28	**Handbook of microstrip antennas, 2 volumes** J.R. James and P.S. Hall (Editors)
Volume 31	**Ionospheric radio** K. Davies
Volume 32	**Electromagnetic waveguides: theory and applications** S.F. Mahmoud
Volume 33	**Radio direction finding and superresolution, 2nd edition** P.J.D. Gething
Volume 34	**Electrodynamic theory of superconductors** S.A. Zhou
Volume 35	**VHF and UHF antennas** R.A. Burberry
Volume 36	**Propagation, scattering and diffraction of electromagnetic waves** A.S. Ilyinski, G. Ya.Slepyan and A. Ya.Slepyan
Volume 37	**Geometrical theory of diffraction** V.A. Borovikov and B.Ye. Kinber
Volume 38	**Analysis of metallic antenna and scatterers** B.D. Popovic and B.M. Kolundzija
Volume 39	**Microwave horns and feeds** A.D. Olver, P.J.B. Clarricoats, A.A. Kishk and L. Shafai
Volume 41	**Approximate boundary conditions in electromagnetics** T.B.A. Senior and J.L. Volakis
Volume 42	**Spectral theory and excitation of open structures** V.P. Shestopalov and Y. Shestopalov
Volume 43	**Open electromagnetic waveguides** T. Rozzi and M. Mongiardo
Volume 44	**Theory of nonuniform waveguides: the cross-section method** B.Z. Katsenelenbaum, L. Mercader Del Rio, M. Pereyaslavets, M. Sorella Ayza and M.K.A. Thumm
Volume 45	**Parabolic equation methods for electromagnetic wave propagation** M. Levy
Volume 46	**Advanced electromagnetic analysis of passive and active planar structures** T. Rozzi and M. Farinai
Volume 47	**Electromagnetic mixing formulas and applications** A. Sihvola
Volume 48	**Theory and design of microwave filters** I.C. Hunter
Volume 49	**Handbook of ridge waveguides and passive components** J. Helszajn
Volume 50	**Channels, propagation and antennas for mobile communications** R. Vaughan and J. Bach-Anderson
Volume 51	**Asymptotic and hybrid methods in electromagnetics** F. Molinet, I. Andronov and D. Bouche
Volume 52	**Thermal microwave radiation: applications for remote sensing** C. Matzler (Editor)
Volume 53	**Principles of planar near-field antenna measurements** S. Gregson, J. McCormick and C. Parini
Volume 54	**Satellite-to-ground radiowave propagation, 2nd edition** J.E. Allnutt
Volume 502	**Propagation of radiowaves, 2nd edition** L.W. Barclay (Editor)
Volume 534	**Emerging Evolutionary Algorithms for Antennas and Wireless Communications** S. Goudos

Bioelectromagnetics in Healthcare

Advanced sensing and communication applications

Edited by
William Whittow

The Institution of Engineering and Technology

Published by SciTech Publishing, an imprint of The Institution of Engineering and Technology, London, United Kingdom

The Institution of Engineering and Technology is registered as a Charity in England & Wales (no. 211014) and Scotland (no. SC038698).

First published 2022

The Institution of Engineering and Technology
Futures Place
Kings Way, Stevenage
Hertfordshire, SG1 2UA, United Kingdom

www.theiet.org

British Library Cataloguing in Publication Data
A catalogue record for this product is available from the British Library

ISBN 978-1-83953-349-5 (hardback)
ISBN 978-1-83953-350-1 (PDF)

Typeset in India by MPS Ltd

Contents

About the editor

Professor William G. Whittow (AFWES, SFHEA, SMIEEE) received the B.Sc. degree in Physics and the Ph.D. degree in Computational Electromagnetics from the University of Sheffield, Sheffield, UK, in 2000 and 2004, respectively. From 2004 to 2012, he was a research associate at Loughborough University, Loughborough, UK. In 2012, he became a lecturer in the Wolfson School of Mechanical, Electrical, and Manufacturing Engineering (WSMEME) at Loughborough University. He became a Senior Lecturer in 2014, a reader (Associate Professor) in 2018, and a Professor of Radiofrequency Materials in 2020. He leads the Wireless Communications Research Group (WiCR) and is the Director of the Connected Infrastructure Research Hub. He is a named investigator on research grants totalling in excess of £12m. He has authored more than 250 peer-reviewed journal and conference papers in topics related to metamaterials, metasurfaces, synthetic and heterogeneous dielectrics, dielectric measurements, 3D-printing, wearable antennas and phantoms, specific absorption rate, embroidered antennas, inkjet printing, and RFID tags. From 2007 to 2011, he was the coordinating chair of the Loughborough Antennas and Propagation Conference (LAPC). He has served as an associate editor of IET's *Electronics Letters and also Microwaves, Antennas, and Propagation*. He serves on the technical programme committees of several IEEE international conferences. He has been asked to give more than 25 invited conference presentations; a 4-day invited workshop on bioelectromagnetics and teaches about dielectric measurements at the European School of Antennas. In 2017, he won the Women in Engineering Society (WES) Men As Allies Award and he is the inaugural male Associate Fellow of WES. He is a Senior Fellow of the Higher Education Academy. More than 90 of his academic journal papers can be freely downloaded here: http://publications.lboro.ac.uk/publications/all/collated/elwgw.html.

Chapter 1

Introduction, overview, and future directions

William Whittow[1]

1.1 Introduction to this book

This book on Bioelectromagnetics in Healthcare is a collection of 12 invited chapters from international experts from UK, Japan, Switzerland, and United States of America. It has been a great pleasure to collate and edit these chapters to form a cohesive architecture that covers the state-of-the-art in terms of sensing and communications with relevance to Bioelectromagnetics in Healthcare. We have endeavoured to write these chapters so that they contain the latest developments but are also accessible to people who do not have a pure specialism in these exact topics. We have strived to include valuable practical advice that other engineers can implement to devise the next generation of healthcare devices.

Electromagnetics, as the name suggests, is the study of varying electric and magnetic fields that form electromagnetic (EM) waves. Bioelectromagnetics means applying these EM waves inside the human body which can be used for healthcare and medical applications.

Although EM waves can only be seen by humans at visible frequencies, we have been aware of the underlying concepts for hundreds of years. The ancient Egyptians were using electric shock therapy from 3,000 BC to treat joint pain.[*] Thales discovered static electric charges by experimenting with amber and silk in ancient Greece in 600 BC. In the nineteenth century, many scientific giants advanced our understanding of electric and magnetic fields via experimentation and drawing conclusions from the measured results. Prominent names include Charles Coulomb, Henry Cavendish, Carl Friedrich Gauss, Alessandro Volta, Hans Christian Oersted, Andre Marie Ampere, George Simon Ohm, Jean-Baptiste Biot, Michael Faraday, Joseph Henry, Gustav Kirchhoff, James Clerk Maxwell, and Heinrich Hertz—all of whom have their names dedicated to units or equations and subsequently etched in history. The experimental work by Oersted[†] and the

[1]Wolfson School of Mechanical, Electrical and Manufacturing Engineering, Loughborough University, Loughborough, UK

[*]https://en.wikipedia.org/wiki/Electric_catfish
[†]https://www.youtube.com/watch?v=oEyLq1QTWyo

mathematical development by Ampere demonstrated that direct currents produce a magnetic field. Faraday[‡] demonstrated that changing magnetic fields can induce a voltage difference. In 1861 and 1862, Maxwell published a set of four equation based on the work by Gauss, Faraday, and Ampere, that together define all of electromagnetics and demonstrates that changing electric and magnetic fields creates EM waves.

In 1887, Heinrich Hertz published his work about sending EM waves a short distance and thus verifying the Maxwell's predicted electromagnetic waves. He is purported to have said "*I do not think that the wireless waves I have discovered will have any practical application.*" In 1901, Guglielmo Marconi further developed the practical aspects to the extent where he was able to send EM waves across the Atlantic Ocean. For this he was awarded the Nobel Prize in 1909.

The theory and practical side of EM waves has therefore been widely known for over one hundred years. Yet electromagnetics is not a simple topic and can still be intimidating to undergraduate electrical engineers choosing their specialisms. The topic is not intuitive, and the behaviour can often not be observed directly but has to be concluded indirectly from observations. Prior to the advancement of modern computers, there were two types of engineers who worked on electromagnetics: (i) the mathematical types who derived everything from first principles, and (ii) the practically minded engineers who physically made radiofrequency (RF) experiments and then altered the physical geometry to observe the results. While analytical techniques can lack the realism and intricate designs, practical fabrication is a slow and expensive technique. Prior to computers becoming mainstream, antennas on mobile phones and vehicles often used a simple whip monopole antenna. The advancement in speed and memory of desktop computers as well as commercial electromagnetic software since the beginning of this century have led to a new way of designing RF devices. Now, (tens of) thousands of iterations of design can be run in a few days, allowing RF engineers to imagine new devices that are smaller, use less power, and have advanced functionality.

To put the healthcare demand in terms of numbers; it is estimated in the United Kingdom that 10 million people alive today may live to be 100 years old; two-thirds of NHS beds are occupied by people over 65; and every night someone spends in a hospital costs the taxpayer several hundred pounds. How can we continue to care for our aging population? Necessity is the mother of invention, and engineers will be required to develop new systems that can be placed on the outside of the body and sense changes on the inside; or be placed inside the body and wirelessly communicate information in/out of the body. Having wires that go through the skin from outside to inside is unacceptable and can lead to infections. If we could monitor people remotely, then there is less need for them to physically visit doctors or hospitals which is expensive, time-inefficient, increases the risk of spreading diseases, and reduces the quality of life of the patient. Hence Bioelectromagnetics is a

[‡]https://www.youtube.com/watch?v=7MTTyWzX1EY

key technology that is a rapidly evolving area where the engineering developments mirrors the societal need.

To demonstrate how the world has changed in a generation, when mobile phone first began to become popular, people wondered why you would ever need a phone as people already had a phone at home and a phone at work. When I started my PhD, I did not at that point own a mobile phone. This was always entertaining as after I had told people the topic of my PhD was to investigate whether mobile phones were dangerous to human health, the second question was always along the lines of "Do you own a mobile phone?" which was promptly followed by a look of concern and fear on their faces. When I recently gave some outreach talks at a local school, I asked the audience of ~1,000 students how many people had ever used a traditional red public phone box; about four students raised their hands.

In the same way that mobile phones used near the human body have completely changed how we live and communicate over the last 25 years, it is envisioned that EM devices in or on the body will change how we perceive healthcare in the next 25+ years. At first, there is hesitancy and concerns, and then the technology advances to the point where the benefits outweigh the initial concerns.

All of electromagnetics is generally scalable in frequency and Maxwell's equations are the same for all frequencies. The main design criteria is the size of the object in wavelengths. The frequency and wavelength are always inversely related. For example, the wavelength at 400 MHz is 75 cm, while the wavelength at 4 GHz is 7.5 cm. Every human tissue has a different relative permittivity that defines the speed of the EM wave in the tissue and the wavelength; and also, a different loss tangent or conductivity which defines the amount of power converted into heat. These values vary with frequency. A rough approximation for the average tissues of the human body is a relative permittivity of 40. This means the wavelength inside the body is 6 or 7 times smaller than in free space. The average conductivity of the human body can be estimated to be of the order of 1 S/m. The electrical skin depth (which is not to be confused with the epidermal skin depth) defines how deep the electric fields penetrate into the human body and also defines how the EM wave propagate out of the body. The fields will penetrate less deeply as the conductivity increases and also as the frequency increases. These factors define the useful frequency range for bioelectromagnetic healthcare applications. For example, high frequencies (>5 GHz) will only be able to sense changes at the surface and in the outer few centimetres or even millimetres of the body. Similarly, if we are trying to communicate from outside the body to inside the body, a lower frequency is desirable where the electrical skin depth is larger. For example, many implanted devices often operate at ~400 MHz. This creates considerable challenges in terms of miniaturization as the implanted device might be of the order of 2 cm in size which must also include the electronics. This is a tiny fraction of the free space wavelength. The communication range could be enhanced by increasing the power; however, this requires larger batteries. Furthermore, there are power absorption compliance limits which limit the power absorbed by the human tissues which are referred to as the specific absorption rate (SAR).

1.2 Overview of each chapter

This section will summarize each chapter. Each chapter is led by a world leading expert on the topic, but it is of course our excellent PhD students and doctoral researchers who have carried out the majority of the design and practical work. Chapter 2 summarizes the theoretical background related to dielectric properties and their measurements. Chapters 3 and 4 summarize the simulation and physical phantom challenges that can be used as the tools to develop bioelectromagnetic devices. Chapters 5 and 6 cover the challenges of communicating from inside to outside the body, with Chapter 5 concerning implanted antennas and Chapter 6 examining nearfield coupling. Chapters 7–11 cover different sensing technologies to monitor changes within the body. Chapter 12 covers therapeutic applications to use electromagnetics to treat cancers and related conditions. Finally, Chapter 13 uses optical frequencies transmitted and reflected through the skin's surface to monitor changes.

Chapter 2 is led by Sami Gabriel, who is a Distinguished Engineer at the Vodafone Group in the United Kingdom. Four of his key papers on this topic have been cited more than 16,000 times. This chapter covers some of the fundamental definitions of dielectric properties including the relative permittivity, the conductivity, and the polarization. Common methods of measuring dielectric properties at a range of frequencies are discussed. The chapter provides several practical tips to carry out the measurements in a robust and repeatable manner. The chapter also provides some useful applications of microwaves inside the body which include tomography for imaging, processing of food, and in a related area the same methodology can be used to sinter ceramics.

Chapter 3 is led by Professor Cynthia Furse, from the University of Utah, USA. This chapter focusses on simulations of EM waves inside the body and discusses the different tissue properties and the range of digital phantoms that are available. It explains the equations required to calculate the EM fields inside the body and develops these in an easy-to-follow approach, starting from Maxwell's equations to the finite-difference time-domain (FDTD) equations that are widely used in commercial software. The chapter discusses the limitations and tips to run simulations accurately and quickly. It is a valuable read for people writing their own code and also those using commercial software. Even if you will never write your own code, it is invaluable to understand exactly what the software is doing when you set it to run and also whether the results are sensible. The chapter concludes by discussing how to simulate SAR inside the body.

Chapter 4 is led by Professor Will Whittow, from Loughborough University, UK. This chapter discusses the challenges of putting an antenna inside the human body. While it is now easy to simulate these designs, the true test of the performance of any device can only be ascertained by measurements. Typically, phantoms which emulate the human body are composed of a human-shaped plastic shell and a single liquid that has the average properties of the human body. This lacks realism to some extent as the homogenous liquid can only represent one average tissue and hence does not consider the reflections and internal resonances that occur due to the boundaries between distinct tissues. This chapter uses some of the fundamental work by Sami Gabriel to create semi-solid phantoms that mimic individual tissue properties. By adjusting the recipes and obtaining feedback from measuring the properties, phantoms can be designed that emulate blood, bone, and muscle over the required frequency range. By using multiple semi-solid materials, heterogeneous phantoms that contain multiple materials can be developed.

Chapter 5 is led by Professor Anja Skrivervik, from École Polytechnique Fédérale de Lausanne, Switzerland. This chapter discusses the design challenges and practical design rules for placing a miniaturized antenna inside the lossy human body. The antennas must be power efficient and use low power electronics. The key parameters used for electrically small antennas in free space, such as gain, do not apply in when placed in the body. "*The losses of an implantable antenna will depend on three main contributions: the near field losses, the propagating field losses, and the reflection at the body/air interface.*" There is a tradeoff in frequency as near field losses will dominate at low frequencies, and the propagation losses will dominate at higher frequencies.

Chapter 6 is led by Professor Chris Stevens, from Oxford University, UK. This chapter examines nearfield inductive coupling using magnetic coils for near field communications and wireless power transfer. The chapter specifically investigates the coupling when the medium is low conductivity human tissues. The chapter develops the equivalent circuit model and the closed form analytical solutions, which are then compared to simulations. The chapter also give insights into the mechanism of wireless inductive power transfer in tissue-equivalent environments, which is of great significance interest to implanted devices.

Chapter 7 is led by Dr Akram Alomainy, from Queen Mary University of London, UK. This chapter reviews metamaterials-based microfluidic sensors with high sensitivity for detection of chemicals and biochemicals. It also explores the various applications of such sensors in the field of biomedical engineering (Wireless Endoscopy, Cancer detection, Wireless Drug Delivery and Cardiovascular healthcare) and the scope of metamaterials and resonators in such sensing devices. The chapter focuses on chemical sensing, and the role of electromagnetics, antennas, metamaterials and resonators for biomedical applications and healthcare.

Chapter 8 is led by Dr Chinthana Panagamuwa, from Loughborough University, UK. It is often challenging to understand how fractured bones heal without resorting to multiple X-rays which are expensive and potentially adverse for human health. Some severe fractures require a metal pin to be placed inside the body to ensure the bone heals correctly. This chapter develops a prototype system where the same implanted pins could be used as antennas and a system of two of these antennas could be used to detect the healing of fractured bones by monitoring the transmission between antennas. This is based on the knowledge that the properties of bone are very different from the properties of blood that tends to fill the fracture. Hence by measuring the power transmitted from one antenna to another, the healing rate can be monitored. This was analysed via simulations in homogenous and realistic phantoms, along with measurements using the heterogeneous phantoms developed in Chapter 4, and a lamb joint.

Chapter 9 is led by Dr Asimina Kiourti, who is an Associate Professor at The Ohio State University, USA. This chapter reviews various types of wearable antennas including RFID tags, antenna arrays, and radar systems to analyse health issues inside the body. Electromagnetic parameters of such wearable antennas, including received power, resonant frequency, and impedance value, are modified as a function of the sensing parameter of interest. They can be applied to monitor and detect conditions that include respiration/breath, cardiac and cardiopulmonary activities, body fluid, and steatotic liver issues. In several cases, they are also advantageous as they do not require batteries.

Chapter 10 is led by Professor John Batchelor, from University of Kent, UK. This chapter also uses RFID tags to monitor and sense different potential health issues inside the body. A tag placed on the throat can detect absences of saliva which could be an indication of infections on a throat prosthesis. The paper includes measurements on different people with different body mass indexes and discusses how the relative permittivity and conductivity values affect the tag performance. The second half of this chapter uses battery-assisted passive RFID tags to transmit higher data rates which could be used in electrocardiography (ECG) applications.

Chapter 11 is led by Dr Dimitris Anagnostou, who is an Associate Professor at Heriot Watt University, UK. This chapter investigates the concept of using radar not located on the body's surface. This can be more comfortable for the patient than having a device physically attached to the skin. The application is to monitor respiration rate, heart rate, and sleep. For example, the chest can move between 4 and 12 mm during breathing that can be detected via radar. The chapter revisits some radar principles and demonstrates how they can be practically used for potential health monitoring.

Chapter 12 is led by Professor Koichi Ito, from Chiba University, Japan. This chapter covers some of the applications of intentionally using EM waves inside the body. This includes hyperthermia and microwave ablation for cancer treatment. Clearly, it is important to have an excellent understanding of the EM waves inside the body. They must be controlled and directed towards only the tumour. Efforts must be taken to minimize the SAR of the nearby tissues. There is considerable work in choosing the optimal electrodes for different applications. The results have been rigorously analysed using extensive simulations and measurements in realistic phantoms.

Chapter 13 is led by Dr Sijung Hu, who is an Associate Professor, and his colleague Dr Yasmin Elsahar, a Biomedical Researcher at Loughborough University, UK. In the chapter, electromagnetic principle is applied for opto-physiological modelling (OPM) reflecting in a real world of optoelectronic sensing application. This chapter uses illumination frequencies to show how light interacts with biological tissue. The contact and non-contact opto-physiological monitoring are illustrated in two aspects: (1) in contact format with the skin is currently popular smart watches and fitness trackers; and (2) non-contact to access human vital signs as an emerging technology. The chapter also describe how these can get these vital signs, i.e., heart rate (HR), oxygen saturation (SpO_2), and respiration rate (RR).

This chapter reviews OPM driven health monitoring research activities and discusses some of the electronics requirements and investigates two case studies.

1.3 Future directions

There are clear financial and socioeconomic drivers for the need to develop intelligent healthcare provision which can be administered with minimal requirements on medical staff or hospital beds or expensive machines such as MRI or X-rays. The market for implanted devices will continue to grow for applications such as intelligent drug delivery or potentially treating depression or epilepsy; all these devices will be capable of being intelligent with improved functionality if wireless communication can be achieved between inside and outside the human body.

The healthcare technology sector is especially challenging, as new devices take many years to be approved, and regulations are different in different countries. Every person will have a different shape and will have different internal ratios of muscle and fat; this makes it challenging to make one system work across the general population. Whereas it is possible for a trained electromagnetic engineer who has worked on the system for several years to achieve repeatable results in the laboratory in controlled conditions, it is a completely different matter for a busy nurse or doctor to make the system function properly on a real person in a real hospital. Whereas the engineer will have built up the intuition about the correct results, a medical professional will regularly use many pieces of equipment and is unlikely to build up that technical understanding to intuitively know if results are slightly misleading. While these challenges are not easy to solve, this is the job of engineers and medical experts, and the direction of future healthcare is clear.

The chapters in this book provide a valuable insight into the current state-of-the-art and also future possibilities. EM engineers will need to keep improving the RF device performance in terms of better efficiency, greater sensitivity, reduced unintended power absorption by the body, small size, and low-power consumption.

In terms of wireless communication, there is potential to use new metamaterial-inspired devices to increase the efficiency or further miniaturize the design in terms of wavelengths. If a lower frequency antenna could be fitted into the same volume, this would improve the propagation losses. Lower frequencies come at the expense of lower data rates and hence intelligent algorithms will be required to only transmit essential data. If a device is only inside the body for a short amount of time, such as a smart pill, then the battery requirements are less critical. However, if the device is to be permanently implanted inside the body, then either wireless charging or energy harvesting will be required.

This book demonstrates the great potential to use a range of RF and microwave devices to monitor the internal structure of the body. To a certain extent, any medical condition will cause some internal change in the body composition which can potentially be monitored if the medical knowledge can be combined with the engineering vision. These can be small and obtrusive devices that can be much less expensive than conventional imaging machines. In some of these techniques, the uncertainty due to the device placement and the behaviour on/in different people may be of a similar magnitude as the expected measurable change. Future devices may need to increase the reliability by incorporating dual measuring techniques, including reconfigurability so that the same system can measure at multiple frequencies, or by including artificial intelligence so that the devices can learn the behaviour on that specific individual.

Detecting health issues and intelligently communicating are both highly valuable capabilities. A further challenge is to use EM waves to treat medical conditions. There is potential to link the imaging and treatment systems into one device.

We have thoroughly enjoyed putting this book together, and we hope that it informs and inspires your future endeavours in the exciting area of Bioelectromagnetics.

Chapter 2

Introduction to dielectric measurement and some common applications

Sami Gabriel[1]

In its simplest form, a dielectric is a material that can store charge when subjected to an electric field. The electric field should penetrate the material, interact with, and rearrange charged entities within it for this to happen. If all the charges are bound as in a perfect conductor, the effect is a slight charge displacement, giving rise to an element of induced polarisation as in a capacitor. Natural materials with free and bound charges would experience charge polarisation and conduction phenomena. These occur over time scales from seconds to picoseconds and manifest over field frequencies from 1 Hz to 100 s of GHz. The complexity of the interaction is commensurate with the structure and composition of the material. The dielectric properties are a measure of such interactions and help explain a material's electrodynamics and related microscopic properties, as well as highlighting the effects of the field exposure on the material.

For this reason, knowledge of the dielectric properties is of practical importance in all fields of science where electromagnetic fields (EMF) are applied. From the processing of oil in refineries to the manufacture of pharmaceuticals to the cultivation of agricultural farmland. Outside of industry use cases, dielectric properties are also central to the optimisation of microwave heating in food processing and to RF welding in manufacturing. Furthermore, despite significant progress, the measurement of dielectrics is still evolving and emerging in new frontiers. One rapidly evolving field is designing and optimising medical devices for diagnostic and therapeutic purposes.

Today EMF has become an essential element in our environment, and people are exposed to low levels all the time. This exposure has required many studies on assessing and quantifying people's exposure. Whilst the exposure assessments often differ in their experimental methodologies according to their frequency range of interest, the dielectric properties of body tissues are fundamental to the understanding of the biological effects resulting from that exposure. Information on the dielectric properties of body tissues is thus vital to these studies for the computation of exposure metrics to provide a mechanistic explanation of any biological effects

[1]Distinguished Engineer, Vodafone Group, London, UK

and the design of biomedical devices where EMFs are used for diagnostic and therapeutic processes.

Dielectric properties of materials are primarily obtained experimentally. For this reason, this chapter will focus on measurement techniques and the practical aspects of carrying out these measurements.

The permittivity of a material is the parameter used to measure the material's dielectric response. In harmonic fields, the permittivity is a complex function originating from the magnitude and phase shift of the polarisation with respect to the polarising field.

The permittivity of a material relative to that of free space is expressed as:

$$\widehat{\varepsilon} = \varepsilon' - j\varepsilon'' = \varepsilon' - \frac{j\sigma}{\omega\varepsilon_0} \tag{2.1}$$

The real part ε' is a measure of the capacitive property due to the induced polarisation per unit field. The imaginary part ε'' is the loss factor or conductive element associated with it. The loss factor can also be represented by a conductivity term $\sigma = \omega\varepsilon_0\varepsilon''$, where ω is the angular frequency and ε_0 the permittivity of free space (8.8542, $\times\ 10^{-12}$ F/m). The SI unit of conductivity is Siemens per metre (S/m), ε' and ε'' are dimensionless relative quantities. Note, in this chapter, the term permittivity will always refer to relative permittivity unless stated otherwise.

In the following sections, we will look at the basic physics of the basic principles of the interaction of EMF with materials. This will be followed by measurement, looking in particular as coaxial probes, as this is particularly suited to biological samples and small samples. Finally, we look at some established applications in material processing and then briefly look at medical diagnostics and treatments.

2.1 How EMFs interact with a range of materials

Understanding the dielectric properties and their intrinsic derivation from the structure and composition of materials requires knowledge of the main possible interaction mechanisms between EMFs and matter.

2.1.1 Electronic polarisation

Consider a solid monomolecular material where the charge configuration is such that the centres of positive and negative charges coincide. Such material is described as non-polar because it possesses no electric dipole. Under the action of an external field, the electron cloud around the nucleus is perturbed to create a slight separation of centres of charge from their equilibrium position and create an induced dipole, see Figure 2.1.

Electronic polarisation occurs in all matter, but when electronic polarisation is the only polarisation mechanism, the effect is small, and so the permittivity is small. Good insulators fall in this category. For example, polytetrafluoroethylene (PTFE) has a permittivity of about 2 and a loss factor of the order 10^{-4}. The

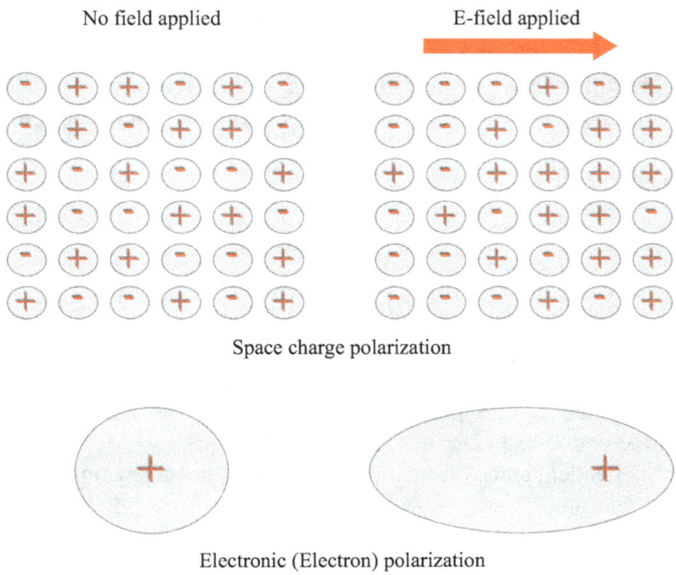

Figure 2.1 Non-polar material electronic structure with and without field

permittivity decreases slightly with temperature and is virtually frequency-independent because the effect is practically instantaneous.

2.1.2 Atomic polarisation

Non-polar materials can also incur atomic polarisation where groups of atoms displace with respect to each other. When present, this too is a small effect and virtually instantaneous and frequency-independent.

2.1.3 Orientation of polar molecules

Some molecules are polar by virtue of their molecular configuration that allows a charge separation with the molecule. The molecule behaves as a dipole with an associated dipole moment. Under the action of an external field, the molecule will tend to orient with respect to the field. When present, this type of dipolar orientation polarisation is significantly larger than electronic and atomic polarisation; it also develops over a measurable time scale of picoseconds to milliseconds depending on the molecular size and viscosity. An ideal monomolecular polar material has an associated polarisation time constant (τ), corresponding to an angular frequency $\omega = 1/\tau$

At frequencies such that $\omega\tau \ll 1$, the polarisation has time to develop fully within the period of the field, and the resulting permittivity is at its highest. Conversely, at high frequencies where $\omega\tau \gg 1$, the polarisation has no time to develop and consequently, the permittivity is at a much lower level. The drop in permittivity as a function of frequency is described as a relaxation phenomenon. It has an associated polarisation mechanism determining it.

The highest permittivity value at the limit of low frequencies is often referred to as static permittivity ε_S. The permittivity at high frequencies is termed ε_∞. The amplitude of the dispersion is $\Delta\varepsilon = \varepsilon_S - \varepsilon_\infty$, the parameters are related by the following expression:

$$\widehat{\varepsilon} = \varepsilon_\infty + \frac{(\varepsilon_S - \varepsilon_\infty)}{1 + j\omega\tau} = \varepsilon' - j\varepsilon'' \tag{2.2}$$

This expression is known as the Debye equation. It assumes a linear and causal rate process where the polarisation rises exponentially, reaching an equilibrium level and decays exponentially as a function of frequency. The corresponding frequency dependence of the permittivity is described as a relaxation, τ is the relaxation time and $f_\tau = 1/2\pi\tau$ is the relaxation frequency. All the parameters are temperature dependent since orientation polarisation occurs against the background of Brownian motion.

There are many ways of deriving this expression and those interested can read much more in textbooks on the subject, Grimmes *et al.* [1] is one such example.

Water is a polar substance; the water molecule possesses a dipole moment due to the configuration of its hydrogen molecules with respect to the oxygen molecules. It displays a near-perfect Debye dispersion, see Figure 2.2. At 20 °C ε_S= 80.1, ε_∞= 5.2, τ = 9.47 ps, which gives a relaxation frequency of 16.8 GHz. For more on water properties, see Liebe *et al.* [2]. These parameters depend on and could be derived from molecular dipole moment, size and viscosity. For more information on the relationship between dielectric and molecular parameters, refer to Böttcher *et al.* [3].

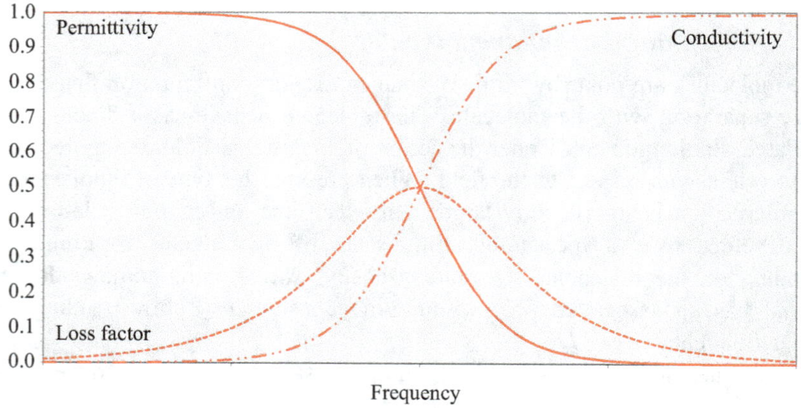

Figure 2.2 A normalised Debye dispersion of a polar material

2.1.4 Ionic conduction

In nature, as in biological materials, water is a solvent for electrolytes and all sorts of organic matter. The dielectric response of low concentration electrolytes is similar to that of pure water with the addition of an ionic conductive element σ_i. The Debye model for electrolytes in solution is thus:

$$\hat{\varepsilon} = \varepsilon_\infty + \frac{(\varepsilon_S - \varepsilon_\infty)}{(1 + j\omega\tau)} - \frac{j\sigma_i}{\varepsilon_0\omega} \tag{2.3}$$

2.1.5 Orientation polarisation of induced dipoles

Many biological molecules in an aqueous solution have polar configurations or have induced dipoles. Under the influence of an external field, water and biological molecules will orient individually according to their own time constants. Because of the difference in size, biological molecules will typically disperse in the kilo and megahertz range, whereas the water dispersion occurs at gigahertz frequencies. The dispersion of a complex system is the superposition of the various polarisations. When it is impossible to analytically separate out two or more dispersions, the resulting spectrum appears wider than a Debye dispersion. It may be described by an expression that considers a distribution of relaxation times. The most commonly used expression is:

$$\hat{\varepsilon} = \varepsilon_\infty + \frac{(\varepsilon_S - \varepsilon_\infty)}{1 + (j\omega\tau)^{1-\alpha}} \tag{2.4}$$

It is generally referred to as the Cole–Cole model in which α is a relaxation time distribution parameter in the range $1 > \alpha \geq 0$; for $\alpha = 0$ the model reverts to the Debye equation. A more general expression:

$$\hat{\varepsilon} = \varepsilon_\infty + \frac{(\varepsilon_S - \varepsilon_\infty)}{(1 + (j\omega\tau)^{1-\alpha})^{\beta}} \tag{2.5}$$

Therefore when $\alpha = 0$, the function is known as a Cole–Davidson, it reverts back to a Cole–Cole for $\beta = 1$ and a Debye for $\alpha = 0$ and $\beta = 1$.

2.1.6 Interfacial polarisations

If dielectric interfaces are introduced in an electrolyte solution, they will act as barriers to ionic current flow. Charges will accumulate at the interfaces creating induced capacitive elements. This is the situation of biological tissue where the cell membrane acts as a dielectric interface. The time constant of the charging of cell membranes is of the order of milliseconds, and the resulting dispersion is in the kilohertz range.

Charged interfaces induce the formation of counterions layers, which perform as a further polarisation. The time constant of this effect is longer than all other polarisations. This is the major contributor to the permittivity of biological tissue.

2.1.6.1 Polarisation mechanism in biological tissues

Water, the main component of biological tissue, is a polar molecule. Most other organic molecules are polar or polarisable. The cell membrane is the site of interfacial polarisation. Consequently, the dielectric spectrum of a biological tissue exhibits several polarisations in the frequency range from extremely low frequencies (ELF) to GHz regions. They manifest in three main dispersion regions referred to as α, β and γ dispersions at low, intermediate and high frequencies, respectively. The α and β in the Hz to MHz range originate from interfacial effects and organic macromolecules. The γ dispersion originates from the dipolar polarisation of tissue water, and it manifests at GHz frequencies.

The three main dispersion regions have been observed experimentally in the dielectric spectrum of tissues in the frequency range 10 Hz–20 GHz [4]. It was also observed that the α dispersion is strongly related to the presence of cell membranes. Destroying the membrane by chemical or mechanical means changes the parameters of the α dispersion significantly. The strong correlation between the water content and the magnitude of the γ dispersion is also well established.

Due to pathological conditions, tissue structure and composition changes may result in specific changes to their dielectric properties. Cancerous and ischemic tissues have been the subject of many investigations to establish differences between healthy and diseased tissue that can serve as a basis for clinical applications.

Figure 2.3 shows a fitted curve line through measured data for human liver. The wide frequency range is the result of multiple open-ended co-axial probe measurements using three vector network analysers (VNA) to achieve this (see

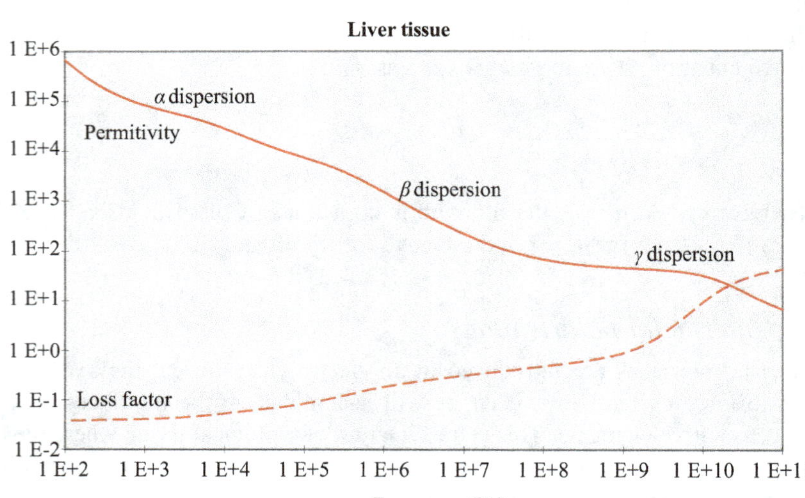

Figure 2.3 The dielectric spectrum of liver tissue – a 4-Cole–Cole fitted curve based on lab measured data. Further information is available from Gabriel et al. [4]

Section 2.2.3). The continuous curve is achieved due to good correlation in the overlapping data points between each VNA.

2.2 Common measurement systems

The dielectric properties of a material may be obtained from numerical modelling of the electrodynamics of all possible interactions within it. While this is theoretically possible and has been done in some instances, most of our knowledge of dielectric properties comes from experimental measurements.

By definition, a dielectric is a material capable of storing charge. Most materials and, more specifically, biological materials have the ability also to conduct charge. They behave as leaky capacitors the complex impedance Z and corresponding complex admittance $Y = 1/Z$ are given by:

$$Z = R + jX \text{ and } Y = G + jB \tag{2.6}$$

Z and its components (resistance R and reactance X) are expressed in ohms (Ω). Y and its components (conductance G and susceptance B) are expressed in Siemens (S). Impedance and admittance carry the same information but, for convenience, one form may be used instead of the other.

The capacitive and conductive elements of the impedance of dielectric material are a function of the permittivity and conductivity. The relationship's complexity depends on the experimental setup, including the geometry of the sample holder and the frequency range. This is because, at extremely low frequencies, it is possible to make simplifying assumptions that do not hold at radio and microwave frequencies.

2.2.1 Measurement and instrumentation at ELF

It is reasonable to assume that quasi-static conditions hold at frequencies below 1 MHz. Under such conditions, the sample holder may be a parallel plate capacitor capable of holding the sample. At its simplest, the experimental setup consists of a signal generator and means of measuring the increase in capacitance of the holder when the sample is placed within it.

For the empty cell $B = \omega C_0$ and $G = 0$, where C_0 is the capacitance which, for a parallel plate capacitor of area A and separation d is $C_0 = \varepsilon_0 A/d$. For a dielectric material of permittivity $\widehat{\varepsilon}$:

$$B = \omega C_0 \varepsilon' \tag{2.7}$$

and

$$G = \omega C_0 \varepsilon'' = \frac{C_0 \sigma}{\varepsilon_0} \tag{2.8}$$

The admittance was traditionally measured with a Wheatstone bridge or equivalent instrumentation. The sample holder could be a constant or variable separation parallel plate capacitor with temperature control sleeves. Examples of such cells are described in numerous textbooks, such as Grant *et al.* [5].

Early measurements were carried out one frequency at a time, and the dielectric spectrum was painstakingly constructed over a period of time. That was until the availability of impedance analysers capable of swept frequency measurements that became the devices of choice for such measurements and are often used in conjunction with coaxial sample holders, including open-ended coaxial probes.

The above analysis assumes perfect circuit components with no stray or inductive elements. In reality, there will be stray capacitance due to the fringing fields at the edge of the sampler and inductive elements from the cables connected to the sampler.

If C_f is the fringing field capacitance, the measured capacitance becomes $C = C_0\varepsilon + C_f$.C_f is assessed by measuring a reference material. The inductance of the cables adds a series element ωL to the measured impedance, which affects the capacitance and conductance at the high-frequency end of the ELF. If present, this effect can be corrected using equivalent circuit analysis.

Another type of unwanted effect is specific to the measurement of ionic samples at frequencies below 100 kHz. It originates from the natural build-up of charges at the electrode–sample interface, which contributes a sizeable additional series capacitance to the system. This phenomenon is known as electrode polarisation, and it is a significant source of error when measuring biological materials at frequencies below 100 kHz. As yet, this problem is not fully resolved. The use of reference materials and/or equivalent circuit analysis does not fully resolve the issue. Reviews by Chassagne *et al.* [6] and Ishai *et al.* [7] describe the problem and methods to reduce its effects.

These effects are not unique to specific sample holder geometry, and they may occur to a greater or lesser extent with all types of sampling devices, including the coaxial probe.

2.2.2 Reference materials

Several well-characterised liquids can be used as a reference for dielectric measurements. For biological materials, water and dilute salt solutions are appropriate references. Their dielectric parameters are available over a wide temperature range and reported in Peyman *et al.* [8] and the references within. Other materials often used include ethanediol, formamide and alcohols such as ethanol and methanol. Reference materials must be high purity and measured under temperature control since their dielectric parameters are temperature dependent.

2.2.3 Measurement and instrumentation at radio and microwave frequencies

The lumped impedance approach used for measurements at ELF is not applicable above 100 MHz. Instead, transmission line theory should be applied. The sample holder is typically a modified section of a transmission line (coaxial or waveguide) connected to a source and detectors capable of measuring the transmission line parameters with and without a sample in place. Transmission line theory is then used to determine the dielectric properties from this data.

For measurement in the frequency domain, right up to the 1980s, researchers used monochromatic or narrow frequency sources and elaborate bespoke sample holders with moving parts to obtain limited data (examples in Grant *et al.* [5]). By the mid-1980s, Hewlett–Packard produced three broadband vector network analysers (VNA), the 8,510, 8,720 and 8,753. The VNAs were developed to simplify the characterisation of RF and microwave components for the fast-developing wireless communication technology. They are 2-port devices incorporating broadband solid-state sources, samplers and microprocessors designed to measure the scattering parameters (reflection and transmission) of components connected to them. It soon became evident that such equipment could be used for dielectric measurements. These VNAs constituted a significant step forward in this field.

In tandem with the use of VNAs for dielectric measurement came the realisation that the sample need not necessarily be enclosed in a coaxial chamber but could be placed at the end of an open-ended coaxial line. The dielectric properties are then derived from an expression for the reflection coefficient at the probe-sample interface. The combination of VNA and open-ended probes was the catalyst that propelled dielectric spectroscopy into mainstream material studies and enabled large-scale measurement of the dielectric properties of biological materials.

2.3 Measurement technique

2.3.1 The probe

The probe is an open-ended section of a coaxial line. If constructed from a high precision air-line, it should be terminated by an impedance-matched PTFE (or equivalent) window. Adding a ground plate at the aperture is desirable to confine the field to the axial direction. Coaxial probes are broadband, enabling measurement over 2–3 frequency decades. The upper-frequency range of interest dictates the size of the probe (e.g. 1–3 mm for the GHz range, 7–10 mm for MHz).

2.3.2 The model

Models for the reflection coefficient of the interface assume a homogeneous medium that is non-magnetic and occupies the entire half-space beyond the ground plate. Samples have to be large enough to comply with this requirement, which adds a practical limit on the size of the probes used to measure biological materials. For specific applications (beyond the scope of this chapter), elaborate models may be derived for laminar structures or thin samples terminated by open or short circuits.

Mosig *et al.* [9] modelled the capacitive and radiating elements of the interface from the propagation of TEM and higher-order modes. Ignoring the effects of higher-order modes leads to a more straightforward expression, originally derived by Marcuvitz [10] and adapted for dielectric measurements by Misra [11]. Gabriel *et al.* [12] compared the performance, limitations, and complexity of the numerical solution of the models. They concluded that provided an appropriate calibration is performed, both models are equally suited to dielectric measurements.

2.3.3 The calibration

For best results when using an open-ended probe, the calibration has to be done at the tip of the probe. This is done to eliminate any error or noise from the probe and its connector. Three standards are needed for a one-port calibration, and these are obtained by shorting the tip of the probe with a metal foil measuring the probe in air and a standard liquid at a known temperature.

Examples of measurements using the Mosig [9] and Marcuvitz [10] admittance models are given in Gabriel *et al.* [12].

2.3.4 Sample preparation

The main advantage of the open-ended probe is that it requires minimal sample preparation. Liquid samples are placed in a container large enough compared to the size of the probe. It is also essential to ensure that no air bubbles are trapped under the probe. Semi-solid samples such as biological tissue need only be placed in contact with the probe for measurement, with no special encasing required. In all cases, the sample is assumed semi-infinite relative to the size of the probe, isotropic and homogeneous. Thus, the sample must be large enough to avoid reflections in all directions, particularly below the probe. Data obtained from measurement on laminar structures (e.g. skin) are challenging to interpret.

This sampling technique is non-destructive, requires minimal handling of the sample and could be used to measure tissue in vivo. One limitation is that the probe's radial geometry does not provide the directionality required for measuring anisotropic material. Temperature control is not easy to maintain over time, but this is mitigated by the short time taken by the measurement and precludes the need for accurate recording of the temperature.

2.3.5 Measurement uncertainty

All biological systems exhibit naturally occurring variability. Biological tissue is non-homogeneous at the level of the organ. The extent of variability depends on the tissue type. Moreover, age and the state of health may affect the structure and composition of a particular tissue, which may, in turn, affect its dielectric properties. Intra and inter-subject differences are to be expected and are indeed observed. In a measurement system free of systematic errors, sample variability and sample handling are the main contributors to all tissue measurement uncertainty [13]. For this reason, it is essential to describe the provenance of tissue samples in as much detail as possible.

2.3.6 In vivo versus in vitro measurement

One of the stated advantages of the open-ended probe is that it enables measurements to be carried out in vivo in preference to excised tissue to ensure their relevance to biological applications. However, it has been established that at high frequencies, the site of the γ dispersion and systematic differences have not been observed [14–16].

It should be noted that in vivo measurements could be subject to additional sources of error. For example, moving the measurement probe and connecting cable is sometimes necessary to reach the tissue sample. The change in position of the connecting cable introduces phase errors depending on the quality of the cable and the frequency range. These are evaluated experimentally and, when present, should be added to the total measurement uncertainty. Limited access to or limited visibility of the target tissues may lead to poor sampling and contamination of the contact surface by bodily fluid.

In vivo measurements are more relevant at ELF, in the range of the α and β dispersions, because of the sensitivity of their causal mechanism on the physiological state of the tissue. This is an invasive procedure fraught with difficulties. It requires specialised tools and instrumentation specific to the application [17]. Three-dimensional microwave tomography (MWT) and electric properties tomography (EPT) are a relatively recent development that captures dielectric data in vivo and non-invasively.

2.3.7 Microwave tomography

MWT was primarily developed as a non-invasive diagnostic procedure for detecting cancerous legions in people [18–20]. It is achieved using arrays of transmitting and receiving antennas around the subject that produce a diffraction pattern from which it is possible to map the dielectric properties numerically. Semenov [21] describes the technique, its successes, limitations, and potential for non-invasive assessment of functional and pathological conditions of soft tissues in vivo. Grzegorczyk *et al.* [22] developed the first clinical 3-D MWT system adequate for breast cancer screening and therapy monitoring. MWT remains an active field of research. The ultimate goal is to detect changes at the precancerous stage prior to their visibility by X-rays and to the emergence of severe clinical symptoms.

2.3.8 Electric properties tomography

EPT is a novel technique that derives the dielectric properties of tissues non-invasively, quantitatively, in vivo using a standard magnetic resonance (MR) system and standard MR measurements. Unlike MWT, EPT does not require transmit and receive antennas. Instead, the conductivity and permittivity are derived by post-processing the measured spatial magnetic field distribution B_1 of the RF transmit coil applied. Katscher *et al.* [23] reviewed the subject comprehensively.

MR uses static fields B_0 up to 7 T and RF magnetic fields B_1. It has been shown that the permittivity and conductivity at the Larmor frequency can be obtained (using non-trivial numerical analysis) from the magnitude and phase B_1.

The Larmor frequency of the MR system used is related to B_0, the static magnetic field strength through the gyromagnetic ratio. The corresponding Larmor frequencies for MR systems operating with static magnetic field B_0 of 0.5, 1.5, 3 and 7 T are 21, 64, 128 and 300 MHz, respectively.

EPT has been evaluated by investigating phantom materials with known electric properties, typically yielding a high correlation between expected and measured values. EPT studies of healthy volunteers yielded electric properties in line with the expected values (1996 database in Gabriel [4,24,25]). The measured electric properties observed across multiple tissue sources can typically be above 10%, while the intra-tissue source variability is typically below 10%.

Clinical EPT studies currently focus on tumours in the breast [26] and brain [27]. These pilot studies indicate a correlation between malignancy and conductivity. On the one hand, the observed tumour conductivity is typically clearly larger than the conductivity of surrounding healthy tissue. On the other hand, conductivity also seems to distinguish between different tumour types or between different malignancy grades. An additional evaluation of EPT is given in Tha *et al.* [27], where in vivo tumour conductivity (measured with EPT) correlates with ex vivo conductivity (of the excised tumour specimen measured with an independent probe).

Furthermore, electric properties obtained with EPT are expected to be useful not only as a biomarker but also as input for patient-individual estimation of the specific absorption rate (SAR) significant in the framework of MR safety management [28] as well as input for hyperthermia treatment planning [29].

2.4 Applications of dielectrics

The following section looks at practical applications of the theory presented above in industrial processing.

Information on the dielectric properties of matter is now either readily available or within reach [4,24,25]. This is thanks to the developments in instrumentation and the associated methodologies of the past three decades. Some of the early work has been reviewed by Duck [30] and Gabriel *et al.* [25]. The Gabriel papers contain experimental data on many tissue types presented in terms of Cole–Cole models. This knowledge is central to current processes in the medical, pharmaceutical, chemical, food and other industries, where microwave heating offers advantages over the conventional process. A few examples are given below, but the list is by no means exhaustive.

This chapter has focused on the dielectric study of biological materials. This feeds directly into several fields related to people's exposure to EMF, either purposefully from medical devices for diagnosis or therapy or incidentally from manufactured sources. The dielectric properties are essential to modelling the exposure to assess its effectiveness or ensure safety.

Knowledge of the dielectric properties of matter is central to numerous applications outside the biomedical field. Some of these applications will be briefly described below.

2.4.1 Optimisation of microwave food processing plants and quality control

One of the first industrial applications of microwave heating was in the food processing industry, such as rendering, drying, pasteurisation and preservation of food materials. The dielectric data primarily obtain to support biomedical applications are equally suited to support the processing of animal tissue in food products. There are also numerous dielectric studies specific to the food industry. Chandrasekaran *et al.* [31] is a review of the subject.

The dielectric properties are used primarily to model the heating process to optimise the parameters in order to achieve uniform heating, not just the high heating rates characteristic of microwaves. Uniform heating is required to obtain better end products.

2.4.2 Microwave-assisted chemical reactions

Microwave dielectric heating has advantages compared to conventional heating for chemical conversions. Strong coupling to one or more components can lead to higher heating rates than conventional heating. This, in turn, may lead to a significantly different temperature profile for the reaction and may lead to an alternative distribution of chemical products from a reaction. Hot spots can be created purposefully to accelerate the reaction and heat the reactants, not the enclosing capsule. The heating can be done under pressure, using Teflon coated or glass containers. For example, ethanol has a conventional boiling point of 79 °C, microwave dielectric heating in a closed vessel can rapidly lead to temperatures of 164 °C [32].

Knowledge of the dielectric properties of chemicals and, more specifically, their variation with temperature is essential to understanding, predicting and controlling the process.

2.4.3 Microwave processing of biomass

Biomass is a by-product of agriculture, forestry and discarded domestic and industrial organic matter. Thermal biomass processing is a form of waste management that produces useful products. The thermal decomposition of biomass, or pyrolysis, entails heating to very high temperatures in the region of 400–500 °C and is well suited to microwave heating. As with chemical reactions, microwave-assisted pyrolysis (MAP) is an effective method to reduce the pyrolysis reaction time and increase the quality of value-added products from different kinds of biomass [33]. The process requires knowledge of the dielectric properties of biomass in its various stages and temperatures.

Identifying biomass by remote sensing of large land and forest areas requires a dielectric model of the area covered. This can be modelled from the dielectric properties of trees (trunk, branches, leaves), other vegetation and soil in different seasons. This identification is an active field of research.

2.4.4 Microwave sintering of ceramics

Ceramics are processed at very high temperatures and benefit from microwave heating in many ways compared to traditional furnaces. A higher heating rate is achieved, leading to a shorter sintering time and lower energy consumption. The process can be manipulated to obtain a finished product with a more refined structure and enhanced mechanical properties [34].

2.5 Concluding remarks

While dielectrics have a long and prestigious history, this science is still evolving. The availability of a new generation of instrumentation and proven methodologies has turned dielectric spectroscopy into a powerful tool for material characterisation benefiting many diverse disciplines. This, in turn, has found applications in numerous fields as varied as medical diagnostics and treatment, industrial processing and remote sensing. The knowledge of dielectrics is also invaluable in the design of many electronic components, from batteries and capacitors to the design of high-efficiency compact RF antennas. The interest and expectation of the scientific community are now focused in the direction of the relatively new and fast-developing areas of RF and MWT.

References

[1] Grimnes S, Martinsen OG. *Bioimpedance and Bioelectricity Basics*. London: Academic Press; 2000.

[2] Liebe H, Hufford G, Manabe T. A model for the complex permittivity of water at frequencies below 1 THz. *International Journal of Infrared Millimeter Waves*. 1991; 12(7): 659–660.

[3] Böttcher CJF, Bordewijk P. *Dielectrics in Time-dependent Fields*. 2nd ed. Amsterdam: Elsevier; 1978.

[4] Gabriel S, Lau R, Gabriel C. The dielectric properties of biological tissues: III. Parametric models for the dielectric spectrum of tissues. *Physics in Medicine & Biology*. 1996; 41(11): 2271–2293.

[5] Middaugh CR. Dieletric behavior or biological molecules in solution (Grant, EH; Sheppard RJ; South GP). *Journal of Chemical Education*. 1979; 56(8): A260.

[6] Chassagne C, Dubois E, Jiménez M, van der Ploeg J, van Turnhout J. Compensating for electrode polarization in dielectric spectroscopy studies of colloidal suspensions: theoretical assessment of existing methods. *Frontiers in Chemistry*. 2016; 4: 30.

[7] Ishai P, Talary M, Caduff A, Levy E, Feldman Y. Electrode polarization in dielectric measurements: a review. *Measurement Science and Technology*. 2013; 24(10): 102001.

[8] Peyman A, Gabriel C, Grant E. Complex permittivity of sodium chloride solutions at microwave frequencies. *Bioelectromagnetics: Journal of the Bioelectromagnetics Society, The Society for Physical Regulation in Biology and Medicine, The European Bioelectromagnetics Association*. 2007; 28(4): 264–274.

[9] Mosig JR, Besson JC, Gex-Fabry M, Gardiol F. Reflection of an open-ended coaxial line and application to nondestructive measurement of materials. *IEEE Transactions on Instrumentation and Measurement*. 1981;1: 46–51.

[10] Marcuvitz N. *Waveguide Handbook*. London: IET; 1951.

[11] Misra D. A quasi-static analysis of open-ended coaxial lines (short paper). *IEEE Transactions on Microwave Theory and Techniques*. 1987; 35(10): 925–928.

[12] Gabriel C, Chan T, Grant E. Admittance models for open ended coaxial probes and their place in dielectric spectroscopy. *Physics in Medicine & Biology*. 1994; 39(12): 2183.

[13] Gabriel C, Peyman A. Dielectric measurement: error analysis and assessment of uncertainty. *Physics in Medicine & Biology*. 2006; 51(23): 6033.

[14] Stauffer PR, Rossetto F, Prakash M, Lee T, Neuman DG. Phantom and animal tissues for modelling the electrical properties of human liver. *International Journal of Hyperthermia*. 2003; 19(1): 89–101.

[15] Peyman S, Holden S, Gabriel C. Department of Health UK. Dielectric properties of biological tissues at microwave frequencies. [Online].; 2005 [cited 2022 1 17. Available from: http://www.mthr.org.uk/research_projects/documents/Rum3FinalReport.pdf.

[16] Farrugia L, Wismayer P, Mangion L, Sammut C. Accurate in vivo dielectric properties of liver from 500 MHz to 40 GHz and their correlation to ex vivo measurements. *Electromagnetic Biology and Medicine*. 2016; 35(4): 365–373.

[17] Latikka J, Kuurne T, Eskola H. Conductivity of living intracranial tissues. *Physics in Medicine & Biology*. 2001; 46(6): 1611.

[18] Hagness S, Taflove A, Bridges J. Three-dimensional FDTD analysis of a pulsed microwave confocal system for breast cancer detection: design of an antenna-array element. *IEEE Transactions on Antennas and Propagation*. 1999; 47(5): 783–791.

[19] Bulyshev A, Semenov S, Souvorov A, *et al.* Computational modeling of three-dimensional microwave tomography of breast cancer. *IEEE Transactions on Biomedical Engineering*. 2001; 48(9): 1053–1056.

[20] Wersebe A, Siegmann K, Krainick U, *et al.* Diagnostic potential of targeted electrical impedance scanning in classifying suspicious breast lesions. *Investigative Radiology*. 2002; 37(2): 65–72.

[21] Semenov S. Microwave tomography: review of the progress towards clinical applications. *Philosophical Transactions of the Royal Society A: Mathematical, Physical and Engineering Sciences*. 2009; 367(1900): 3021–3042.

[22] Grzegorczyk T, Meaney P, Kaufman P, Paulsen K. Fast 3-D tomographic microwave imaging for breast cancer detection. *IEEE Transactions on Medical Imaging*. 2012; 31(8): 1584–1592.

[23] Katscher U, van den Berg C. Electric properties tomography: biochemical, physical and technical background, evaluation and clinical applications. *NMR in Biomedicine*. 2017; 30(8): e3729.

[24] Gabriel S, Lau RW, Gabriel C. The dielectric properties of biological tissues: II. Measurements in the frequency range 10 Hz to 20 GHz. *Physics in Medicine & Biology*. 1996; 41(11): 2251–2269.

[25] Gabriel S, Corthout E, Gabriel C. The dielectric properties of biological tissues: I. Measurements in the frequency range 10 Hz to 20 GHz. *Physics in Medicine & Biology*. 1996; 41(11): 2231–2249.

[26] Kim SY, Shin J, Kim DH, *et al.* Correlation between conductivity and prognostic factors in invasive breast cancer using magnetic resonance electric properties tomography (MREPT). *European Radiology*. 2016; 26(7): 2317–2326.
[27] Tha KK, Katscher U, Yamaguchi S, *et al.* Noninvasive electrical conductivity measurement by MRI: a test of its validity and the electrical conductivity characteristics of glioma. *European Radiology*. 2018; 28(1): 348–355.
[28] Liu J, Wang Y, Katscher U, He B. Electrical properties tomography based on maps in MRI: principles, applications, and challenges. *IEEE Transactions on Biomedical Engineering*. 2017; 64(11): 2515–2530.
[29] Balidemaj E, Kok HP, Schooneveldt G, *et al.* Hyperthermia treatment planning for cervical cancer patients based on electrical conductivity tissue properties acquired in vivo with EPT at 3 T MRI. *International Journal of Hyperthermia*. 2016; 32(5): 558–568.
[30] Duck F. *Physical Properties of Tissues: A Comprehensive Reference Book*. London: Academic Press; 2013.
[31] Chandrasekaran S, Ramanathan S, Basak T. Microwave food processing—a review. *Food Research International*. 2013; 52(1): 243–261.
[32] Gabriel C, Gabriel S, Grant E, Halstead B, Mingos M. Dielectric parameters relevant to microwave dielectric heating. *Chemical Society Reviews*. 1998; 27(3): 213–224.
[33] Motasemi F, Afzal M. A review on the microwave-assisted pyrolysis technique. *Renewable and Sustainable Energy Reviews*. 2013; 28: 317–330.
[34] Borrell A, Salvador MD. Chapter 1: Advanced ceramic materials sintered by microwave technology. In: *Sintering Technology – Method and Application*. London: IntechOPen; 2018; pp. 3–24.

Chapter 3

Dosimetry for bioelectromagnetics

Cynthia M. Furse[1]

Bioelectromagnetic applications rely on the ability to control the interaction between electromagnetic fields and the body. Designing electrical sensors and medical devices therefore requires the ability to determine the fields in the body that are caused by these devices, or to determine how much the body impacts the fields, creating an image or perturbation that can be received by a device, such as an electrical sensor. Dosimetry is the process of determining these fields, and is the topic of this chapter. Dosimetry requires three major components—a sufficiently precise model of the physical structure of the part(s) of the body the fields will traverse, a model of the electrical properties of these structures, and a numerical method to compute the fields within this model. We will cover all of these topics in this chapter, with a focus on the use of the finite-difference time-domain (FDTD) method for bioelectromagnetic simulations.

3.1 Electrical properties of biological tissues

Biological tissues have electrical properties that are similar to salt water, with both dielectric and conductivity effects. The properties result in attenuation, dispersion, reflection, diffraction, guiding, and other electromagnetic field effects within the tissues.

The electrical properties are strongly dependent on the type of tissue. Tissues with large water and salt content such as blood and cerebral spinal fluid have the highest conductivities (see Chapter 2 in this book). Those with less water such as fat, skin, and bone have lower conductivities, and other tissues tend to fall between these two. Table 3.1 gives examples of tissues and their relative permittivity (ε_r) and conductivity (σ (S/m)) at 915 MHz. Values at several frequencies were measured using an open-ended coaxial probe. A four-term Cole–Cole representation was then derived and used to calculate the properties in Table 3.1. Many other types of tissues are included in Gabriel (1996), and a handy calculator for the properties is available at [1].

For the purposes of dosimetry, most tissues are considered to be uniform throughout the organ of interest. Values for these tissues are calculated at the

[1]Department of Electrical and Computer Engineering, University of Utah, Salt Lake City, UT, USA

Table 3.1 Electrical properties of tissue at 915 MHz (calculated using [1])

Tissue	ε_r	σ (S/m)
Cerebral spinal fluid	68.6	2.42
Blood	61.3	1.54
Brain (grey matter)	52.6	0.95
Brain (white matter)	38.8	0.60
Kidney	58.6	1.40
Muscle	55.0	0.95
Bone (cortical)	12.4	0.14
Bone (cancellous)	20.8	0.34
Skin (wet)	46.0	0.85
Skin (dry)	41.3	0.87
Fat (infiltrated)	11.3	0.11
Fat (not infiltrated)	5.5	0.05

frequency(s) of interest and used throughout the simulations. However, this is only a rough approximation. Tissues vary significantly between individuals and species, vary with location within a single organ, and vary with age, health, temperature, hydration, frequency, etc. There is substantial variation between healthy and diseased tissues, and between in vivo and ex vivo tissues. Particularly for tissues with elongated cell structures such as bone and nerve, there is substantial anisotropy in the properties. As we will discuss in Section 3.6, it may be important to include this variation in dosimetry calculations in order to calculate the worst/best case scenarios and expected variability in the results and outcomes.

Electrical properties (ε_r and σ) are dependent on frequency, temperature, and the biological properties including tissue type [2], age [3], individual, in vivo/in vitro, health (e.g. cancers/non-cancerous) [4], species [5], etc. Magnetic relative permeability $\mu_r \sim 1$ for biological tissues and air, so this is treated as a constant in most bioelectromagnetic simulations.

3.2 Anatomical models

In order to calculate the field distribution in and around the body, an anatomical model with sufficient detail to allow simulation of all relevant field effects must be used. Models may be as simple as layered models or spheroids or as complex as millimeter or sub-millimeter scale anatomical models based on MRI, CT, or X-ray scans, or visual images. Anatomical models are used in applications beyond bioelectromagnetics including dosimetry for nuclear radiation and mechanical stressors. Numerous anatomical models exist [6], made from scans of adults and children of several different ages, men and women, including pregnant women. Their postures can be adjusted to be in different positions, and their sizes may be scaled to represent ages or natural variation in sizes of people. High-resolution models of specific organs or parts of the body are available. Many types of animals have been modeled as well.

Anatomical models can be created from any type of scan that gives a visual representation of the body, where dark/light portions of the image correspond to different tissue densities. The challenge is that many tissues share the same density, so there is not a one-to-one correspondence between density and tissue type. Thus, anatomical models have typically been created semi-automatically by identifying regions containing a specific density, intelligently associating that with a specific organ or tissue, and allowing an automated algorithm to pick out surrounding areas of similar density, assuming they are also part of the same organ. This method does a generally good job of identifying major organs; however, there are many regions of the body that are simply filled with fluid, connective tissue, or air, and these require knowledge of the underlying anatomy in order to identify and characterize them. An example of an MRI-derived anatomical model is the Utah Man Model shown in Figure 3.1 along with the specific absorption rate (SAR) in the head from a nearby cellular telephone.

These are several challenges in creating good anatomical models for electromagnetic dosimetry that are related to limitations of the scans themselves. For example, the difference in electrical properties of the tissues creates an apparent shift in location in MR-derived images. This is most apparent between fat and water-based tissues (blood, muscle, major organs). This can make the fat on one side of the body appear as much as 0.5 cm thicker than on the other side of the

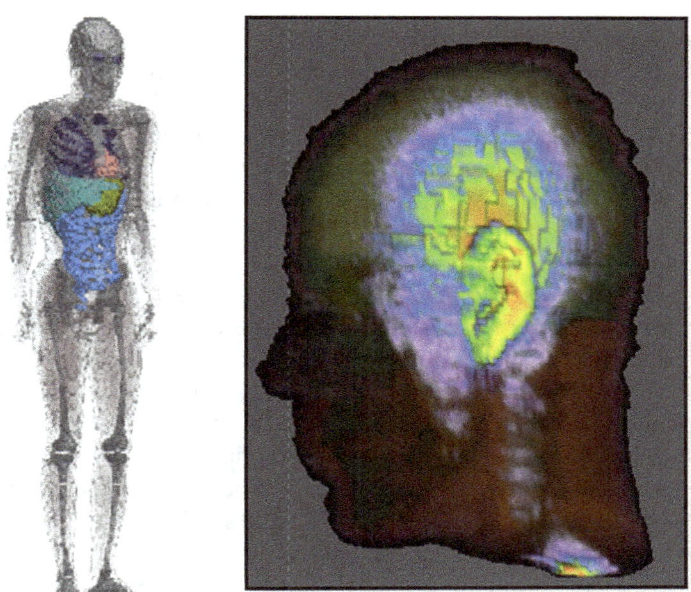

Figure 3.1 The Utah Man Model (left) is derived from MRI scans. It includes 31 tissue types and is $2 \times 2 \times 3$ mm^3 in resolution [7]. The head from this model has been used to calculate the SAR from a cellular telephone at 835 MHz (right) [8]

body, and must often be corrected semi-manually. Skin is often also not seen well in images and must be added semi-manually after the fact. In some applications, the thickness of skin is important, and this varies with age, gender, location on the body, and between individuals.

Other challenges with creating dosimetric models are related to the physical orientation of the person or animal when it is scanned. If the body is deceased, the lungs, heart and other organs very quickly deflate, and all tissues sink to the back of the body (assuming the body is laying face-up), where they appear in the scans. The relaxed position of the body in the scanner also causes the arms to relax to the back and the toes to relax forward. When the body is then digitally "stood up" after scanning, the arms are in an unnatural position for a standing body, and the body may appear to be standing on tip-toe. These orientations are often corrected after the fact. Algorithms have also been created to convert a scan of a body lying prone into models in a variety of poses (standing, sitting, working, etc.). These algorithms typically try to keep the inner tissues (e.g. bones) in places as well. Blur from moving organs (such as the beating heart or expanding lungs) is another artifact that must often be adjusted for. In order to reduce the artifacts from heartbeat and breathing, the images are often time-gated to take scans in sync with the heartbeat or breathing, trying to evaluate them when the organ is in as similar a state/location as possible. Still, dosimetric models of these organs still often require semi-manual manipulation to make accurate models of these organs. For instance, walls and vessels may not properly line up, and may need to be manually moved in the model to make them align.

When considering which model to use, or how to adapt an existing model, it is important to consider what effects may be apparent in the electromagnetic results. Details can make a big difference here. For instance, when analyzing how well fields penetrate into the body from a localized source, the thickness of skin may be important, whereas when considering a less localized source, this may not matter. Height and size may make a big difference when considering field resonances but may have less effect when considering more localized effects. Anything highly conductive in or near the body (medical implants, cell phones, wearables, jewelry, etc.) can have significant impact on the fields, so their exact placement and orientation may be important. Modeling the source of the fields (cell phones or medical implants, for instance) can be very critical, as even a millimeter or two can profoundly affect the very near fields in the vicinity of the source.

Understanding the theoretical aspects of the electromagnetics in the application is essential to choosing and applying an appropriate dosimetry model. It can be helpful to start with a simple model to evaluate these effects and move to progressively more detailed models. For instance, remembering that fields concentrate on tips and edges of highly conductive objects or gaps between them can help explain unreasonably high field strengths at the corners of square or rectangular cells (voxel models). Field diffraction and reflection may also not be correct in these models, due to inaccuracies modeling curved structures. On the other hand, attenuation in conductive tissues often dampens these effects so much that they produce minimal error. Understanding and evaluating these and other electromagnetic effects is key to successful electromagnetic simulations.

The resolution (Δx, Δy, Δz) of an anatomical model for FDTD simulation is controlled by two things. The first is the accuracy of the derivatives in the FDTD simulation. Using a central difference formula, as described in Section 3.3.1, the difference equations have an error on the order of (Δ^2). Typical resolutions are chosen to be at least a tenth of the minimum wavelength in the simulation ($\Delta \leq \lambda_{min}/10$). The minimum wavelength λ_{min} will be found in the tissue with the largest values of relative permittivity (ε_r) and conductivity (σ (S/m), and will occur at the highest frequency used in the model. This voxel size Δ in the simulation is then generally set so $\Delta \leq \lambda_{min}/10$. The other aspect of the model that controls the resolution is the size of the physical structures to be modeled. For instance, ultra-low frequency (e.g. 60 Hz) simulations may be used to calculate current or power distributions in individual tissues, though the entire body is less than 0.002 λ_{min}. Although this gives a very small error in the theoretical differences, it can create a significant error in numerical roundoff error when the differences are calculated, and care must be taken to manage this error [9]. Most notably, the simulation cannot be run "too long," as will be described in Section 3.3.1.

3.3 The FDTD method

Computational methods use the anatomical models as described in Section 3.2, with the tissue properties as described in Section 3.1 to approximate the propagation of the electromagnetic waves from internal and/or external sources. Through approximations to Maxwell's equations, field effects including attenuation, propagation, reflection, diffraction, etc. are calculated. This can be done in either the time and/or frequency domain, and it is common to transfer information back and forth using the Fourier transform. Frequency domain methods include the impedance method, method of moments [10], and finite-element method (FEM) [10]. The FDTD method [11] computes the fields in the time domain. Both frequency- and time-domain methods should give identical results if simulating exactly the same conditions; however, the computational efficiency, flexibility, and ease of use of the FDTD method has led to its being the method of choice for most bioelectromagnetic simulations.

3.3.1 Overview of the FDTD method

The FDTD method was originally developed by [12] and was first applied to bioelectromagnetic simulations in [13]. The FDTD method solves the differential form of Faraday's and Ampere's laws:

$$\nabla \times \overline{E} = -\mu \frac{\partial \overline{H}}{\partial t} \tag{3.1}$$

$$\nabla \times \overline{H} = \sigma \overline{E} + \varepsilon \frac{\partial \overline{E}}{\partial t} \tag{3.2}$$

where the permittivity (ε), conductivity (σ), and permeability (μ) are isotropic, frequency-independent, and constant, respectively, over the small region where the equation is being solved. We can convert these vector equations to scalar equations by equating vector components, giving six coupled partial differential equations of the form [14]:

$$\frac{\partial H_x}{\partial t} = \frac{-1}{\mu}\left(\frac{\partial E_z}{\partial y} - \frac{\partial E_y}{\partial z}\right) \tag{3.3}$$

$$\frac{\partial E_x}{\partial t} = \frac{1}{\varepsilon}\left(\frac{\partial H_z}{\partial y} - \frac{\partial H_y}{\partial z} - \sigma E_x\right) \tag{3.4}$$

The FDTD model space is divided into voxels ($i\Delta x$, $j\Delta y$, $k\Delta z$), and the electric and magnetic field vector components are distributed throughout each of these unit cells as shown in Figure 3.2. Electric fields are defined at time step $n\Delta t$, and magnetic fields are defined at half time steps, $(n + 1/2)\Delta t$. Therefore, the electric and magnetic fields are offset by half a voxel, and by half a time step $\Delta t/2$. This "leap frog" scheme allows the space and time derivatives of the fields to be approximated by the central difference formulas for space (x) and time (t):

$$\frac{\partial F^n}{\partial x} = \frac{F^n\left(i + \frac{1}{2}, j, k\right) - F^n\left(i - \frac{1}{2}, j, k\right)}{\Delta x} \tag{3.5}$$

and

$$\frac{\partial F^n}{\partial t} = \frac{F^{n+1/2}(i, j, k) - F^{n-1/2}(i, j, k)}{\Delta t}. \tag{3.6}$$

For square voxels, the error for these equations is $(\Delta x)^2$ for the spatial derivatives and $(\Delta t)^2$ for the time derivatives. Thus, using the first-order central difference formula (producing the first-order difference equations) gives the second-order accuracy. This advantage is lost if non-uniform voxels are used, so most often FDTD models are made from uniform, cubical voxels.

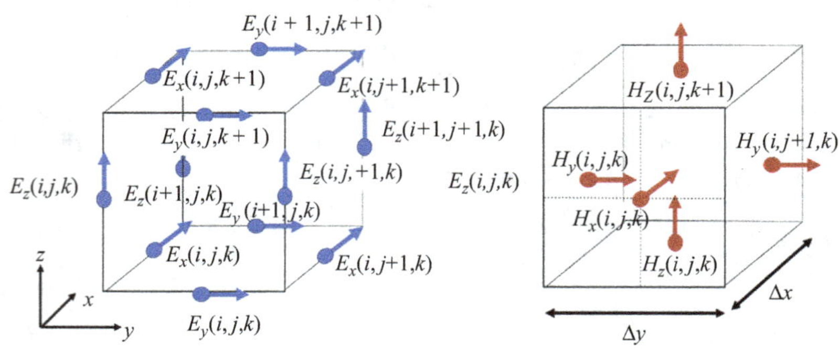

Figure 3.2 The FDTD unit cell has electric fields parallel to the edges and magnetic fields perpendicular to the faces of a voxel of size (Δx, Δy, Δz)

The differential electric and magnetic field equations in (3.3) and (3.4) can then be converted to difference equations of the form [14]:

$$H_x^{n+\frac{1}{2}}(i,j,k) = H_x^{n-\frac{1}{2}}(i,j,k) + C_H(i,j,k)\{[E_x^n(i,j,k) - E_x^n(i,j+1,k)]/\Delta y$$

$$+\left[E_y^n(i,j,k+1) - E_y^n(i,j,k)\right]/\Delta z\} \tag{3.7}$$

$$E_x^{n+1}(i,j,k) = C_{E1}(i,j,k)E_x^n(i,j,k)$$

$$+C_{E2}(i,j,k)\left\{\left[H_x^{n+1/2}(i,j,k) - H_x^{n+1/2}(i,j-1,k)\right]/\Delta y\right.$$

$$\left.+\left[H_y^{n+1/2}(i,j,k) - H_y^{n+1/2}(i,j,k-1)\right]/\Delta z\right\} \tag{3.8}$$

The time domain sampling step size Δt is chosen so that the fields cannot pass any point in the simulation ($i\Delta x$, $j\Delta y$, $k\Delta z$) without being properly sampled. For a square voxel, the longest distance in any voxel is diagonally from corner to corner ($\Delta x/(\sqrt{3})$, and the fastest it can go is the speed of light (c_o), so Δt must be $\leq \Delta x/(\sqrt{3}c_o)$. A common simplifying approximation is that $\Delta t = \Delta x/(2c_o)$, although this results in the equations being slightly less accurate. Another numerical simplification is to normalize $\overline{H}$, which is typically much smaller than $\overline{E}$, by a normalization factor:

$$H_{\text{programmed}} = \sqrt{\mu_o/\varepsilon_o}\, H_{\text{physical}} \tag{3.9}$$

to make $\overline{E}$ and $\overline{H}$ approximately the same order of magnitude. Using this normalization, the fact that for biological tissues, $\mu_r \sim 1$, and the time step resolution $\Delta t = \Delta x/(2c_o)$, the constants in (3.7) and (3.8) become

$$C_{Hy}(i,j,k) = \frac{c_o\Delta t}{\mu_r(i,j,k)} \tag{3.10a}$$

$$C_{E1}(i,j,k) = \frac{2\varepsilon_o\varepsilon_r(i,j,k) - \Delta t\sigma(i,j,k)}{2\varepsilon_o\varepsilon_r(i,j,k) + \Delta t\sigma(i,j,k)} \tag{3.10b}$$

$$C_{E2}(i,j,k) = \frac{2\Delta t\sqrt{\varepsilon_o/\mu_o}}{2\varepsilon_o\varepsilon_r(i,j,k) + \Delta t\sigma(i,j,k)} \tag{3.10c}$$

Similar equations and constants are obtained for the remaining four coupled differential equations from (3.1) and (3.2) (similar to (3.3) and (3.4)). Details can be found in [14].

FDTD is a time-domain simulation, which means that the fields are calculated at every voxel in the model, at every time step. The steps in the simulation are as follows:

1. Set up the model:
   (a) Define the voxel model
   (b) Define the constants associated with the model.

2. Define the source
   (a) Geometry of the source
   (b) Time-domain signals in the source
3. Define initial conditions
4. For each time step, $n\Delta t$
   (a) Specify fields at source.
   (b) Calculate E^n for all voxels.
   (c) Calculate $H^{n+1/2}$ for all voxels.
   (d) Apply numerical boundary conditions to terminate the model.
5. Stop when the solution has converged.
6. Perform post processing.

There are several nuances associated with each step when applying it to bioelectromagnetic applications, some of which are described below:

Step 1(a): Define the voxel model:

Models of the body that are commonly used in bioelectromagnetic simulations are described in Section 3.2. In addition to the model of the body or body section, the full FDTD model also includes air surrounding the body. Generally, minimizing the size of this air space (number of voxels between the body and the outer boundary model) is important, to minimize storage and computation time requirements. Numerical boundary conditions that computationally absorb the electromagnetic fields are applied at the boundary. Typically, the entire model is divided into cubical voxels, where the tissue/material type is defined in each voxel ($i\Delta x$, $j\Delta y$, $k\Delta z$). This requires less memory than storing, for example, the electrical properties (ε_r and σ) in every voxel in the model.

It is important to correctly model any aspect of the body that will impact the electromagnetic fields. For instance, even though skin is often too thin to be modeled by a full voxel, it may have significant impacts on the electromagnetic fields from devices that would be in contact with the skin. Some methods for handling thin layers, or voxels that include multiple tissue types, are to average the properties of the two or more tissues in the voxel [3]. Other potential issues include applications where surface waves are important, where the pixelated voxel model may impede the surface waves. Models that include divisions between two very dissimilar materials (such as metal, often used for sources) must be particularly carefully considered.

Step 1(b): Define the constants:

The voxel model described above will typically have information on which tissue or material is in each voxel. Next, we consider how the properties of the tissues (relative permittivity ε_r and conductivity σ) and FDTD parameters ($\Delta x, \Delta y, \Delta z$ and Δt) are incorporated into the model to calculate the constants in (3.10).

Electrical properties (ε_r and σ) are dependent on frequency, temperature, and the biological properties including tissue type [2], age [3], individual, in vivo/in vitro, health (e.g. cancers/non-cancerous) [4], species [5], etc. Magnetic relative permeability $\mu_r \sim 1$ for biological tissues and air, so this is treated as a constant in most bioelectromagnetic simulations. There are several nuances of the properties of biological materials that impact how the FDTD simulation must be run. Properties may also be anisotropic (as in the case of the long bones in the body and cardiac tissue), necessitating the use of different constants in each direction *x, y, z*. Biological tissues vary with frequency, so if a pulsed or other multi-frequency course is used, a frequency-dependent FDTD method, as described in Section 3.4, must be used. Electrical properties of tissues vary statistically, both because of variability within the tissues themselves, and because of measurement variability. Most often, average properties are used, and the FDTD simulation gives average fields. If the variability of these fields is important (as it may be in the case of detuning of antennas or devices, accuracy for field certification, or evaluation of sensor perturbation), a stochastic FDTD method, described in Section 3.5, can be used to calculate the effect of this variation. Biological properties also vary with temperature, which means if the source being simulated has sufficiently high power that the temperature may change throughout the simulation, this will also need to be considered.

Step 2: Define the source: The geometry of the source and its time domain signature are two aspects of the source that must be carefully simulated.

Step 2(a): Geometry of the source: The physical nature of the source may be distributed such as a plane wave or localized such as a cell phone or implantable medical device. Plane waves are best handled with special plane wave boundaries that remove the plane wave from the fields before they reach the outer boundaries of the simulation [15]. This improves the absorption of the outer boundary conditions and allows for calculations involving the scattered field only (such as radiation pattern, radar cross section, or imaging with scattered fields).

Localized sources must be meticulously modeled, considering the location of the fields within the FDTD grid as shown in Figure 3.2. Any small difference is the location or size of a localized source will translate to very large changes in the near fields of this source, which are typically significant to the outcome of the FDTD simulation. Metallic objects are generally modeled by setting tangential electric fields on their surfaces equal to zero. This requires careful modeling of where these surfaces actually are in the model. For example, modeling a metal box, such as the metal casing of a cell phone or battery pack, that is 1-voxel thick actually requires the tangential electric fields in two adjacent voxels (the voxel where the box resides, and the next voxel) to be set to zero. Metallic objects are more likely than dielectric ones to have problems with "staircasing," where rough surfaces caused by the cubical voxels can corrupt the surface waves on these devices. An example of the staircasing (roughness) in a voxel model can be seen in the Utah head model as shown in Figure 3.3. To model a tilted telephone (where the phone is a metal box with a protruding antenna), if the box had been tilted rather than the head, the voxelization would have made a very rough metallic surface on the phone and antenna, significantly perturbing the results. Tilting the head was a better option in this case.

Step 2(b): Time domain signals in the source:

FDTD can be run with any type of time domain source, including pulses and sine waves. Applications that naturally use time-dependent signatures such as radar

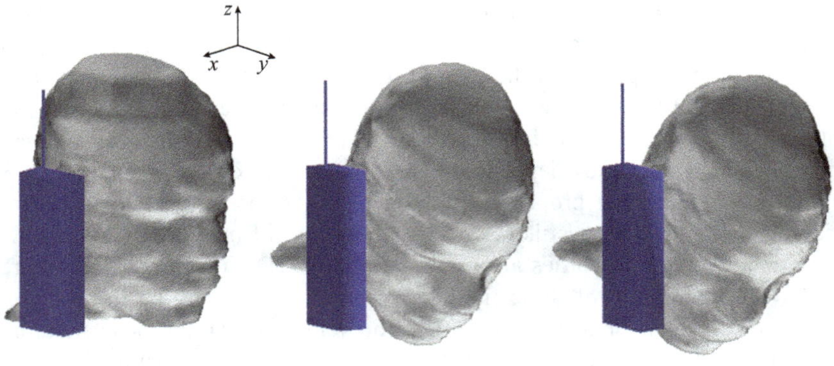

Figure 3.3 Cellular telephones near the human head. Modeling the telephone held in a realistic tilt against the head was best done by leaving the phone straight and tilting the head [16]

are natural fits for pulsed source simulations. Pulses can also be used for applications where output properties are needed at multiple frequencies (such as the impedance of an antenna or sensor), by converting the pulsed responses from a single FDTD simulation from the time to frequency domain. Both of these applications of pulsed FDTD are problematic for bioelectromagnetic simulations, because the properties of biological materials (anything with water in it) vary with frequency. If the frequency spectrum in the pulse is not too large, the average properties may be used, with some reasonable error. Alternatively, a frequency-dependent FDTD simulation (such as described in Section 3.5) can be used to more accurately model these properties.

An interesting nuance of pulsed FDTD simulations is that it might seem as if the pulse would need to be band-limited to the frequencies supportable by the FDTD sampling rate Δt in order to prevent aliasing. However, the FDTD grid prevents propagation of waves that are outside this range. Typical cell sizes (Δx, Δy, Δz) are chosen to $\leq \lambda_{min}/10$), as described in Section 3.3.1. This keeps numerical dispersion (variation of velocity of propagation) $<1\%$ for most angles of propagation within the model [14]. The highest frequency in the simulation has the smallest wavelength. Frequencies that are higher than the grid can support are not propagated, and therefore do not cause problems with aliasing [17].

A sine wave is a single-frequency source, so single-frequency tissue properties can be used. Multiple frequency simulations then require multiple FDTD simulations. If pulsed responses are desired, signal processing can be used to combine the output from these multiple simulations into a pulsed time domain output. A simple sine wave seems like it should be a single-frequency source, but in an FDTD simulation, it is not that simple. At the start of the FDTD simulation, the fields start at zero, and the sine wave is turned on from there. This sudden turn on means the signal that is actually simulated is a step function ($u(t)$) multiplied by the sine wave. The ideal step function ($u(t)$) has infinite bandwidth, so the resultant signal does too. This can lead to unexpected results, such as DC offsets within the simulation [18]. It is better to use a ramped sine wave with a smoother turn-on, rather than a step-turn-on sine wave.

Multi-frequency FDTD simulations:

As described above, results for multiple frequencies can be obtained using either a single frequency-dependent pulsed FDTD simulation or multiple sine wave simulations. It is tempting to think that the pulsed approach (which requires only one FDTD simulation) would be more efficient than running multiple sinusoidal FDTD simulations. It turns out, however, that this is not the case [19]. If output results are needed at many locations, such as the fields or power absorption patterns throughout the body, the time-to-frequency domain transformation using the FFT or DFT (described in Section 3.4) takes more time than running a new sinusoidal FDTD simulation with the more efficient two-equations two-unknowns (2E2U) transformation, as will be described in Section 3.4.2. If results are desired at two or more frequencies, another advantage of using a sinusoidal simulation with the 2E2U method is that each simulation can be run with electrical properties of body tissues that are chosen for that particular frequency. The alternative is to use a single FDTD simulation with a pulsed or other multi- or broad-band source. If this is done, $(FD)^2TD$ should be used to ensure that the electrical properties of the tissues are accurate at each frequency in the simulation.

Step 3. Define initial conditions:

Usually, the source and fields in all the voxels start out at zero, however an efficient way to run multiple similar simulations (such as making small changes in the model and rerunning it, changing the frequency slightly, etc.) can be to start with the converged fields from a similar simulation, rather than restarting the simulations with fields equal to zero.

Step 4: For each time step, $n\Delta t$:

(a) Specify fields at source.
(b) Calculate E^n for all voxels.
(c) Calculate $H^{n+1/2}$ for all voxels.
(d) Apply numerical boundary conditions. A full discussion of possible boundary conditions is out of the scope of this chapter. There are three major classes of boundary conditions.
    (i) Absorbing boundary conditions (ABC) absorb electromagnetic fields that come in contact with the boundary. The most common boundary condition used today is a perfectly matched layer (PML) boundary condition [20], of which there are many variations.
    (ii) Symmetry boundary conditions represent symmetry within the model in terms of either electric or magnetic field symmetry planes [21].
    (iii) Floquet boundary conditions represent arrays of identical elements [22].

Step 5: Stop when the solution has converged:

For transient fields, this means all of the fields have died away to zero. For sinusoidal fields, this means that all of the fields have converged to a steady-state sinusoid. Convergence in the simulation is essentially the same as physical convergence of a wave impinging on the system. If a system has a high quality factor (Q), many reflections will occur within the system. Typically, all of these reflections would also need to be simulated before convergence will be reached. Bioelectromagnetics examples of high-Q systems include reverberation chambers used to measure absorbed power and simulation of the body at its resonant frequencies. Most bioelectromagnetic simulations are not high-Q, however, because the lossy tissues in the body absorb the fields.

Convergence is often reached after one or a few periods of the sine wave. In very low-frequency simulations, however, where a mm-resolution model of the body is used, the time step is so small compared to the period that convergence is reached when only a very small fraction of the sine wave period has propagated into the body. Evaluating convergence is best done using a time-to-frequency domain conversion, evaluating the fields at one time step, $n\Delta t$, in the simulation. allowing the simulation to run a little longer (to $(n+m)\Delta t$), and checking the fields again. If they are the same, within an acceptable margin of error, the simulation is sufficiently converged. It is important to stop the simulation when convergence has been reached, and not keep running it significantly longer, as numerical round off error in the difference equation calculations can cause late-time instabilities [9].

3.4 Conversion between time- and frequency-domains

Many parameters of interest in bioelectromagnetic simulations are evaluated in the frequency domain. These include field or current magnitude and phase, SAR, heating, and antenna radiation patterns and scattering (S-) parameters. The conversion from the FDTD time-domain signals to the frequency-domain can be done using the Fourier transform from either pulsed [23] or sinusoidal data [24]. Although it might seem that running FDTD with pulsed signals that carry multiple frequencies and using the fast Fourier transform (FFT) to convert them to the frequency domain would be the most efficient way to obtain data for multiple frequencies, it is actually more efficient to run multiple single-frequency (sinusoidal) FDTD simulations and use the discrete Fourier transform (DFT) to convert to the frequency domain [24]. However, even with this efficiency, and others [23], applications that require a large number of time-to-frequency domain conversions, such as finding current, field, or SAR distributions in the human body, this conversion can take more time and memory than the FDTD simulation itself. Another, more efficient method based on solving for the magnitude and phase (two unknowns) using sinusoidal equations at two time steps is described in the next section. It was originally developed for low frequency FDTD simulations, but is also more efficient than either the DFT or FFT at higher frequencies.

3.4.1 FDTD for very low frequencies

Another challenge of using the Fourier-transform-based time-to-frequency conversion methods (and one of these reasons they are so computationally expensive) is that they require integration over at least ¼ of a sinusoidal cycle after the simulation has converged. While this is manageable for many bioelectromagnetic simulations, this is prohibitively expensive for low-frequency simulations such as calculation of currents or fields in the body under or near 60 Hz power lines [25]. For example, in a model of the body with 6 mm resolution, a typical FDTD time step would be 10 ps. The 17 ms period of the 60 Hz sine wave would require 417 M steps to run ¼ of a period to complete the DFT. Not only would this be impractical from the point of view of computation time, but it also would raise the very real likelihood of late-time instabilities caused by propagation of numerical round off errors in the FDTD difference calculations. Running a 60 Hz simulation on a mm-scale resolution model of the human body is impractical or impossible.

A better alternative is to use two methods to change how the FDTD simulation is run. The first is to use a quasi-static approximation and use frequency scaling to convert the simulation from a 60 Hz simulation to one at, say, 10 MHz as in [25]. Then it is important to realize that the body is much too small to be resonant at these frequencies, so the FDTD simulation converges in a very small fraction of a period. Needing to compute the DFT or FFT over ¼ of a period would make this conversion method untenable. Instead, a different method is used to calculate the magnitude and phase (frequency-domain parameters) of the sinusoidal fields. This method, called the 2-equations 2-unknowns (2E2U) method, is more efficient than the DFT or FFT at all frequencies [19]. It can also be extended to N-equations N-unknowns (NENU) for multiple frequency FDTD simulations.

3.4.2 The two-equations two-unknowns (2E2U) method

The 2E2U method assumes the FDTD simulation has been run with a single frequency sine wave as the source, and has been run to convergence. The time to reach convergence depends on the nature of the reflections in the body. In general, the body is sufficiently lossy that fields are attenuated before they resonate strongly. There are some exceptions to this, when the body or part of the body (head, torso, etc.) is resonant at the frequency of interest, and in these cases, convergence takes longer. One way or another, once convergence has been reached, the fields at every voxel in the FDTD simulation are sine waves with a specific magnitude and phase that we want to extract using a time-to-frequency domain conversion. The sine waves have converged to their specific magnitude and phase, even if the sine wave has not yet actually reached the peak (magnitude), and even if the sine wave may have only been run a very small fraction of its period before convergence has been reached (as is the case for low frequency simulations).

Peak detection (literally recording the maximum value of the field) can be used to find the magnitude (peak) and phase (the location of the peak in time), but requires at least ¼ of the period of the sine wave after convergence. The 2E2U method is a more efficient way of determining the unknown magnitude (A) and phase (φ) of the sine wave of angular frequency ω ($= 2\pi \times$ frequency (Hz)).

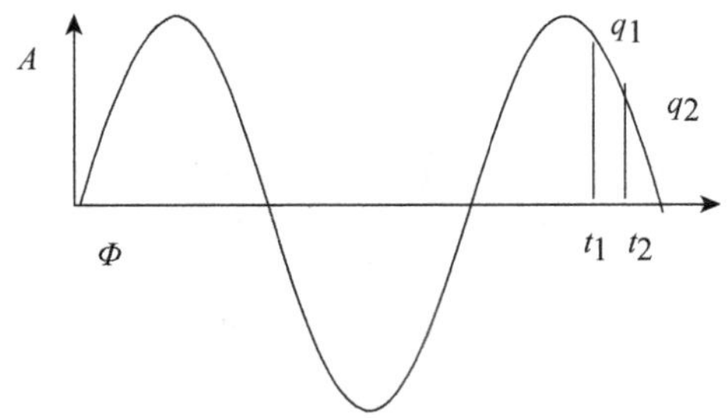

Figure 3.4 *Sine wave with amplitude A and phase* φ. *The two equations two unknowns (2E2U) method can be used to find these values using samples* q_1 *and* q_2 *at times* t_1 *and* t_2 *[19]*

FDTD is used to compute the values of the field at a given location in space at two times, t_1 and t_2, giving values q_1 and q_2.

$$\begin{matrix} A\sin(\omega t_1 + \varphi) = q_1 \\ A\sin(\omega t_2 + \varphi) = q_2 \end{matrix} \quad (2) \tag{3.11}$$

This is shown in Figure 3.4. These two equations can be solved for the two unknowns, magnitude (A) and phase (φ), to give direct relationships for these values:

$$\varphi = \tan^{-1}\left[\frac{q_2\sin(\omega t_1) - q_1\sin(\omega t_2)}{q_1\cos(\omega t_2) - q_2\cos(\omega t_1)}\right] \tag{3.12a}$$

$$A = \left|\frac{q_1}{\sin(\omega t_1 + \varphi)}\right| \tag{3.12b}$$

The choice of t_1 and t_2 depends on the simulation. For most FDTD simulations, the spatial resolution Δ, is on the order of $\lambda/10$ to $\lambda/100$. Then the last two time steps of the simulation (immediately after convergence) can be used for t_1 and t_2. For higher-resolution simulations, such as mm-scale low frequency simulations, t_1 and t_2 are so close together in time that they are nearly equal, so q_1 and q_2 are also nearly equal. This results in errors due to numerical round-off when calculating A and φ. For these simulations, it is better to choose t_1 to be the time step at convergence and t_2 to be a few (say, 50) time steps later. The 2E2U method is a substantial saving in time over the DFT or FFT, and it can also provide a significant savings in memory if the time-to-frequency domain must be computed for very many voxels in the model.

It is also important to note that this method provides accurate magnitude and phase calculations only for clean, sine wave output. Noise, computational round off error, and DC offsets will cause errors. Ramped sine excitations known not to cause a DC offset are generally advisable [18,26].

Using these methods, the current densities in the Utah 3D human body model were calculated for a frontally incident 60 Hz plane wave [25]. Figure 3.5 shows the current densities in a layer of the torso containing the lungs. The current density is highest in the muscles of the back. The peak current density in each layer of the

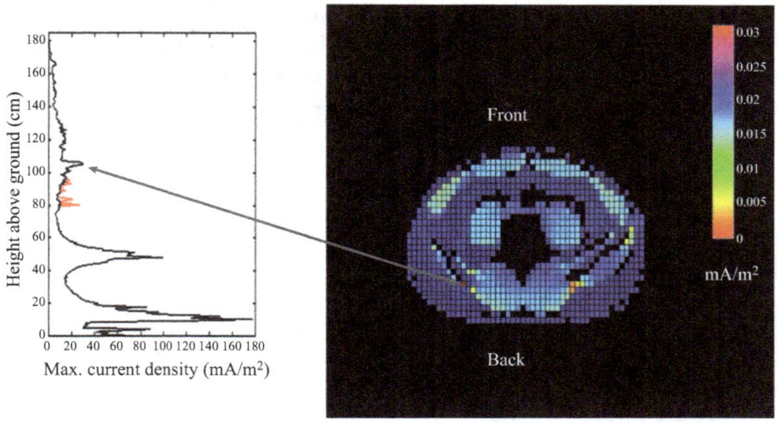

Figure 3.5 (Right) Current density in the body from a frontally-incident 60 Hz plane wave. (Left) Peak current density in each layer is shown on the left, and was originally published in [25]. The plot on the right shows the current density distribution and the location of the peak value in this layer of the body. The layer contains the lungs, and the peak values are seen in the muscles in the back

body is shown in the left plot, and the location for this layer is indicated. The impact of voxelization is seen clearly in this figure.

3.5 The $(FD)^2TD$ method

The frequency-dependent finite difference time domain ($(FD)^2TD$) method is used to evaluate multiple frequencies from a single FDTD run. The electrical properties (ε_r and σ) of biological tissue vary with frequency. If an FDTD simulation is to be run with multiple frequencies, such as using a pulsed source, this frequency variation must be considered. The complex permittivity is typically modeled with a Debye [17,27,28] or Lorentz [29] equation. Then one of two methods is used to calculate the fields. One method [27,28] is to convert the complex permittivity from the frequency-domain to the time-domain and recursively convolve it with the electric field FDTD equations. A second approach [17] is to add a differential equation relating the electric flux density $\boldsymbol{D}$ to the electric field $\boldsymbol{E}$ and solve this additional equation along with the standard FDTD equations. We will describe this second approach here.

We begin with Maxwell's equations in the form:

$$\nabla \times \overline{E} = -\mu \frac{\partial \overline{H}}{\partial t} \tag{3.13}$$

$$\nabla \times \overline{H} = \frac{\partial \overline{D}}{\partial t} \tag{3.14}$$

The flux density is related to the electric field according to $\overline{D} = \varepsilon^*(\omega)\overline{E}$, where $\varepsilon^*(\omega)$is the frequency-dependent complex permittivity. To convert this relationship to the time domain, we choose a rational function for $\varepsilon^*(\omega)$ such as a Debye equation with two relaxation constants (τ_1 and τ_2):

$$\varepsilon^*(\omega) = \varepsilon_o\left[\varepsilon_\infty + \frac{\varepsilon_{s1}-\varepsilon_\infty}{1+j\omega\tau_1} + \frac{\varepsilon_{s2}-\varepsilon_\infty}{1+j\omega\tau_2}\right] \tag{3.15}$$

Rearranging this equation, and assuming an $e^{j\omega t}$time dependence gives:

$$\tau_1\tau_2\frac{\partial^2\overline{D}}{\partial t^2} + (\tau_1+\tau_2)\frac{\partial\overline{D}}{\partial t} + \overline{D} = \varepsilon_o\left[\varepsilon_s\overline{E} + (\varepsilon_{s1}\tau_2 + \varepsilon_{s2}\tau_1)\frac{\partial\overline{E}}{\partial t} + \varepsilon_\infty\tau_1\tau_2\frac{\partial^2\overline{E}}{\partial t^2}\right] \tag{3.16}$$

where the dc (zero frequency) dielectric constant is

$$\varepsilon_s = \varepsilon_{s1} + \varepsilon_{s2} - \varepsilon_\infty \tag{3.17}$$

The second order differential equation in (3.16) is then converted to a second-order difference equations and solved along with the first-order differential equations for (3.13) and (3.14) in the iterative FDTD solution process. The second-order differences in time require use of two previous time steps. This roughly doubles the memory and computation time requirements for the FDTD simulation.

If strictly time domain information is desired, such as the pulsed response of a frequency dispersive system (all/most biological systems are frequency dispersive), the $(FD)^2TD$ method is the most efficient way to obtain this time-domain response. If frequency-domain information is required for only a small number of voxels, requiring minimal time-to-frequency domain transformations, this method is also idea. However, if frequency-domain information is required for a large number of voxels (for instance finding field, power or SAR, or current distributions throughout the body), using a set of single-frequency sinusoidal simulations with the 2E2U time-to-frequency domain conversion as described in Section 3.4.1 is more efficient for both memory and computational time.

3.6 The S-FDTD and GS-FDTD methods

As previously discussed, the electrical properties of tissues vary with age, individual, in vivo/in vitro status, species, etc. They also have normal variation across a single organ, and variation due to errors in measurement. When a source impinges on a body with these statistical variations, this will cause statistical variation in the electric fields, as well. These variations can be evaluated using the FDTD method in one of two ways. In a non-intrusive approach, multiple FDTD simulations are used, each with properties selected from a statistical pool. The original algorithm is unchanged, making it easy to apply these methods, however since they require multiple runs, this can be computationally time consuming. The gold standard for uncertainty analysis is the Monte Carlo method (MC-FDTD) [30], which is a non-invasive method. Other non-intrusive methods include polynomial chaos (PC-FDTD) [31] and stochastic collocation (SC-FDTD) [32]. They are more efficient than the FDTD method, requiring multiple FDTD runs, but converging faster than Monte Carlo methods, and giving up only a little accuracy compared to Monte Carlo. Even greater efficiency can be gained by using an intrusive method, which adds equations that represent the statistical variation of parameters to the FDTD algorithm and evaluates the resultant variation in fields using a single FDTD run [33,34]. Accuracy may be slightly reduced, however the tremendous gains in efficiency can enable statistical evaluations that might otherwise be neglected. In this section, we will describe a very efficient intrusive method, stochastic FDTD (S-FDTD).

The S-FDTD method estimates the standard deviation of fields caused by variation in the electrical properties [34,35]. The mean of the electric and magnetic fields is calculated with the standard FDTD method, and results are very similar to Monte Carlo evaluations. Variation in the fields is calculated with an additional set of stochastic equations. The S-FDTD approximations are very close to Monte Carlo for standard deviation of the fields [35,36]. The full 3D S-FDTD formulations and their derivations can be found in [37]. Although too extensive to include here, an example 3D S-FDTD equation for the standard deviation of the x-component of the electric field is:

$$
\begin{aligned}
\sigma\left\{E_x^{n+1/2}\left(i,j+\frac{1}{2},k+\frac{1}{2}\right)\right\} &= \left(\frac{1-\frac{\widehat{\sigma}\Delta t}{2\varepsilon}}{1+\frac{\widehat{\sigma}\Delta t}{2\varepsilon}}\right)\sigma\left\{E_x^{n-1/2}\left(i,j+\frac{1}{2},k+\frac{1}{2}\right)\right\} \\
&+\frac{4\Delta t\left(\widehat{\sigma}\rho_{\varepsilon,E}\sigma\{\varepsilon\}-\varepsilon\rho_{\widehat{\sigma},E}\sigma\{\widehat{\sigma}\}\right)}{(2\varepsilon+\Delta t\widehat{\sigma})^2}E_x^{n-1/2}\left(i,j+\frac{1}{2},k+\frac{1}{2}\right) \\
&+\frac{2\Delta t}{(2\varepsilon+\Delta t\widehat{\sigma})}\left[\frac{\sigma\{H_z^n(i,j+1,k+\frac{1}{2})\}-\sigma\{H_z^n(i,j,k+\frac{1}{2})\}}{\Delta y}\right. \\
&\left.-\frac{\sigma\left\{H_y^n(i,j+\frac{1}{2},k+1)\right\}-\sigma\left\{H_y^n(i,j+\frac{1}{2},k)\right\}}{\Delta z}\right] \\
&-\frac{2\Delta t\left(2\rho_{\varepsilon,E}\sigma\{\varepsilon\}-\Delta t\rho_{\widehat{\sigma},H}\sigma\{\widehat{\sigma}\}\right)}{(2\varepsilon+\Delta t\widehat{\sigma})^2}\left[\frac{H_z^n(i,j+1,k+\frac{1}{2})-H_z^n(i,j,k+\frac{1}{2})}{\Delta y}\right. \\
&\left.+\frac{H_y^n(i,j+\frac{1}{2},k+1)-H_y^n(i,j+\frac{1}{2},k)}{\Delta z}\right]
\end{aligned}
\tag{3.18}
$$

where σ represents standard deviation, and $\widehat{\sigma}$, ε, and μ are the tissue conductivity, permittivity, and permeability, respectively. Δ is the spatial or temporal resolution, where the prefix shows the type and direction. The spatial indices in the *x, y,* and *z* directions are shown using *i, j, k.* The temporal index is *n.* The correlation

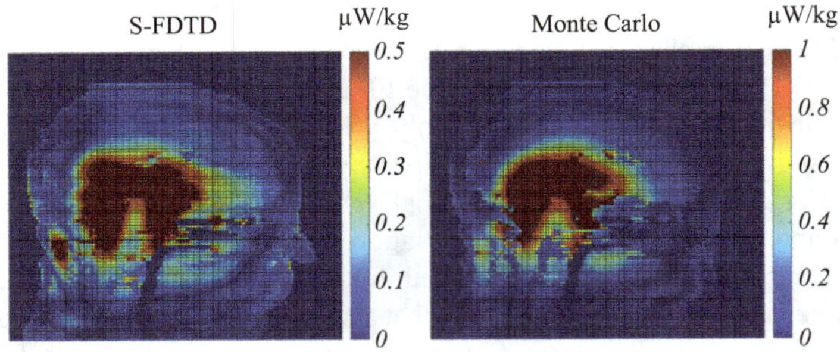

Figure 3.6 Standard deviation of the SAR for an 835 MHz half-wave dipole antenna next to the head, calculated using S-FDTD with a single run (left) and Monte Carlo (right) with 1,000 FDTD runs. From [39]

coefficient ρ is between two items specified by its subscript. The correlation coefficients vary between $-1 \leq \rho \leq 1$ and are approximated in the S-FDTD derivation. These approximations effect the results, and can be improved with appropriate choices of ρ [36,38]. An example comparison of standard deviation between S-FDTD and Monte Carlo is shown in Figure 3.6 [39] for an 835 MHz half-wave dipole antenna next to the head, using a correlation coefficient of 1 (which typically slightly over-estimates the variation).

The S-FDTD method has also been extended to geometrical variation [40,41], where it was used to estimate variation due to variation in skin thickness when evaluating the SAR for cell phone studies.

3.7 Dosimetry—What are we measuring?

In this chapter, we have described computational methods for dosimetry—determining the dose of electromagnetic radiation. This could be electric or magnetic field strength in or around a body from any variety of source(s), current or current density, absorbed or reflected power, SAR, antenna radiation pattern, etc. One overall consideration with computational models is the difference between localized values in individual voxels, and those averaged over a region or volume. Values in individual voxels can be very large, particularly very near localized sources. The smaller the voxel, the larger the value. Voxel values are, therefore, somewhat problematic. Instead, values averaged over a region or volume are used.

One example of values averaged over a region is the SAR. SAR is a measure of power per weight, and is given by:

$$\text{SAR} = \frac{\sigma|\overline{E}|^2}{2m_d}(\text{W/kg}) \tag{3.19}$$

where σ is the electrical conductivity, and m_d is the mass density. $|\overline{E}|$ is the magnitude of the electric field (typically found from converting time domain data

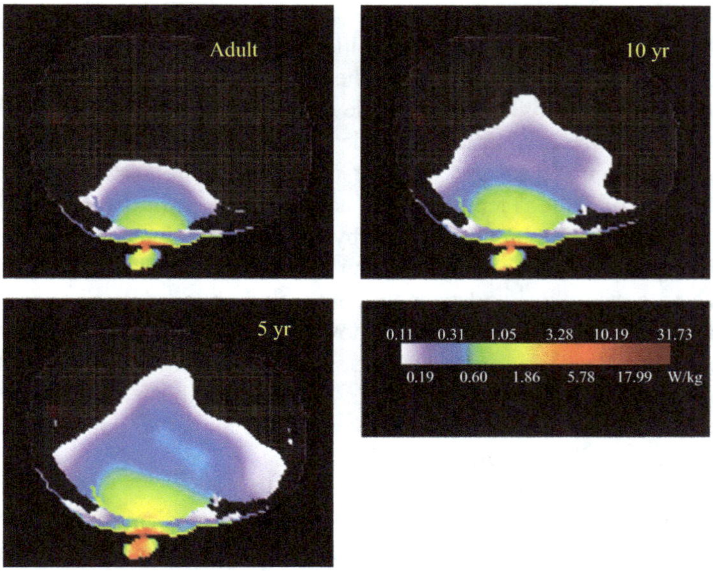

Figure 3.7 SAR in the human head of an adult, 10-year old, and 5-year old from an 835 MHz cellular telephone. The black and white versions of these figures were originally published in [42]

to the frequency domain using a Fourier transform or 2E2U method as described in Section 3.4.1). $|\overline{E}|^2/2$ is the root-mean-square (RMS) value of the electric field. Values of SAR can be found in each (*i, j, k*) voxel in a computational model. An example of voxel SAR values is shown in Figure 3.7 for a cellular telephone at 835 MHz [42]. Guidelines and regulations related to SAR consider the SAR averaged over 1-g or 10-g of tissue [43].

3.8 Dosimetry—future work and applications

Today's dosimetry relies on the computational electromagnetics research of the past. Each new application brings with it challenges that have to be assessed and overcome. These may be improvements in computational efficiency for running variational analyses, more efficient memory use or parallel processing for voxel models that are more and more detailed (and hence, more and more voxels), decisions and global agreements in what we are actually trying to measure and how we will agree to measure it, ever larger frequencies and broad/multi-frequency applications (such as 5G and on to 6G cellular applications), and more accurate models of the most electromagnetically or biologically significant aspects of the computational models. These are some of the aspects of computational research that continue to challenge and inspire creative new ideas and improvements to existing methods.

Most dosimetry calculations today are done with commercial software packages, many of which also allow multiphysics assessments such as combinations with mechanical or heat stresses. These packages are constantly evolving to include the new methods and models developed in the computational research community. They are easy to use, provide a great deal of flexibility in modeling and simulation approaches, and provide outputs that are readily viewable and immediately useful. But as tempting as it can be to use and trust these codes, they are only as good as the input provided by the user. Having a strong theoretical understanding of what the software is doing is essential to obtaining reliable results. Having a strong theoretical understanding of the underlying electromagnetics of the system, and what aspects are critical to how it functions, is critical to choosing a correct model and method for simulations, being sure that it models the effects of this physics correctly. Comparisons and validations with theoretical and experimental models, and comparisons between different modeling approaches, remain important aspects of computational dosimetry.

Bioelectromagnetic dosimetry remains challenged by the fact that we are often evaluating field levels that are too low to be of medical significance, evaluating guidelines that are based on heating rather than other more localized biological effects [44]. Maintaining relevant evaluations when field levels are medically insignificant can be difficult. In other cases, bioelectromagnetic dosimetry may be used to evaluate exposure conditions that produce observable effects, such as tissue healing, pain relief, hyperthermia, or nerve/muscle/brain stimulation. In these cases, the pathways that produce the expected effects may not be well understood, medically, so it can be difficult to know what parts of the model need to be modeled especially carefully. Each new medical application comes with its own unique challenges for assessment and evaluation of the biological significance of the results. This is one of the most interesting aspects of dosimetry, and often requires significant collaboration with biological and medical experts.

Bioelectromagnetic dosimetry is a relatively mature field, but as it is applied to new and interesting biological and medical applications, new challenges and nuances emerge, keeping dosimetric methods constantly new and evolving.

References

[1] V. Anderson and J. Rowley, "Tissue Dielectric Properties Calculator." 1998. [Online]. Available: https://www.arpansa.gov.au/sites/default/files/legacy/pubs/rps/tissues5.xls

[2] S. Gabriel, R. Lau, and C. Gabriel, "The dielectric properties of biological tissues: II. Measurements in the frequency range 10 Hz to 20 GHz," *Physics in Medicine and Biology*, vol. 41, no. 11, p. 2251, 1996.

[3] C. M. Furse, C. Waterman, and L. Griffiths, "To-average or not-to-average in FDTD modeling of dielectric interfaces," *Applied Computational Electromagnetics Society Journal*, vol. 21, no. 2, pp. 173–176, 2006.

[4] S. Chaudhary, R. Mishra, A. Swarup, and J. M. Thomas, "Dielectric properties of normal & malignant human breast tissues at radiowave & microwave frequencies," *Indian Journal of Biochemistry & Biophysics*, vol. 21, no. 1, pp. 76–79, 1984.

[5] C. Gabriel, S. Gabriel, and E. Corthout, "The dielectric properties of biological tissues: I. Literature survey," *Physics in Medicine and Biology*, vol. 41, no. 11, pp. 2231–2249, 1996.

[6] X. G. Xu and K. F. Eckerman, "Handbook of anatomical models for radiation dosimetry," in *Series in Medical Physics and Biomedical Engineering*, CRC Press LLC, London, 2009.

[7] O. P. Gandhi, "Millimeter-resolution MRI-based models of the human body for electromagnetic dosimetry from ELF to microwave frequencies," in Proceedings of the International Workshop on Voxel Phantom Development, 1995, 1995, pp. 24–31.

[8] V. Pandit, R. McDermott, G. Lazzi, C. Furse, and O. Gandhi, "Electrical energy absorption in the human head from a cellular telephone," in Visualization '96. Proceedings, 1996, pp. 371–374.

[9] C.-T. Hwang and R.-B. Wu, "Treating late-time instability of hybrid finite-element/finite-difference time-domain method," *IEEE Transactions on Antennas and Propagation*, vol. 47, no. 2, pp. 227–232, 1999.

[10] F. J. Meyer, D. B. Davidson, U. Jakobus, and M. A. Stuchly, "Human exposure assessment in the near field of GSM base-station antennas using a hybrid finite element/method of moments technique," *IEEE Transactions on Biomedical Engineering*, vol. 50, no. 2, pp. 224–233.

[11] A. Taflove, A. Oskooi, and S. G. Johnson, *Advances in FDTD Computational Electrodynamics: Photonics and Nanotechnology*, Artech House, London, 2013.

[12] K. Yee, "Numerical solution of initial boundary value problems involving Maxwell's equations in isotropic media," *IEEE Transactions on Antennas and Propagation*, vol. 14, no. 3, pp. 302–307, 1966.

[13] D. M. Sullivan, D. T. Borup, and O. P. Gandhi, "Use of the finite-difference time-domain method in calculating EM absorption in human tissues," *IEEE Transactions on Biomedical Engineering*, vol. BME-34, no. 2, pp. 148–157, 1987, doi: 10.1109/TBME.1987.326039.

[14] A. Taflove and S. C. Hagness, *Computational Electrodynamics*, Artech House, London, 2005.

[15] I. R. Capoglu and G. S. Smith, "A total-field/scattered-field plane-wave source for the FDTD analysis of layered media," *IEEE Transactions on Antennas and Propagation*, vol. 56, no. 1, pp. 158–169, 2008.

[16] G. Lazzi and O. P. Gandhi, "Realistically tilted and truncated anatomically based models of the human head for dosimetry of mobile telephones," *IEEE Transactions on Electromagnetic Compatibility*, vol. 39, no. 1, pp. 55–61, 1997, doi: 10.1109/15.554695.

[17] C. M. Furse, J.-Y. Chen, and O. P. Gandhi, "The use of the frequency-dependent finite-difference time-domain method for induced current and

SAR calculations for a heterogeneous model of the human body," *IEEE Transactions on Electromagnetic Compatibility*, vol. 36, no. 2, pp. 128–133, 1994.

[18] C. M. Furse, D. H. Roper, D. N. Buechler, D. A. Christensen, and C. H. Durney, "The problem and treatment of DC offsets in FDTD simulations," *IEEE Transactions on Antennas and Propagation*, vol. 48, no. 8, pp. 1198–1201, 2000.

[19] C. M. Furse, "Faster than Fourier: ultra-efficient time-to-frequency-domain conversions for FDTD simulations," *IEEE Antennas and Propagation Magazine*, vol. 42, no. 6, pp. 24–34, 2000, doi: 10.1109/74.894179.

[20] J.-P. Berenger, "Improved PML for the FDTD solution of wave-structure interaction problems," *IEEE Transactions on Antennas and Propagation*, vol. 45, no. 3, pp. 466–473, 1997.

[21] D. Crouch, "On the use of symmetry to reduce the computational requirements for FDTD analyses of finite phased arrays," *Microwave and Optical Technology Letters*, vol. 13, no. 3, pp. 123–128, 1996.

[22] J. Ren, O. P. Gandhi, L. R. Walker, J. Fraschilla, and C. R. Boerman, "Floquet-based FDTD analysis of two-dimensional phased array antennas," *IEEE Microwave and Guided Wave Letters*, vol. 4, no. 4, pp. 109–111, 1994.

[23] C. M. Furse, S. P. Mathur, and O. P. Gandhi, "Improvements to the finite-difference time-domain method for calculating the radar cross section of a perfectly conducting target," *IEEE Transactions on Microwave Theory and Techniques*, vol. 38, no. 7, pp. 919–927, 1990.

[24] C. M. Furse and O. P. Gandhi, "Why the DFT is faster than the FFT for FDTD time-to-frequency domain conversions," *IEEE Microwave and Guided Wave Letters*, vol. 5, no. 10, pp. 326–328, 1995.

[25] C. M. Furse and O. P. Gandhi, "Calculation of electric fields and currents induced in a millimeter-resolution human model at 60 Hz using the FDTD method," *Bioelectromagnetics*, vol. 19, pp. 293–299, 1998.

[26] D. N. Buechler, D. H. Roper, C. H. Durney, and D. A. Christensen, "Modeling sources in the FDTD formulation and their use in quantifying source and boundary condition errors," *IEEE Transactions on Microwave Theory and Techniques*, vol. 43, no. 4, pp. 810–814, 1995.

[27] R. Luebbers, F. P. Hunsberger, K. S. Kunz, R. B. Standler, and M. Schneider, "A frequency-dependent finite-difference time-domain formulation for dispersive materials," *IEEE Transactions on Electromagnetic Compatibility*, vol. 32, no. 3, pp. 222–227, 1990.

[28] D. M. Sullivan, "Frequency-dependent FDTD methods using Z transforms," *IEEE Transactions on Antennas and Propagation*, vol. 40, no. 10, pp. 1223–1230, 1992, doi: 10.1109/8.182455.

[29] R. J. Luebbers and F. Hunsberger, "FDTD for Nth-order dispersive media," *IEEE Transactions on Antennas and Propagation*, vol. 40, no. 11, pp. 1297–1301, 1992.

[30] M. N. O. Sadiku, *Monte Carlo Methods for Electromagnetics*, CRC Press, London, 2018.

[31] J. Wiart, E. Conil, A. Hadjem, M. Jala, P. Kersaudy, and N. Varsier, "Handle variability in numerical exposure assessment: The Challenge of the Stochastic Dosimetry," in 2013 7th European Conference on Antennas and Propagation (EuCAP), April 2013, pp. 1979–1981.

[32] O. Aiouaz, D. Lautru, M.F. Wong, *et al.*, "Uncertainty analysis of the specific absorption rate induced in a phantom using a stochastic spectral collocation method," *Annals of Telecommunications-annales des télécommunications*, vol. 66, no. 7, pp. 409–418, 2011.

[33] A. C. M. Austin and C. D. Sarris, "Efficient analysis of geometrical uncertainty in the FDTD method using polynomial chaos with application to microwave circuits," *IEEE Transactions on Microwave Theory and Techniques*, vol. 61, no. 12, pp. 4293–4301, 2013, doi: 10.1109/TMTT.2013.2281777.

[34] S. M. Smith and C. Furse, "A stochastic FDTD method for statistically varying biological tissues," in 2011 IEEE International Symposium on Antennas and Propagation (APSURSI), 2011, pp. 2274–2277.

[35] S. M. Smith and C. Furse, "Stochastic FDTD for analysis of statistical variation in electromagnetic fields," *IEEE Transactions on Antennas and Propagation*, vol. 60, no. 7, pp. 3343–3350, 2012.

[36] K. Masumnia-Bisheh, M. Ghaffari-Miab, and B. Zakeri, "Evaluation of different approximations for correlation coefficients in stochastic FDTD to estimate SAR variance in a human head model," *IEEE Transactions on Electromagnetic Compatibility*, vol. 59, no. 2, pp. 509–517, 2016.

[37] S. M. Smith, "Stochastic Finite-Difference Time-Domain," Ph.D. Dissertation, University of Utah, 2011.

[38] K. M. Bisheh, B. Z. Gatabi, and S. M. H. Andargoli, "Stochastic FDTD accuracy improvement through correlation coefficient estimation," *Waves in Random and Complex Media*, vol. 25, no. 2, pp. 154–169, 2015.

[39] K. Masumnia-Bisheh and C. Furse, "Estimating the variance of SAR in a 3D human head model using stochastic FDTD," in 2020 IEEE International Symposium on Antennas and Propagation and North American Radio Science Meeting, Montreal, QC, Canada, Jul. 2020, pp. 1989–1990. Accessed: Dec. 01, 2021. [Online]. Available: https://ieeexplore.ieee.org/document/9330224/

[40] K. Masumnia-Bisheh and C. Furse, "Bioelectromagnetic uncertainty analysis using geometrically stochastic FDFD method," *IEEE Transactions on Antennas and Propagation*, vol. 69, no. 4, pp. 2433–2436, 2020.

[41] K. Masumnia-Bisheh, K. Forooraghi, M. Ghaffari-Miab, and C. Furse, "Geometrically stochastic FDTD method for uncertainty quantification of EM fields and SAR in biological tissues," *IEEE Transactions on Antennas and Propagation*, vol. 67, no. 12, pp. 7466–7475, 2019.

[42] O. P. Gandhi, G. Lazzi, and C. M. Furse, "Electromagnetic absorption in the human head and neck for mobile telephones at 835 and 1900 MHz," *IEEE Transactions on Microwave Theory and Techniques*, vol. 44, no. 10, pp. 1884–1897, 1996.

[43] IEC/IEEE International Standard – Determining the Peak Spatial-Average Specific Absorption Rate (SAR) in the Human Body from Wireless Communications Devices, 30 MHz to 6 GHz – Part 1: General Requirements for using the Finite-Difference Time-Domain (FDTD) Method for SAR Calculations. 2017, pp. 1–86. [Online]. Available: https://ieeexplore-ieee-org.ezproxy.lib.utah.edu/stamp/stamp.jsp?tp=&arnumber=8088404

[44] O. P. Gandhi, "The ANSI radio frequency safety standard: its rationale and some problems," *IEEE Engineering in Medicine and Biology Magazine*, vol. 6, no. 1, pp. 22–25, 1987, doi: 10.1109/MEMB.1987.5006370.

Chapter 4
Phantom recipes for implanted and wearable antenna applications

Symeon Symeonidis[1], Chinthana J. Panagamuwa[1] and William G. Whittow[1]

4.1 Design principles of antennas in/on lossy media

The human body is a lossy medium; therefore, implanted antennas that operate at lower frequencies are desirable to engineers due to their propagation capabilities through the lossy tissues. This typically would require bigger antenna geometries, which are challenging to implement due to the limited available space inside the human body, hence miniaturisation techniques must be developed and measured in phantoms. Over the last 10 years, multiple authors have addressed these challenges in the literature and have proposed a variety of optimised antenna geometries for various applications. A good trade-off between frequency of operation, antenna efficiency and size were established in the Medical Devise Radiocommunications Service (MedRadio) frequency band at the 401–406 MHz range and the Industrial, Scientific and Medical (ISM) radio bands at the 2.44–2.46 GHz range [1]. Most of the proposed applications in the literature include incorporation of the antenna under the skin, fat or muscle tissues [2].

Many phantoms use a plastic shell which is then filled with a liquid where the electrical properties (relative permittivity and conductivity) resemble the average of the human body. Typical values are relative permittivity ~40 and conductivity ~1 S/m. The relative permittivity tends to decrease with frequency while the loss tangent increases with frequency. Hence different liquids with different electrical properties may be required for different frequencies. For implanted antennas, it is especially important to consider heterogenous phantoms which contain multiple materials to represent different tissues. The fields penetrate deeper at lower frequencies, hence for wearable antennas, it becomes more important to consider heterogenous phantoms compared to conventional homogeneous phantoms at these frequencies. The plastic shell can be problematic for measuring wearable antennas as this lacks realism of the quasi-conductive human skin, however it could emulate a layer of clothing.

[1]Wolfson School of Mechanical, Electrical and Manufacturing Engineering, Loughborough University, Loughborough, UK

Other studies include implants beneath the skull for monitoring brainwave signals [3], intraocular implementation of antennas for retinal prosthesis [4], ingestible capsule antennas for the monitoring human intestines [5] and antennas operating in the Global Positioning System (GPS) frequency range for the localization of people suffering from Alzheimer's disease [6].

The principles of consideration when designing phantoms for implanted and wearable antennas are presented in this Chapter. Section 4.2 discusses the electromagnetic properties of human body phantoms. Section 4.3 reports the tissue-mimicking phantoms found in the literature, and details nine phantom recipes consisting of off-the-shelf biocompatible ingredients and provides a test-bed for validation in terms of relative permittivity and conductivity. Finally, in Section 4.4, the concluding remarks of this chapter can be found.

4.2 The electromagnetic properties of the medium

Permittivity ε, permeability μ and conductivity σ of homogenous phantoms are complex quantities defined by the following formulas [7]:

$$\varepsilon = \varepsilon_0(\varepsilon'_r - j\varepsilon''_r)\mathrm{F/m} \tag{4.1}$$

$$\mu = \mu_0(\mu'_r - j\mu''_r)\mathrm{H/m} \tag{4.2}$$

$$\sigma = \sigma' - j\sigma''\mathrm{S/m} \tag{4.3}$$

where:

$$\varepsilon_0 = 8.85 \times 10^{-12}\mathrm{F \cdot m^{-1}} \tag{4.4}$$

$$\mu_0 = 4\pi \times 10^{-7}\mathrm{H/m} \tag{4.5}$$

ε_0 and μ_0 correspond to the permittivity and permeability of vacuum. ε''_r and μ''_r account for the losses due to the damping of the vibrating dipole moments. ε'_r and μ'_r are the relative permittivity and relative permeability of a medium with respect to the permittivity and permeability of a vacuum.

The values of ε, μ, and σ are frequency dependent. Therefore $\varepsilon = \varepsilon(\omega)$, $\sigma = \sigma(\omega)$ and $\mu = \mu(\omega)$ for $\omega = 2\pi f$.

The definition of real permittivity ε'_e and conductivity σ'_e can be given as follows:

$$\varepsilon'_e = \varepsilon'_r - \frac{\sigma''}{\omega\varepsilon_0} \tag{4.6}$$

$$\sigma'_e = \sigma' + \omega\varepsilon_0\varepsilon''_r \tag{4.7}$$

The definition of the complex effective permittivity ε_e is given by the substitution of the above formulas as follows:

$$\varepsilon_e = \varepsilon_0(\varepsilon'_e - \mathrm{j}\varepsilon''_e) = \varepsilon_0\left(\varepsilon'_e - \mathrm{j}\frac{\sigma'_e}{\omega\varepsilon_0}\right) \tag{4.8}$$

The dissipation of the EM power inside a medium can be given by the loss tangent tanδ [8], as shown below:

$$\tan\delta = -\frac{\mathrm{Re}\{\varepsilon_e\}}{\mathrm{Im}\{\varepsilon_e\}} = \frac{\sigma'_e}{\omega\varepsilon_0\varepsilon'_e}\delta = \sqrt{\frac{2}{\mu\omega\sigma'_e}} \tag{4.9}$$

δ corresponds to the skin depth for $\sigma'_e \gg \omega\varepsilon_0\varepsilon'_e$

The spatial frequency of the propagated wave inside a lossy medium is given by the wave number k_e:

$$k_e = \omega \tag{4.10}$$

The wavelength λ_ε inside the medium is given by:

$$\lambda_\varepsilon = \frac{2\pi}{\omega\sqrt{\mu_0\varepsilon_0}}\mathrm{Re}\left\{\sqrt{\frac{\omega\varepsilon_0}{\omega\varepsilon_0\varepsilon'_e - \sigma'_e j}}\right\} \tag{4.11}$$

4.2.1 Efficiency and radiated power of implanted antennas

The radiation efficiency of an antenna is the ratio between the power radiated from the antenna to the power delivered at the feed of the antenna. Highly efficient antennas have most of the total power delivered from the source radiated to their environment [9].

The radiation efficiency of the antenna is given by the equation below:

$$e_{\mathrm{rad}} = \frac{P_{\mathrm{rad}}}{P_{\mathrm{source}}} \tag{4.12}$$

The average power, P_{rad}, flowing through a spherical surface S_r of volume V is given by the following equations found in [10]:

$$P_{\mathrm{source}} = P_{\mathrm{rad}} + P_{\mathrm{abs}} = \oint_{S_r} Re\ \{S * \widehat{r}\}dS + \frac{\omega}{2}\int_V \varepsilon_0\varepsilon''_e|E|^2 dV \tag{4.13}$$

Where P_{abs} is the power absorption and S the pointing vector equal to $(\frac{1}{2}\overline{E}x\overline{H})$. Further expanding the equation above:

$$\begin{aligned} P_{\mathrm{source}} &= \oint_{S_r} Re\ \left\{S(1/r^2) + S(1/r^n)\right\} * \widehat{r}dS \\ &\quad + \frac{\omega}{2}\int_V \varepsilon_0\varepsilon''_e\left[|E|^2(1/r^2) + +|E|^2(1/r^n)\right]dV \\ &= P_{\mathrm{rad-FF}} + P_{\mathrm{rad-NF}} + P_{\mathrm{abs-FF}} + P_{\mathrm{abs-NF}} \end{aligned} \tag{4.14}$$

The $1/r^2$ dependency identifies exclusively the far-field component, while $1/r^n$, with $n > 2$, refers to the near-field components of the electromagnetic radiation. The power terms $P_{\mathrm{rad-FF}}$, $P_{\mathrm{rad-NF}}$, $P_{\mathrm{abs-FF}}$ and $P_{\mathrm{abs-NF}}$ can be defined to separate the radiated and absorbed power with respect of the nature ($1/r^n$ dependency) of the EM field. Obviously, $P_{\text{rad-NF}}$ is only appreciable within a small distance from the source.

Implanted and wearable antennas that are prone to near-field coupling due to the surrounding tissue materials have as a result low radiation efficiencies and increased power loss [11]. A way to reduce this coupling effect is by using insulators between the tissue and the antenna.

As observed in Figure 4.1, as the insulation increases in thickness, the efficiency of the implanted antenna increases accordingly. All e_{rad} values were measured for the same antenna depth inside the body phantom.

For implanted antennas, radiation from different parts of the conductor is not attenuated evenly. This is a result of the variable dielectric properties of the tissues around the antenna and the uneven geometry of the human body [13]. Therefore, the precise representation of the properties and shape of the human body is important for accurate prediction of the radiation patterns, and this is crucial in terms of designing an efficient radiator.

4.2.2 Bandwidth

The bandwidth in antenna theory describes the frequency range in which the antenna can properly radiate or receive energy and it is directly depended on the propagation medium [14]. For lossy media like the human body, a wider bandwidth phenomenon is typically observed. This phenomenon is a result of the dissipation of the near field of the radiator into the lossy medium. No matter how efficient a radiator may be, most of the provided power will be absorbed by the medium resulting in low efficiency antennas with wide band characteristics [15]. Usually, the wide bandwidth is generally an indication of losses and hence lower antenna efficiency for implanted antennas. Note, the antennas will have the same efficiency and bandwidth if they are transmitting or receiving.

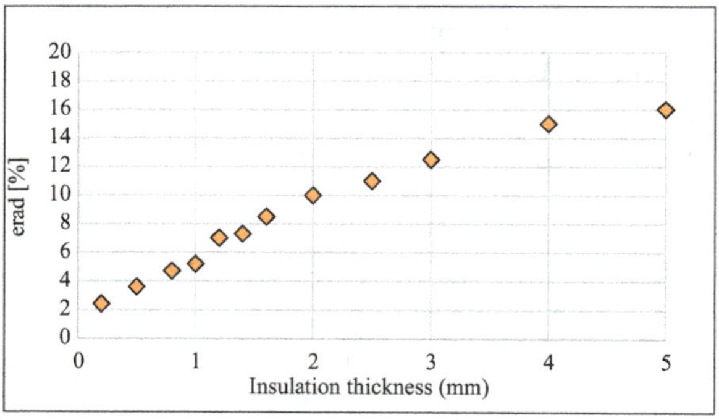

Figure 4.1 e_{rad} *values for different insulation thicknesses inside a body phantom, redrawn from [12]*

4.3 Review of tissue-mimicking phantoms

Since the earliest research regarding the interaction between the electromagnetic radiation and the biological tissues, a large amount of body phantoms has been proposed, designed, and developed [1,16,17].

The dielectric properties of the human tissues are well known over a wide frequency spectrum (2 Hz–100 GHz). Most of the research in the microwave range is well established in academia and industry due to the work of Gabriel [18].

According to investigations on the applications conducted in [19], the electromagnetic fields radiated out of the body is highly dependent to the geometry and composition of the test-bed phantom. Therefore, when designing a body phantom, it is important to reproduce close approximations of the dielectric properties of the lossy tissues of the human body and geometries for the investigated body section.

In order to sufficiently report the available phantoms proposed in the literature, it is necessary to divide them into two major categories. The first category is geometrical shaped phantoms while the second is anatomical shaped phantoms. Additionally, each category is followed by a sub-section of simulation based and measurement-based phantoms.

4.3.1 Geometrical phantoms

The most common geometries used for simulations and measurements of geometrical phantoms are canonical: cubical [1], cylindrical [16] or spherical [12]. These types are frequently used to emulate parts of the human body in order to validate the electromagnetic propagation of a radiator through a lossy medium.

4.3.2 Simulations of geometrical phantoms

In simulations, geometrical phantoms offer fast results by providing easy to interpret conditions for the measurements of implanted devices. This is due to the simplicity of their geometries. Examples of reported canonical geometries include a cylindrical phantom that was used for the validations of EM wave propagation through sea water [20] and spherical phantoms for the modelling of the human body [21], and the human head [17]. Another example shown in Figure 4.2 is a

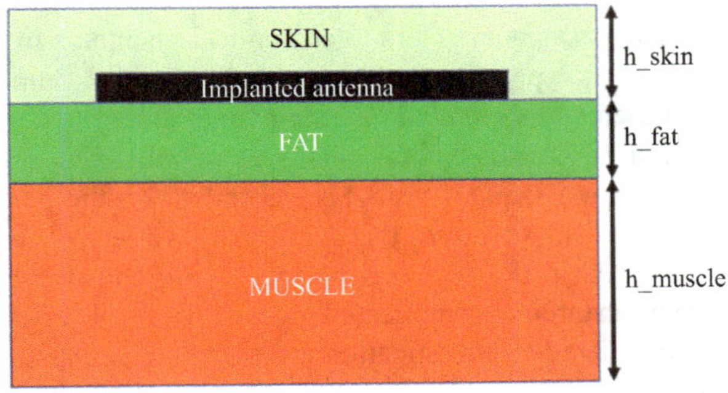

Figure 4.2 A three-layer canonical geometry redrawn from [22]

simulated cuboid phantom [22] used for the representation of the human body. It is also possible to use a combination of canonical geometries. This can offer a better imitation of the human body. Additional combinations of canonical geometries based on the human body emulation can be found in [23] for MedRadio band implants and at 2.4 GHz for on-body applications. Moreover, simulations of canonical geometries offer an easier transition to validations via measurements than anatomical phantom geometries. Each layer takes the shape of the container used, ensuring minimum dimensional error between the simulated and constructed phantoms, along with high repeatability for each experiment.

4.3.3 Measurements of geometrical phantoms

Phantoms with canonical geometries usually consist of homogenized liquids or gels with averaged dielectric properties according to the investigated area of the human head or body. They can be manufactured by mixing deionized water and specific amounts of sugar and/or salt to match the desired relative permittivity and conductivity [2,24].

It is worth mentioning that phantoms should be refrigerated to avoid evaporation or bacteria growing. The measurements should be carried out at room temperature as the dielectric properties change with temperature. Bactericides are often used in order to avoid bacterial growth and increase the lifespan of phantoms [25]. A liquid tissue phantom has been reported to reach over one-year long lifetime [26] and gel type phantoms has shown sufficient dielectric consistency for over 40 days [27,28]. In [29], solutions at 22 °C are reported to possess similar dielectric properties to real muscle tissue at 37 °C. Finally, in [28], recipes with more robust temperature tolerance are presented. The main disadvantage of geometrical homogeneous phantoms is the frequent unrealistic approach in terms of dimensions compared to the real human body tissue. The complex geometry of the human body can contribute to the creation of standing waves in the channel of propagation. This effect can cause discrepancies in comparisons of simulations of a realistic human model and the measurements in a geometrical phantom [30].

4.3.4 Phantoms with realistic dimensions (anatomical phantoms)

The realistic dimensions category refers to anatomical phantoms in relation to the targeted body tissue. This type of phantoms possesses complex geometry attributes and are mainly represented by simulations using voxel models.

Selection of realistic dimensions for a body phantom is very important as the efficiency of an implanted radiator is highly dependent on the position that it is placed inside the body [7]. Adding to that is the large variety of different human body types in nature that differ in terms of age, height and weight, making the appropriate choice of phantom a complicated process. These observations highlight the importance of a detailed investigation of anatomical phantoms [31] with designs that will be adjustable to the body type of the person about to receive an implant, such as the phantom presented in this thesis.

4.3.5 Simulations of anatomical phantoms

One of the most frequently reported realistic dimensions model is the visible human project (VHP) Hugo that has been studied thoroughly in the literature [32–34]. Hugo was an adult male who was frozen, sliced and photographed after execution. The MRI data and photographs were then combined to produce a detailed 3D model including his internal body structure. The final computer model of Hugo has more than 30 tissue types identified, with each one allocated a frequency dependant relative permittivity and conductivity. The advantage of this model, and anatomical phantoms in general, is that the electromagnetic propagation and behaviour of the implants inside the body can be studied in more detail. The main disadvantage compared to homogenous phantoms is increased simulation time as a consequence of the phantom's level of detail. Extended examples of realistic models can be found in [35–37]. Results on performance comparisons of implanted radiators found in [3,12,36,37]. Other regularly used models are the Japanese male and female body phantoms developed in [38], that allow modification of their postures [39]. These phantoms are also compared with the VHP model in [40] using implanted antennas.

An example of a voxel model can be seen in Figure 4.3. Similar types of voxel models like the VHP models found in [41] can feature up to 300 tissues and organs per model and a resolution of $0.5 \times 0.5 \times 0.5$ mm^3 throughout the entire body with the ability to adjust the posture. These models incorporate a complete human family based on magnetic resonance images allowing investigations of different body constitutions according to age and gender.

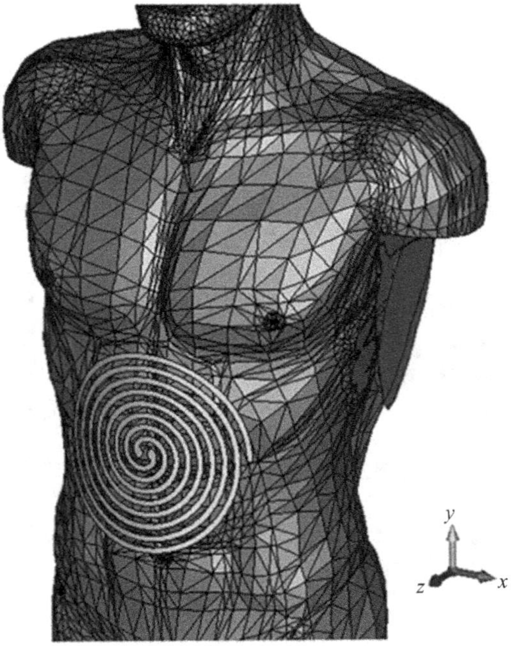

Figure 4.3 An example of a voxel model

4.3.6 Measurements of anatomical phantoms

In order to match the dimensions of the investigated real life tissues, specific animal parts, such as lamb, pork or rabbit joints are often used in literature [42–44]. These animal joints can provide good geometric detail when measuring implanted devices. Due to the process that commercially available meat undergoes (maturation in low humidity and blood draining), a known issue is that the dielectric properties of the investigated part may exhibit minor variations. This usually results in a drop of the average relative permittivity and conductivity of the animal part due to the lack of blood [45]. This may lead to differences between the simulated and measured properties of investigated phantoms [16].

4.3.7 Composition of human body phantoms

The composition of tissue-mimicking phantoms is highly reliant on the body part under investigation. For simulations and measurements of body parts in the microwave spectrum where one tissue layer takes up most of the available space, like the brain inside the human head, the most common phantoms used are homogenous [12,46]. In both simulations and measurements, multilayer phantoms are used when an investigation of a radiator implanted into a tissue layer with respect to the surrounding tissues is required [47]. Multilayer phantoms usually consist of bone, fat, and muscle layers [28].

While there is a large variety of simulated body phantoms, there are several limitations in the realisation of a human body phantom in measurements [48]. Several solid phantoms have been reported, as the example shown in Figure 4.4 that

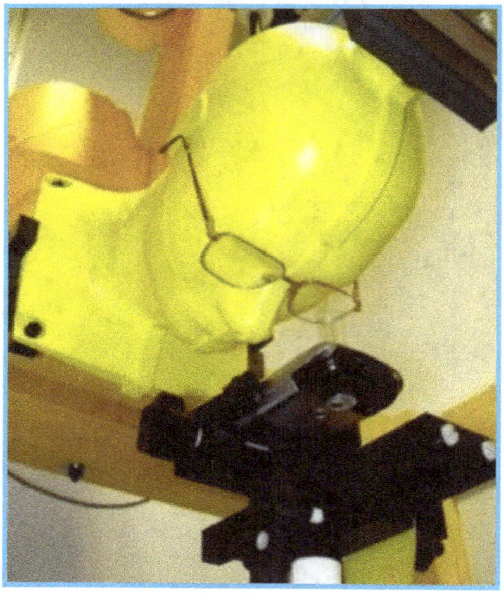

Figure 4.4 An example of realistic shape phantom

are anatomical and homogenous. The cast used for these measurements is given a human shape and filled with liquids that represent an approximation on the average dielectric properties of the body part.

Due to the large variation of the dielectric properties of human tissues in the microwave frequency range, developing multilayer phantom test-beds with dielectric properties similar to the desired human tissues in a wide frequency band is challenging. Detailed inhomogeneous phantoms with exact human body anatomy is almost impossible to achieve with liquids, and very hard with gels due to the complexity of the geometric characteristics and the multilayer nature of real life tissues [49].

In the literature, an ultra-wideband multilayer phantom with head equivalent properties can be found in [50] for the 0.2–20 GHz frequency range and in [51] featuring blood, fat, wet skin and muscle layers at 0.3–20 GHz for the monitoring of blood glucose levels. Such phantoms require the combination of multiple layers, each with a specific recipe and thickness according to its equivalent tissue. The recipes are either gels for high permittivity and conductivity tissues or flour based for lower values. Combining multiple recipes into a phantom can often result to undesirable mixing among the contact area of the layers [49,51,52]. This can affect the properties of each layer. Therefore, the use of high amounts of gelatine for high permittivity and conductivity recipes that require deionized water was followed as a process for the creation of the recipes of this chapter. The phantoms presented throughout the rest of this chapter are narrowband and wideband inhomogeneous geometric phantoms. The geometries of the proposed measured in-house developed phantoms consist of bone marrow and bone cortical with realistic dimensions and muscle with canonical dimensions. The dimensions of bone marrow and bone cortical materials match their real-life equivalent geometries of bone marrow and bone cortical and can be adjusted to match specific bone sizes and types. This is due to the ductile nature of their constituting materials. All the recipes for each tissue layer of the in-house developed phantoms presented in this chapter are frequency dispersive and wide band according to the IEEE specific absorption rate (SAR) measurement standards for the 0.3–2 GHz and the 2–3 GHz frequency spectrum [53].

4.3.8 Methodology and design of the in-house developed phantom recipes

Combining results from the literature along with experimental investigations for specific frequencies, bone marrow, bone cortical, blood and muscle phantom recipes were developed in-house. The dielectric behaviour of each ingredient based on experimentation can be summarized as: Deionized water has the highest permittivity and conductivity values. By adding sugar ($C_{12}H_{22}O_{11}$), the permittivity of the mixture reduces. Adding salt (NaCl) will increase its permittivity and conductivity; adding oil, either vegetable or olive, will decrease both these parameters over the desired frequency spectrum. It is important to mention that oil will not dissolve in water because water molecules form a polar covalent bond and oil molecules have non-covalent bonds. In order to mix oil and water, a detergent containing surfactants is required for the reduction of the tension between molecules of oil and water [54]. The detergent used for these measurements was a Fairy Lemon™ washing-up liquid containing 15–30% anionic surfactants and 5–15% non-anionic surfactants. Gelatine was used as a solidifying agent for the mixture.

4.3.9 Measurement test-bed for the dielectric parameters of the phantom recipes

For the measurements of the dielectric parameters of the recipes presented in this section, an open ended 50 Ohm RG-402 semi-rigid coaxial cable probe with an SMA connector was used (see Figure 4.5). This technique was chosen due to the low cost coaxial cable required and the accuracy of the results when measuring body tissues [55]. Measurements were taken with an Anritsu MS46524A vector network analyzer that was calibrated over the 0.5–4 GHz frequency range. The coaxial probe's dimensions were: 3 mm diameter and 9.5 cm length. As detailed in [56], the relative permittivity and conductivity of a liquid can be calculated by measuring the complex reflection coefficient Γ^* of three different calibration standards: open, short and deionised water. The open complex reflection coefficient Γ^* was recorded with the probe left open-ended in free space. The short-circuited Γ^* was recorded with an aluminium foil pushed against the cleaved end. The deionized water immersed Γ^* was measured with the probe dipped 3 cm into the liquid to ensure the majority of the radiated energy is absorbed in the water [57]. A detailed analysis on the operation principles of the probe can be found in [56]. In order to minimize any phase-related errors due to the flexible coaxial cable that was used, the probe was calibrated at the exact same position that the materials under measurement were placed. In addition to that, a soldering cable stand was used as the probe holder in close proximity to the materials under measurement.

The complex permittivity of deionized water was calculated using the Debye equation (3.1) [56]:

$$\varepsilon_w^* = \varepsilon' + j\varepsilon'' = \varepsilon_\infty + \frac{\varepsilon_S - \varepsilon_\infty}{1 + j\omega\tau} \tag{4.15}$$

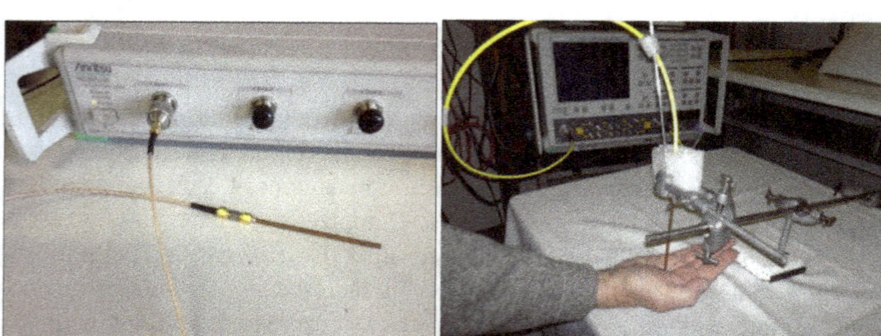

Figure 4.5 The probe used for the dielectric measurements

ε_w^*: complex relative permittivity of deionized water

ε', ε'': real and imaginary parts of the complex relative permittivity respectively

ε_∞: relative permittivity at high frequencies

ε_s: static relative permittivity at low frequencies

τ: relaxation time

According to the literature measurements of deionized water at room temperature from [58,59], $\varepsilon_\infty = 4.6$, $\varepsilon_s = 78.3$ and $\tau = 8.08$ ps. The complex relative permittivity ε_r^* and the effective conductivity σ_r can be calculated as:

$$\varepsilon_r^* = -\frac{\Delta_{r2} \times \Delta_{13}}{\Delta_{r1} \times \Delta_{32}} \varepsilon_w^* - \frac{\Delta_{r3} \times \Delta_{21}}{\Delta_{r1} \times \Delta_{32}} \tag{4.16}$$

$$\sigma_r = \omega \varepsilon_0 \varepsilon'' \tag{4.17}$$

$$\Delta_{ij} = \Gamma_i^* - \Gamma_j^* \tag{4.18}$$

Γ_1^*, Γ_2^*, Γ_3^*, Γ_r^*: Measured complex reflection coefficients for the short, open, deionized water and material under test respectively.

4.3.10 Measurements of homogenous tissue emulating phantom recipes

Refection coefficient measurements were taken with 300 mg of sample (this was applied for both liquid and solid samples) held in a 350 ml cylindrical plastic cup of 3.5 cm radius. Each measurement was taken by immersing the coaxial probe into the middle of the sample found at 3 cm depth from the surface, thus a good contact between the probe and the sample was achieved. The size of the container (350 ml, 3.5 cm radius) and the depth that the probe was immersed was selected according to [55,60,61] in order to ensure all of the radiated waves were confined inside the sample. A Microsoft© Excel spreadsheet produced by the authors of [57] that implemented the formulas (4.15) to (4.18) was used. The real and imaginary values of the probe's S_{11} for the open, short and deionised water were used as inputs and the relative permittivity and conductivity values of the sample under test were extracted and plotted as a function of frequency. The procedures given in the flowcharts of Figures 4.6 and 4.7 were followed to produce recipes for bone marrow, bone cortical, blood and muscle respectively. In these flowcharts, the recipes were compared with the real life values of the equivalent body tissues as measured by Gabriel *et al.* [18]. A similar process of ingredient mixing for producing representations of different body tissues, such as the human skin, can be found in [51].

A small amount of blue food colouring (1mg) was used in order to provide the blood-emulating liquid a distinct colour. This made it easier to observe the liquid's spread inside the phantom during experiments that will be detailed later in Chapter 8. Once the mixing was completed, the material was allowed to rest for 10 minutes. Then, the S_{11} was measured from 0.5 to 4 GHz three times using the probe. The three measurements were input into the Microsoft Excel spreadsheet

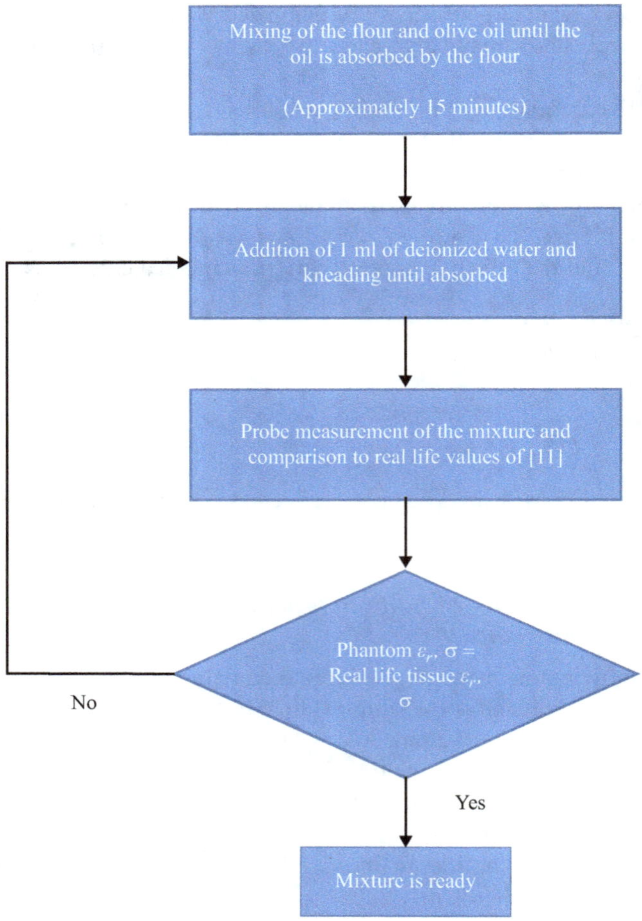

Figure 4.6 Bone marrow and bone cortical recipes development process

separately, yielding three sets of relative permittivity and conductivity curves for the material. The average of the three measurements was calculated and compared to the target curves taken from the published literature [18]. Following extensive testing, when the curve finally showed good agreement with the target values, the recipe was complete. The material was then reproduced twice more using that recipe and the S_{11} measurement procedure and proceeding Microsoft© Excel calculations were conducted in the same way. This yielded nine sets of relative permittivity and conductivity curves for the material. The average of all nine sets of relative permittivity and conductivity curves were then calculated and are henceforth referred to as the measured relative permittivity and measured conductivity. These are the values used in all proceeding phantom models of Chapter 8. This

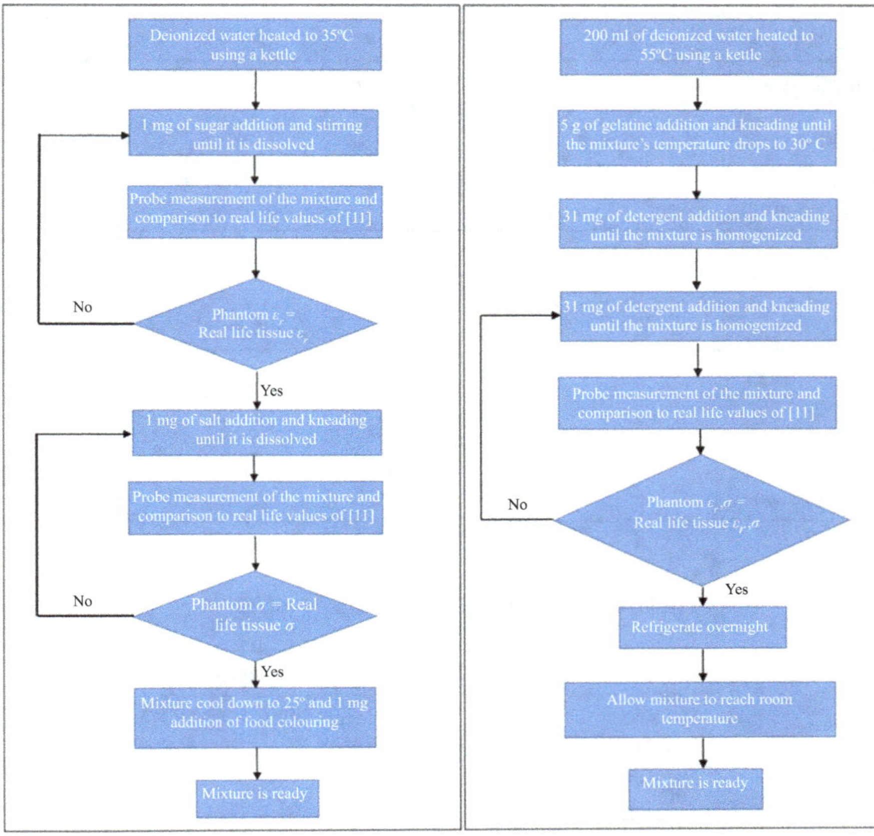

Figure 4.7 Blood emulating liquid recipe development process (left); broadband muscle recipe development process (right)

Table 4.1 Bone marrow recipes for 0.5–4 GHz

Ingredient (g)	**0.5–0.8 GHz**	**0.5–4 GHz**	**2–4 GHz**
Flour	80	80	80
Oil	33	33	33
Deionized water	–	1.51	31

detailed procedure was used to produce measured relative permittivity and conductivity curves for blood, muscle, bone cortical and bone marrow emulating material. To cover the full 0.5–4 GHz frequency band, two or three recipes were needed for each material (see Tables 4.1–4.4).

Table 4.2 Bone cortical recipes for 0.5–4 GHz

Ingredient (g)	0.5–1.5 GHz	0.5–4 GHz
Flour	80	80
Oil	30	30
Deionized water	9.6	10.5
Food colouring	7.2	7.2

Table 4.3 Blood recipes for 0.5–4 GHz

Ingredient (g)	0.5–1.5	0.5–4	3–4
Flour	114	91	70
Oil	2	1	–
Deionized water	164.5	164.5	151.5
Food colouring	1	1	1

Table 4.4 Muscle recipes for 0.5–4 GHz

Ingredient (g)	0.5–4 GHz
Deionized water	200
Gelatine	5
Detergent	31
Oil	31

Figure 4.8 Flour-based bone cortical recipe (left); deionized water-based muscle recipe held in a 350 ml cylindrical plastic cup of 3.5 cm radius (right)

The measured dielectric parameters (both narrowband and wideband) of blood, muscle, bone cortical and bone marrow are presented in the next series of graphs (see Figures 4.9–4.17). Data found in literature for the relevant body tissue [62] is also plotted for comparison purposes. In several of the following plots a green

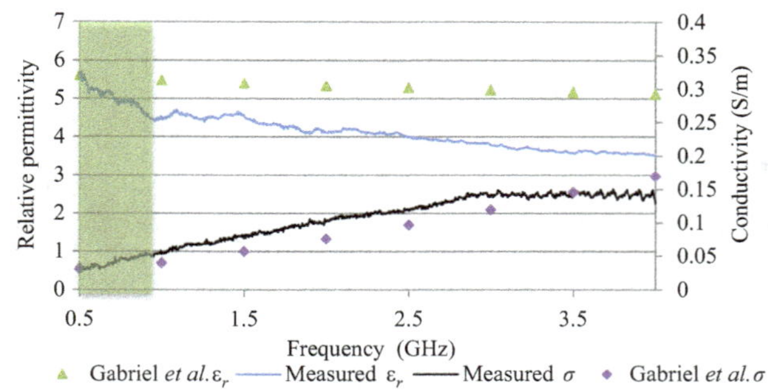

Figure 4.9 Relative permittivity and conductivity measurements compared to literature data for bone marrow at 0.5–0.8 GHz

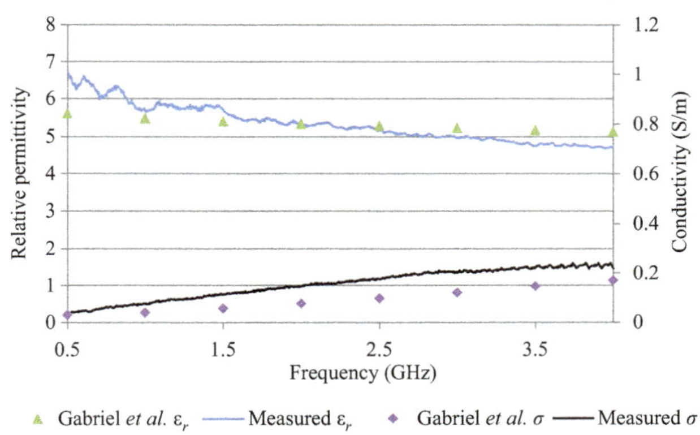

Figure 4.10 Relative permittivity and conductivity measurements compared to literature data for bone marrow at 0.5–4 GHz

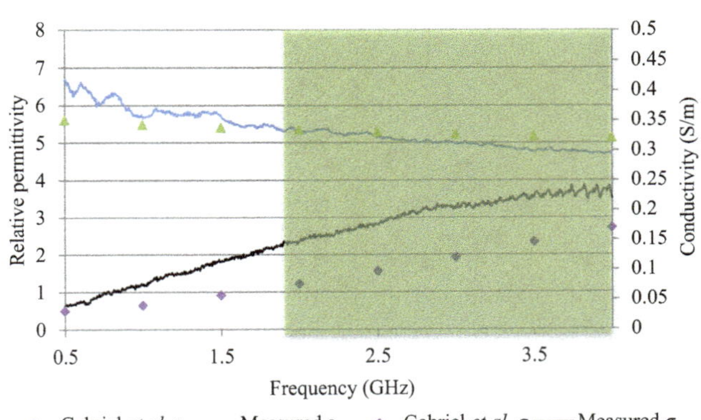

Figure 4.11 Relative permittivity and conductivity measurements compared to literature data for bone marrow at 2–4 GHz

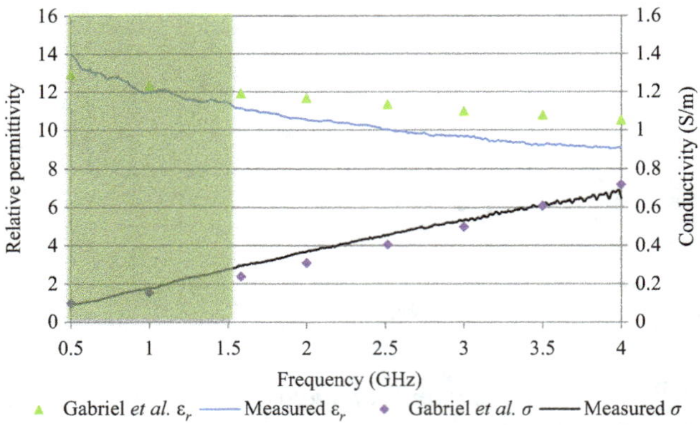

Figure 4.12 Relative permittivity and conductivity measurements compared to literature data for bone cortical at 0.5–1.5 GHz

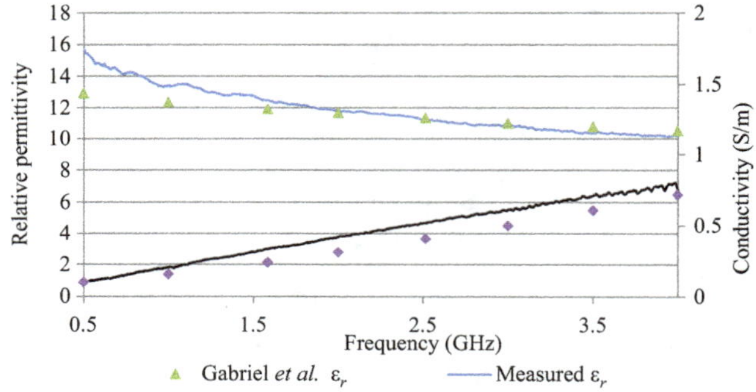

Figure 4.13 Relative permittivity and conductivity measurements compared to literature data for bone cortical at 0.5–4 GHz

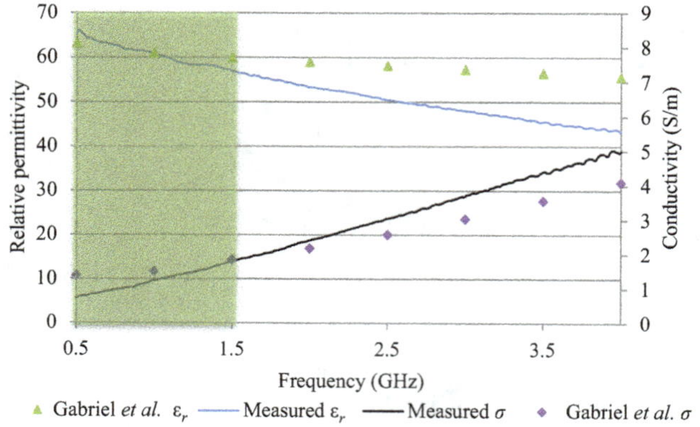

Figure 4.14 Relative permittivity and conductivity measurements compared to literature data for blood at 0.5–1.5 GHz

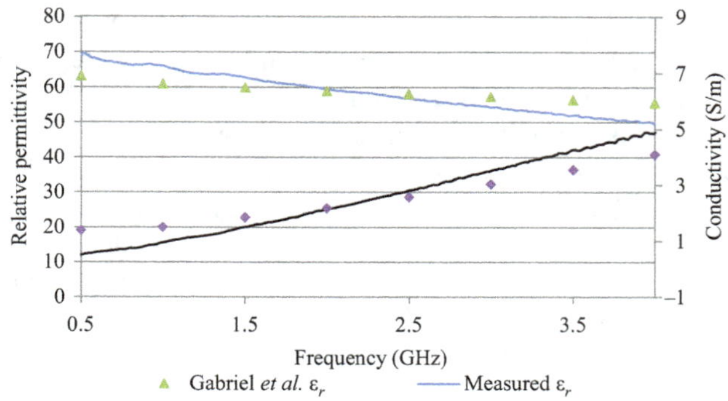

Figure 4.15 *Relative permittivity and conductivity measurements compared to literature data for blood at 0.5–4 GHz*

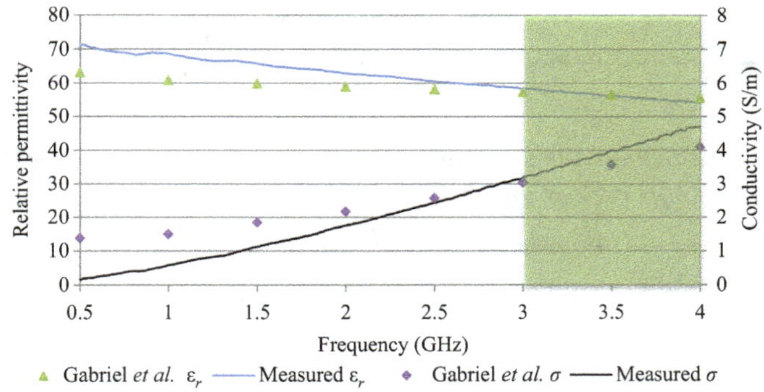

Figure 4.16 *Relative permittivity and conductivity measurements compared to literature data for blood at 3–4 GHz*

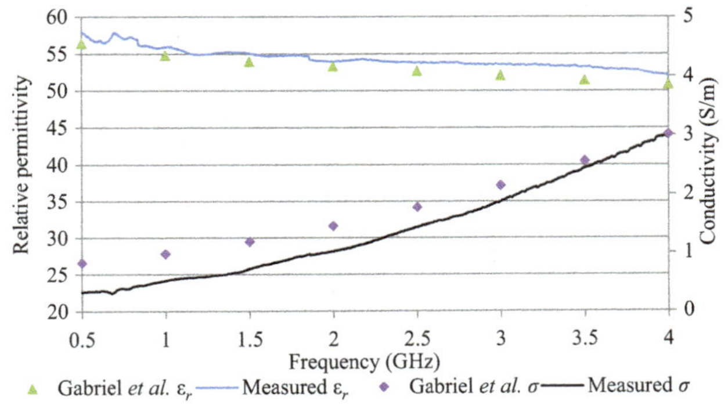

Figure 4.17 *Relative permittivity and conductivity measurements compared to literature data for muscle at 0.5–4 GHz*

shade is used to highlight the narrowband area of interest inside the 0.5–4 GHz frequency spectrum. In cases of where broadband recipes were investigated, the complete frequency spectrum was targeted.

Tables 4.5–4.8 describe the mean square error between the nine phantom recipes that were presented in this section and the measurements of Gabriel *et al.* [18] for their real-life equivalents. According to the IEEE SAR measurement standard for broadband phantoms [53], the measured conductivity and relative permittivity of recipes targeted between the 0.3 and 2 GHz frequency spectrum, should be within ± 5% of their real-life tissue equivalent values. For the 2–3 GHz,

Table 4.5 The calculated mean square error of the bone marrow phantom versus bone marrow data from [18]

Bone marrow phantom	**0.5–0.8 GHz**	**0.5–4 GHz**	**2–4 GHz**
Relative permittivity	0.09	0.03	0.04
Conductivity (S/m)	0.01	0.01	0.02

Table 4.6 The calculated mean square error of the bone cortical phantom versus bone cortical data from [32]

Bone cortical phantom	**0.5–1.5 GHz**	**0.5–4 GHz**
Relative permittivity	0.24	0.14
Conductivity (S/m)	0.02	0.01

Table 4.7 The calculated mean square error of the blood phantom versus blood data from [32]

Blood phantom	**0.5–1.5 GHz**	**1.5–3 GHz**	**0.5–4 GHz**
Relative permittivity	2.77	3.41	1.29
Conductivity (S/m)	0.12	0.13	0.45

Table 4.8 The calculated mean square error of the muscle phantom versus muscle data from [32]

Muscle phantom	**0.5–4 GHz**
Relative permittivity	2.39
Conductivity (S/m)	0.11

frequency spectrum measured conductivity should be within ±5% and relative permittivity cannot be more than ±10% from the target values. The mean square error for ε_r and σ of the measured tissue-mimicking phantoms compared to [18] that are found at Tables 4.5–4.8, shows that the recipes of Figures 4.10, 4.13, 4.15 and 4.17, that correspond to bone marrow, bone cortical, blood and muscle accordingly, exhibit a broadband behaviour at the frequency spectrum of 0.5–4 GHz [49,51,52]. The rest of the recipes found in Figures 4.11, 4.12, 4.14, and 4.16 can be used in measurements where targeting a specific frequency is required.

The ingredients of the presented recipes are easy to source and the procedure for producing the materials is straight forward. The mean-square error calculation has shown that the dielectric properties of four of the proposed phantom recipes exhibited broadband behaviour over the frequency spectrum of 0.5–4 GHz according to the IEEE standards for phantom development [53]. The narrowband recipes can be used in measurements where targeting a specific frequency is required. Finally, the phantoms' dielectric properties can be maintained for a maximum amount of 12 h in room temperature. Single material phantoms can be maintained for a maximum of five days within a refrigerator. At least 30 minutes are necessary for the refrigerated phantoms to return to room temperature before initiating the measurements.

4.4 Conclusions

This chapter has provided details on the most important aspects for consideration when designing antennas that operate inside lossy media and extensively referenced the relevant sources found in literature. Furthermore, a literature overview has been provided on the types of frequency dependent tissue-mimicking phantoms that are currently available. It has been highlighted that multi-material phantoms with broadband behaviour, without the undesirable mixing of their constituting ingredients is necessary for implanted antenna applications. Following this, nine different tissue-mimicking phantom recipes have been developed, each approximating the dielectric properties of bone marrow, bone cortical, blood and muscle at different frequencies constituting of off-the-shelf biocompatible ingredients. It has been shown that four of the proposed phantom recipes exhibited broadband behaviour at the frequency spectrum of 0.5–4 GHz according to the IEEE standards for phantom development [53]. The rest of the recipes are narrowband and can be used in measurements where targeting a specific frequency is required.

References

[1] W. Xia, K. Saito, M. Takahashi, and K. Ito, "Performances of an implanted cavity slot antenna embedded in the human arm," *IEEE Trans. Antennas Propag.*, vol. 57, no. 4, pp. 894–899, 2009, doi: 10.1109/TAP.2009.2014579.

[2] J. Abadia, F. Merli, J.-F. Zürcher, J.R. Mosig, and A.K. Skrivervik, "3D-spiral small antenna design and realization for biomedical telemetry in the MICS band," *Radioengineering*, vol. 18, no. 4, pp. 359–367, 2009.

[3] Z. N. Chen, G. C. Liu, and T. S. P. See, "Transmission of RF signals between MICS loop antennas in free space and implanted in the human head," *IEEE Trans. Antennas Propag.*, vol. 57, no. 6, pp. 1850–1854, 2009, doi: 10.1109/TAP.2009.2019933.

[4] K. Gosalia, M. S. Humayun, and G. Lazzi, "Impedance matching and implementation of planar space-filling dipoles as intraocular implanted antennas in a retinal prosthesis," *IEEE Trans. Antennas Propag.*, vol. 53, no. 8, pp. 2365–2373, 2005, doi: 10.1109/TAP.2005.852514.

[5] P. M. Izdebski, H. Rajagopalan, and Y. Rahmat-Samii, "Conformal ingestible capsule antenna: a novel chandelier meandered design," *IEEE Trans. Antennas Propag.*, vol. 57, no. 4, pp. 900–909, 2009, doi: 10.1109/TAP.2009.2014598.

[6] M. Z. Azad and M. Ali, "A miniature implanted inverted-F antenna for GPS application," *IEEE Trans. Antennas Propag.*, vol. 57, pp. 1854–1858, 2009, doi: 10.1109/TAP.2009.2016805.

[7] F. Merli, "Implantable antennas for biomedical applications," Ph.D. Thesis, pages 232, 2011, doi:10.5075/epfl-thesis-5110.

[8] S. Y. Huang, "Loss tangent," *Notes*, no. 5, pp. 8–9, 2012.

[9] W. G. Scanlon, J. B. Burns, and N. E. Evans, "Radiowave propagation from a tissue-implanted source at 418 MHz and 916.5 MHz," *IEEE Trans. Biomed. Eng.*, vol. 47, no. 4, pp. 527–34, 2000, doi: 10.1109/10.828152.

[10] C. A. Balanis, *Modern Antenna Handbook.* New York, NY: Wiley-Interscience, 2008.

[11] Y. Hao, A. Alomainy, P. S. Hall, Y. I. Nechayev, C. G. Parini, and C. C. Constantinou, "Antennas and propagation for body centric wireless communications," *2005 IEEE/ACES International Conference on Wireless Communications and Applied Computational Electromagnetics*, pp. 588–591, 2005, doi: 10.1109/WCACEM.2005.1469656.

[12] J. Kim and Y. Rahmat-Samii, "Implanted antennas inside a human body: simulations, designs, and characterizations," *IEEE Trans. Microw. Theory Tech.*, vol. 52, no. 8, pp. 1934–1943, 2004, doi: 10.1109/TMTT.2004.832018.

[13] R. Moore, "Effects of a surrounding conducting medium on antenna analysis," *IEEE Trans. Antennas Propag.*, vol. 11, no. 3, pp. 216–225, 1963, doi: 10.1109/TAP.1963.1138043.

[14] A. D. Yaghjian and S. R. Best, "Impedance, bandwidth, and Q of antennas," *IEEE Trans. Antennas Propag.*, vol. 53, no. 4, pp. 1298–1324, 2005, doi: 10.1109/TAP.2005.844443.

[15] T. Alves, B. Poussot, and L.-M. Laheurte, "A simple method to estimate the antenna efficiency over a lossy medium," *7th Eur. Conf. Antennas Propag.*, vol. 1, no. 3, pp. 3750–3753, 2013.

[16] T. Dissanayake, K. P. Esselle, and M. R. Yuce, "Dielectric loaded impedance matching for wideband implanted antennas," *IEEE Trans. Microw. Theory Tech.*, vol. 57, no. 10, pp. 2480–2487, 2009, doi: 10.1109/TMTT.2009.2029664.

[17] K. S. Nikita, G. S. Stamatakos, N. K. Uzunoglu, and A. Karafotias, "Analysis of the interaction between a layered spherical human head model

and a finite-length dipole," *IEEE Trans. Microw. Theory Techn.*, vol. 48, no. 11. pp. 2003–2013, 2000, doi: 10.1109/22.884189.

[18] C. Gabriel, "Compilation of the dielectric properties of body tissues at RF and microwave frequencies," *Environ. Heal.*, vol. 1, pp. 1–21, 1996, doi: Report N.AL/OE-TR- 1996-0037.

[19] P. S. Hall, Y. Hao, Y. I. Nechayev, *et al.*, "Antennas and propagation for on-body communication systems," *IEEE Antennas Propag. Mag.*, vol. 49, no. 3. pp. 41–58, 2007, doi: 10.1109/MAP.2007.4293935.

[20] W. Blair, "Experimental verification of dipole radiation in a conducting half-space," *IEEE Trans. Antennas Propag.*, vol. 11, no. 3. pp. 269–275, 1963, doi: 10.1109/TAP.1963.1138047.

[21] J. C. Lin, A. W. Guy, and C. C. Johnson, "Power deposition in a spherical model of man exposed to I-20-MHz electromagnetic fields," *IEEE Trans. Microwave Theory Techn.*, vol. 21, no. 12. pp. 791–797, 1973, doi: 10.1109/ TMTT.1973.1128133.

[22] C. J. Sanchez-Fernandez, O. Quevedo-Teruel, J. Requena-Carrión, L. Inclán-Sánchez, E. Rajo-Iglesias, and M. Kehn, "Dual-band implantable antenna based on short-circuit SRR," in *2010 Proceedings of the Fourth European Conference on Antennas and Propagation*, pp. 1–4, 2010.

[23] C. Miry, T. Alves, R. Gillard, J. M. Laheurte, R. Loison, and B. Poussot, "Analysis of the transmission between on-body devices using the bilateral dual-grid FDTD technique," *IEEE Antennas Wireless Propag. Lett.*, vol. 9. pp. 1073–1075, 2010, doi: 10.1109/LAWP.2010.2090645.

[24] D. Valderas, C. Schmidt, and X. Chen, "Broadband implanted UHF RFID antenna," *2010 IEEE Antennas and Propagation Society International Symposium*, no. 1, pp. 1–4, 2010.

[25] A. T. Mobashsher and A. M. Abbosh, "Artificial human phantoms: human proxy in testing microwave apparatuses that have electromagnetic interaction with the human body," *IEEE Microw. Mag.*, vol. 16, no. 6, pp. 42–62, 2015, doi: 10.1109/MMM.2015.2419772.

[26] G. Hartsgrove, A. Kraszewski, and A. Surowiec, "Simulated biological materials for electromagnetic radiation absorption studies," *Bioelectromagnetics*, vol. 8, no. 1, pp. 29–36, 1987.

[27] K. Ito, K. Furuya, Y. Okano, and L. Hamada, "Development and characteristics of a biological tissue-equivalent phantom for microwaves," *Electron. Commun. Japan (Part I Commun.)*, vol. 84, no. 4, pp. 67–77, 2001, doi: 10.1002/1520-6424(200104)84:4<67::AID-ECJA8>3.0.CO;2-D.

[28] K. Fukunaga, S. Watanabe, H. Asou, and K. Sato, "Dielectric properties of non-toxic tissue-equivalent liquids for radiowave safety tests," in *IEEE International Conference on Dielectric Liquids, 2005. ICDL 2005.* pp. 425–428, 2005, doi: 10.1109/ICDL.2005.1490116.

[29] C. K. A Chou, G. W. Chen, A. W. Guy, and K. H. Luk, "Formulas for preparing phantom muscle tissue at various radiofrequencies," *Bioelectromagnetics*, vol. 5, no. 4, pp. 435–441, 1984.

[30] S. Symeonidis, W. G. Whittow, and C. Panagamuwa, "Design and characterization of a three material anatomical bone phantom for implanted antenna applications," in *Loughborough Antennas & Propagation Conference (LAPC)*, 2017, pp. 4–8, doi: 10.1049/cp.2017.0253.
[31] P. B. Johnson, S. R. Whalen, M. Wayson, B. Juneja, C. Lee, and W. E. Bolch, "Hybrid patient-dependent phantoms covering statistical distributions of body morphometry in the U.S. adult and pediatric population," *Proceedings of the IEEE*, vol. 97, no. 12, pp. 2060–2075, 2009, doi: 10.1109/JPROC.2009.2032855.
[32] A. Alomainy and Y. Hao, "Modeling and characterization of biotelemetric radio channel from ingested implants considering organ contents," *IEEE Trans. Antennas Propag.*, vol. 57, no. 4, pp. 999–1005, 2009, doi: 10.1109/TAP.2009.2014531.
[33] J. H. Jung, S. W. Kim, Y. S. Kim, and S. Y. Kim, "Electromagnetic propagation from the intestine-ingested source in the human body model," *IEEE Trans. Antennas Propag.*, vol. 58, no. 5, pp. 1683–1688, 2010, doi: 10.1109/TAP.2010.2044338.
[34] L. Xu and M. Q. H. Meng, "Effects of dielectric parameters of human body on specific absorption rate for Ingestible Wireless Device at operating frequency of 430 MHz," *IEEE/ASME International Conference on Advanced Intelligent Mechatronics, AIM*, vol. 56, no. 8, pp. 215–220, 2009, doi: 10.1109/AIM.2009.5230011.
[35] P. Dimbylow, T. W. Dawson, K. Caputa, and M. A. Stuchly, "Current densities in a 2 mm resolution anatomically realistic model of the body induced by low frequency electric fields," *Phys. Med. Biol.*, vol. 45, pp. 1013–1022, 2000.
[36] I. G. Zubal, C. R. Harrell, E. O. Smith, Z. Rattner, G. Gindi, and P. B. Hoffer, "Computerized three-dimensional segmented human anatomy," *Med. Phys.*, vol. 21, no. 2, pp. 299–302, 1994, doi: 10.1118/1.597290.
[37] "Center for NMR Research – Penn State College of Medicine Research".
[38] T. Nagaoka, S. Watanabe, K. Sakurai, *et al.* "Development of realistic high-resolution whole-body voxel models of Japanese adult males and females of average height and weight, and application of models to radio-frequency electromagnetic-field dosimetry," *Phys. Med. Biol.*, vol. 49, no. 1, pp. 1–15, 2004, doi: 10.1088/0031-9155/49/1/001.
[39] T. Nagaoka and S. Watanabe, "Postured voxel-based human models for electromagnetic dosimetry," *Phys. Med. Biol.*, vol. 53, no. 24, pp. 7047–61, 2008, doi: 10.1088/0031-9155/53/24/003.
[40] A. Sani, A. Alomainy, and Y. Hao, "Numerical characterization and link budget evaluation of wireless implants considering different digital human phantoms," *IEEE Trans. Microw. Theory Techn.*, vol. 57, no. 10. pp. 2605–2613, 2009, doi: 10.1109/TMTT.2009.2029770.
[41] A. Christ, W. Kainz, E. G. Hahn, *et al.*, "The virtual family—development of surface-based anatomical models of two adults and two children for

dosimetric simulations," *Phys. Med. Biol.*, vol. 55, no. 2, pp. N23–N38, 2010, doi: 10.1088/0031-9155/55/2/N01.

[42] M. Khan, S. P. S. M. A. Sirdeshmukh, and K. Javed, "Evaluation of bone fracture in animal model using bio-electrical impedance analysis," *Perspect. Sci.*, vol. 8, pp. 567–569, 2016, doi: 10.1016/j.pisc.2016.06.022.

[43] T. Yoshida, W.-C. Kim, K. Kawamoto, T. Hirashima, Y. Oka, and T. Kubo, "Measurement of bone electrical impedance in fracture healing," *J. Orthop. Sci.*, vol. 14, no. 3, pp. 320–329, 2009, doi: 10.1007/s00776-009-1333-z.

[44] T. Yoshida, W.-C. Kim, Y. Oka, N. Yamada, and T. Kubo, "Assessment of distraction callus in rabbits by monitoring of the electrical impedance of bone," *Acta Orthop.*, vol. 81, no. 5, pp. 628–633, 2010, doi: 10.3109/17453674.2010.519910.

[45] J. L. Damez, S. Clerjon, S. Abouelkaram, and J. Lepetit, "Beef meat electrical impedance spectroscopy and anisotropy sensing for non-invasive early assessment of meat ageing," *J. Food Eng.*, vol. 85, no. 1, pp. 116–122, 2008, doi: 10.1016/j.jfoodeng.2007.07.026.

[46] T. Karacolak, A. Z. Hood, and E. Topsakal, "Design of a dual-band implantable antenna and development of skin mimicking gels for continuous glucose monitoring," *IEEE Trans. Microw. Theory Tech.*, vol. 56, no. 4, pp. 1001–1008, 2008, doi: 10.1109/TMTT.2008.919373.

[47] R. King, S. Prasad, and B. Sandler, "Transponder antennas in and near a three-layered body," *Microw. Theory Tech.*, vol. 28, no. 6, pp. 586–596, 1980.

[48] K. Ito, "Human body phantoms for evaluation of wearable and implantable antennas," *The Second European Conference on Antennas and Propagation, EuCAP 2007*. pp. 1–6, 2007, doi: 10.1049/ic.2007.1117.

[49] A. Sani, M. Rajab, R. Foster, and Y. Hao, "Antennas and propagation of implanted RFIDs for pervasive healthcare applications," *Proc. IEEE*, vol. 98, no. 9, pp. 1648–1655, 2010, doi: 10.1109/JPROC.2010.2051010.

[50] Y. Okano, K. Ito, I. Ida, and M. Takahashi, "The SAR evaluation method by a combination of thermographic experiments and biological," *IEEE Trans. Microw. Theory Tech.*, vol. 48, no. 11, pp. 2094–2103, 2000, doi: 10.1109/22.884200.

[51] T. Yilmaz, R. Foster, and Y. Hao, "Broadband tissue mimicking phantoms and a patch resonator for evaluating noninvasive monitoring of blood glucose levels," *IEEE Trans. Antennas Propag.*, vol. 62, no. 6, pp. 3064–3075, 2014.

[52] R. Augustine, S. Raman, and A. Rydberg, "Microwave phantoms for craniotomy follow-up probe development," *2014 IEEE Conference on Antenna Measurements & Applications (CAMA)*, pp. 1–2, 2014, doi: 10.1109/CAMA.2014.7003308.

[53] IEEE Standards Coordinating Committee 34, IEEE *Recommended Practice for Determining the Peak Spatial-Average Specific Absorption Rate (SAR) in the Human Head from Wireless Communications Devices: Measurement*

Techniques, February. The Institute of Electrical and Electronics Engineers, Inc., 2003.

[54] J. L. Salager, A. M. Forgiarini, and J. Bullón, "How to attain ultralow interfacial tension and three-phase behavior with surfactant formulation for enhanced oil recovery: a review. Part 1. *Optimum formulation for simple surfactant-oil-water ternary systems,"* *J. Surfactants Deterg.*, vol. 16, no. 4, pp. 449–472, 2013, doi: 10.1007/s11743-013-1470-4.

[55] D. M. Hagl, D. Popovic, S. C. Hagness, J. H. Booske, and M. Okoniewski, "Sensing volume of open-ended coaxial probes for dielectric characterization of breast tissue at microwave frequencies," *IEEE Trans. Microw. Theory Tech.*, vol. 51, no. 4 I, pp. 1194–1206, 2003, doi: 10.1109/TMTT.2003.809626.

[56] T. P. Marsland and S. Evans, "Dielectric measurements with an open-ended coaxial probe," *IEEE Proc. Microw. Antennas Propag.*, vol. 134, no. 4, p. 341, 1987, doi: 10.1049/ip-h-2.1987.0068.

[57] C. J. Panagamuwa, I. Howells, and W. G. Whittow, "Conductivity and permittivity measurements of children and adult's hands covering mobile communications frequency bands," in *Progress in Electromagnetics Research Symposium (PIERS)*, 2013, pp. 1–5.

[58] H. P. Schwan, R. J. Sheppard, and E. H. Grant, "Complex permittivity of water at 25°C," *J. Chem. Phys.*, vol. 64, no. 5, pp. 2257–2259, 1976, doi: 10.1063/1.432416.

[59] V. Komarov, S. Wang, and J. Tang, "Permittivity and measurements," *Encycl. RF Microw. Eng.*, pp. 1–19, 2005, doi: 10.1002/0471654507.eme308.

[60] M. Wolf, R. Gulich, P. Lunkenheimer, and A. Loidl, "Broadband dielectric spectroscopy on human blood," *Biochim. Biophys. Acta*, vol. 1810, pp. 727–740, 2011, doi: 10.1016/j.bbagen.2011.05.012.

[61] M. H. Moussa, M. A. Abu-Khousa, and N. N. Qaddoumi, "Measurement of liquids dielectric property using monopole probes operating at microwave frequencies," in *Proceedings of the IEEE International Conference on Electronics, Circuits and Systems*, vol. 3, pp. 1204–1207, 2003, doi: 10.1109/ICECS.2003.1301729.

[62] R. W. Lau and C. Gabriel, "The dielectric properties of biological tissues: III. Parameteric models for the dielectric spectrum of tissues," *Phys. Med. Biol.*, vol. 41, pp. 2271–2293, 1996, doi: 10.1088/0031-9155/41/11/003.

Chapter 5

Antennas for implants: design and limitations

Anja K. Skrivervik[1], Marko Bosiljevac[2] and Zvonimir Sipus[2]

Wireless communication with an implanted sensor represents a huge challenge due to the large losses in the link. A key element in such a system is the implanted antenna, which has an important impact on the global efficiency of the communication link. The design of implantable antennas presents two main challenges: first, the miniaturization, as the available space in an implantable capsule is usually much smaller than the wavelength. Second, the antenna radiates into a lossy environment, which is very different from the lossless free space environment we know from classic antenna theory. This has a deep impact not only on the link performance but also more fundamentally on the way antenna radiation performance is described and on the bounds on the performance of these antennas. In this chapter, we will gain insight into the main fundamental differences between classic electrically small antennas and implantable antennas. We will then study some canonical cases to understand the loss mechanisms and propose certain physical bounds on the efficiency of implantable antennas. Based on these results, we will propose design rules, illustrated by realistic examples. Finally, we will discuss issues linked to the measurement of implantable antennas.

5.1 Introduction

Diagnosis and treatment of a wide range of medical conditions is possible today using real-time biotelemetric data, which is gathered using various implantable, injectable or ingestible wireless medical devices. In recent years in particular, the field of use of wireless medical devices has been continuously enlarged providing new methods and devices with the aim of maintaining mobility and quality of life for patients [1]. This was made possible through advances in microelectromechanical and microfluidic systems, which combined with continuous miniaturization of electronic devices, have generated a wide range of innovations in biotelemetry. Transmission of various diagnostic data (endoscopic imagery,

[1]Microwave and Antenna Group, Ecole Polytechnique Fédérale de Lausanne, Lausanne, Switzerland
[2]Faculty of Electrical Engineering and Computing, University of Zagreb, Zagreb, Croatia

temperature, pressure, pH, glucose levels, etc.) is the most common application; however, typical implanted medical devices also include pacemakers, defibrillators, glucometers, cochlear implants and drug delivery systems. This list is continuously expanded due to advances in miniaturization of electronic devices and new fields are being opened, such as neural implants and brain–machine interfaces [2,3]. A review on implantable antennas can be found in [4–8] for classic applications, and in [9,10], for neural and brain implant applications. In addition, there is also an increased interest in single-cell identification through RF-ID, leading to extremely small antennas implanted in living cells [11,12].

To achieve their full potential, in-body devices require a reliable and efficient communication link with an external monitoring or controlling device. The most common technique is based on magnetic induction; however, novel applications require communication systems with improved characteristics. In particular, this means increased bandwidth, improved range for the external receiver, additional link security and even more advanced functions, which has given rise to fully wireless RF communication systems for implanted devices. The first standard frequency band dedicated to these applications was the Medical Implant Communication Service band (402–405 MHz). The Medical Body Area Network band (2,360–2,400 MHz) was introduced when the requirements increased. Today, these bands are primary when considering the longest range and sufficient data rates for implanted wireless devices [8].

Developing an active wireless implant is very challenging. One of the major problems is the energy efficiency, since many implanted devices are battery powered. The power consumption of the transceiver needs to be minimal to ensure that during the lifetime of the device no surgery for battery replacement is needed. This implies that the used communication protocols and electronic circuits must consume very low power and that the antenna radiation efficiency should be carefully taken into account in the overall system design of the implant. Related to the antenna efficiency is the requirement of the minimal volume for the device, since the implant needs to be safe and unobtrusive for the patient, and allow a comfortable implantation process.

Wireless power transfer (WPT) systems can solve the energy efficiency problem of the implanted device; however, WPT techniques, which have reasonable efficiency in most applications in lossless media, can exhibit highly deteriorated performances when implantable devices are considered. A careful design of the WPT system dedicated to in-body scenarios is needed to reach the required performance and efficiency levels. Due to the properties of the body tissue, which is very lossy and heterogeneous, this is very challenging and requires an application specific design of the implant transceiver and the access point on the body surface.

In general, sensor devices with antennas in living bodies can be either injected, ingested or implanted [4,8,13], but as the way of getting inside the body does not have an impact on the main radiating characteristics of the antenna, we will refer to all as "implantable." Note that the antenna's surrounding will depend on the means by which the antenna got into the host body, as ingestible antennas will be found in the digestive tracks, injectable will be subcutaneous and implantable will be found anywhere in the body.

Antennas implanted in living tissues date back as far as the 60s [1–14] and since then many designs have been proposed in literature. Early work on implantable antennas concerned mostly antennas for therapeutic or sensing applications [15–18], where propagation over a certain distance is not an issue as antennas work in their near field. However, further designs of small implantable sensor capsules faced the problem that an implant antenna's performance is strongly dependent on the tissue immediately surrounding it [19] with potential radiation pattern fragmentation, polarization distortion, reduced radiation efficiency, and changes in antenna input impedance [20].

Depending on the application and technique of implantation, implants will have dimensions going from a few cm^3 to a few mm^3 and even μm^3 for cellular implants. This of course affects the dimension of the antennas on those devices, which in the vast majority of cases will be electrically small to electrically very small. It is known since many decades that decreasing the dimensions of an antenna will have a negative impact on its radiating performances and that a good design will always be the best compromise between size and performance for a specific application [21]. Good designs require benchmarks to compare their performances to, which can be obtained through physical bounds to the radiation performance of an electrically small antenna (ESA) with respect to its size. This topic has been studied since the 1940s, in the seminal works of Chu and Wheeler [22,23], followed by many others [24–28]. It is noticeable that most of the results obtained in these studies are based on a spherical wave expansion (SWE) of the electromagnetic fields radiated by the antennas. The boom of wireless communication triggered by the boom of mobile phones in the 1990s of last century has induced a renewal in studies of bounds for ESAs, first with a reformulation of the results obtained earlier and thus still based on SWE (see for instance [29]). This was followed in the first decade of this new millennium by new work taking into account the form factor of the antenna [30], and the work of Gustafsson enabling the computation of physical bounds of antennas based on their static characteristics [31–33].

In all these studies on classic ESAs, the antennas radiate into free space, or at least, a lossless medium. The bounds achieved, and thus the experience and the rules obtained in the design of ESAs over all these decades do not apply in the case of implants, which radiate into a high loss medium, biological tissues. More importantly, not even the key performances indicators (quality factor and gain) used for classic ESAs apply to implantable antennas, for which the key radiation characteristic is the amount of power, which reaches outside the host body. It appears thus obvious that new physical bounds are required for implanted antennas in order to obtain efficient design rules.

Moore was among the first to consider the impact of radiation into lossy media on antenna performance [34], demonstrating the effect of seawater on submarine antenna radiation. He showed that fundamental antenna characteristics like the far field, the antenna radiation pattern, or the bandwidth do not apply when the antenna radiates into an infinite lossy medium. More recently, the optimal frequency for an implant depending on size, depth of implant, and type of antennas has been studied [35–38]. An initial attempt to provide a physical bound on the radiated power reaching the lossless space outside the host body of an implantable antenna was proposed in [39]. It is noticeable that this initial work on bounds for implantable antennas is based on a SWE of the radiated fields, as were the founding studies on bounds for ESAs [22–29]. As these pioneering works not only gave bounds to the performance of ESAs in free space but also enabled the understanding of the physics underlying their behaviors, the question arises if parallels can be drawn with implantable antennas, and thus physical insight gained. This is what we will address in this chapter. In Section 5.2, we remind the key performance indicators on ESAs, and the physical bounds obtained for these antennas using SWE. In Section 5.3, we will use SWEs to explain the loss mechanisms for implantable antennas and differentiate between unavoidable propagation losses through the lossy medium, and avoidable losses due to near field coupling between the electromagnetic fields near the implanted antenna and the lossy medium hosting it. Initial bounds for efficiency will be proposed, based on the work presented in [39,40]. In Section 5.4, these results will be used to explain the performances of the implantable antennas presented in [41,42]. In Section 5.5, the difficulties of properly measuring the radiation performances of ESAs will be summarized and extended to the case of implantable antennas. Finally, Section 5.6 presents some conclusions.

5.2 Key performance indicators: implantable antennas versus classic electrically small antennas

Implantable antennas are in nearly all cases also electrically small antennas (ESA). However, their key performance indicators (KPI) differ. Indeed, for ESAs the aim of a good design is often to minimize the antenna quality factor (thus maximizing the bandwidth), maximize the antenna gain, the antenna efficiency, the gain-bandwidth product or a combination of these features. The case is different for implantable antennas as they radiate into a lossy medium, making the meaning of characteristics as gain or impedance bandwidth less meaningful. The gain, for instance, is linked to the radiation pattern in the far field. In a lossy medium, the radiation will never reach far field, as was pointed out by Moore in 1963 [34]. Thus, the gain or radiation pattern of an implantable antenna can only be considered in conjunction with its host body: it is defined according to the far field reaching outside the host body. As an example, let us consider an implantable antenna, slightly modified from an antenna found in literature [43]. A sketch of the antenna is depicted in Figure 5.1a, and the radiator was modified, removing the superstrate and including the presence of a 1-mm thick plastic insulation that required minor

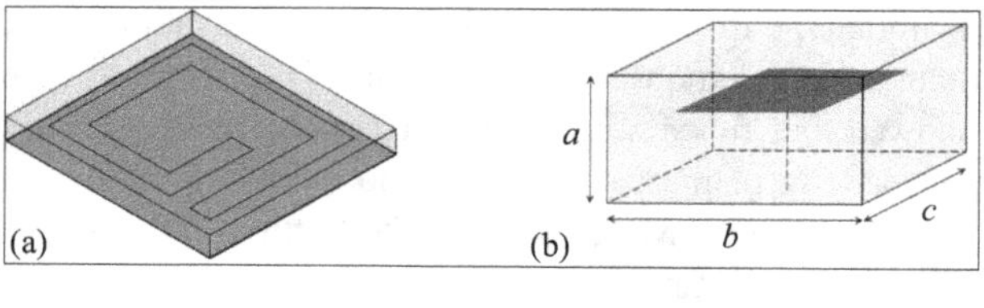

Figure 5.1 Implantable antenna (a) and phantom (b)

changes of the design. The antenna, intended to work in the MedRadio band at 404 MHz, is implanted in a human body model made of a homogeneous rectangular cuboid having the characteristics of muscle tissue surrounded by air, see Figure 5.1. Two different dimensions have been considered for this phantom in order to assess its effect on the radiation pattern of the antenna: $72 \times 80 \times 50$ mm^3 and $216 \times 240 \times 50$ mm^3. These phantoms were selected arbitrarily, keeping a simple shape, in order to assess their influence on the antenna result. Figure 5.2 depicts the radiation pattern in the plane perpendicular to the antenna considering both phantoms, clearly demonstrating the influence of the phantom on the antenna radiation.

As there are no standard phantom models that could be used to compare the radiation characteristics of two different antennas found in literature, gain and pattern are not useful KPIs for implantable antennas, due to their dependency on the used phantom.

The impedance bandwidth, a key characteristic of ESAs, becomes less meaningful when comparing two implantable antennas. Figure 5.3 shows the input reflection coefficient of the antenna depicted in Figure 5.1 for two cases. In the first, the antenna is implanted in the lossy muscle phantom of Figure 5.1b, in the second, we consider a lossless phantom with the same real dielectric constant. As can be seen, most of the s_{11} bandwidth stems from the losses in the phantom, and is thus not a useful measure of the real radiation bandwidth of the antenna.

Clearly, KPIs that are more meaningful need to be defined for implantable antennas. Moreover, we see that unlike the case of an antenna radiating into the free space, we cannot separate the implanted antenna from its surroundings,

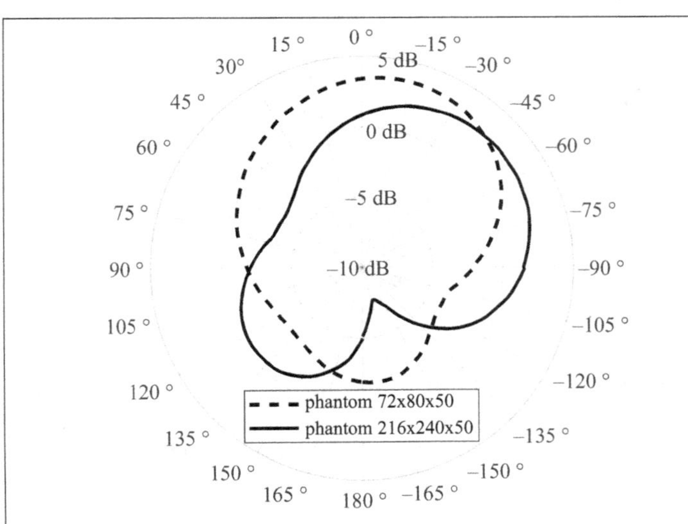

Figure 5.2 Radiation pattern of the antenna depicted in Figure 5.1 for two different phantom sizes

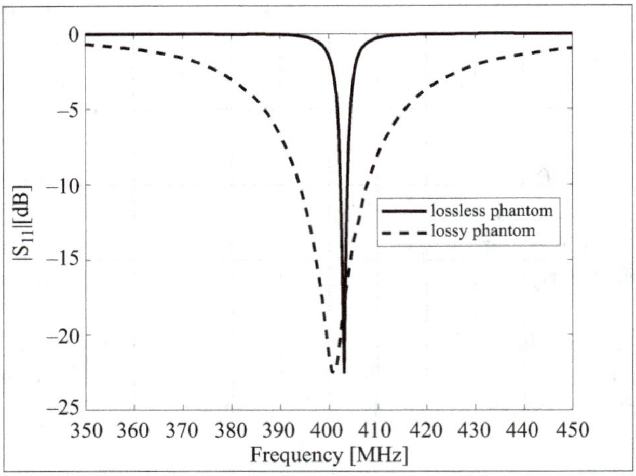

Figure 5.3 Reflection coefficient for the antenna depicted on Figure 5.1, considering a lossy or lossless body phantom

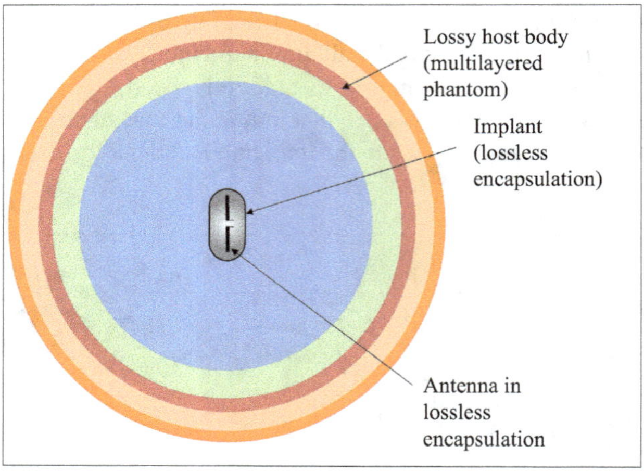

Figure 5.4 Classical example of an implant in a spherical phantom

and the KPIs will depend on the latter. To this aim, let us consider a classic example of an implant, as depicted in Figure 5.4. It consists of an antenna in a lossless encapsulation placed in a body phantom. The latter can have any shape, be homogeneous or inhomogeneous. In the case pictured here, we have a multilayered spherical phantom.

A possible KPI for this implantable antenna is its radiation efficiency, defined as the power reaching outside the lossy host phantom normalized with the power leaving the lossless encapsulation. Clearly, this KPI will depend on the phantom shape, volume, and material, that all affect the losses. However, the encapsulation will also affect the KPI, as it influences the amount of near field leaking into the lossy host phantom, and thus dissipated. Thus, a fair comparison between implantable antennas requests to place the antennas in the same encapsulation and the same phantom.

Another possible KPI, closely related to the radiation efficiency, is the maximum power density reaching free space, i.e. just outside the phantom or host body (where we could put an on-body antenna to establish the communication link). The advantage of this KPI is that it in principle does not depend on the geometry of the phantom, but on the distance between the implanted antenna and the body surface. This KPI enables us to model the body with phantoms of different shapes leading to similar results when designing e.g., the link between the implanted sensor and access point outside the body.

Once the KPIs are defined, having access to simple physical bounds is of great interest. In the case of ESAs, these bounds will depend mainly on the antenna electrical size, defined as the radius of the smallest sphere encapsulating the antenna divided by the free space wavelength. In the case of implantable antennas, we have seen that they will depend on the size of the encapsulation, the characteristics of the phantom (or host body) and the location of the implant within the latter. Physical bounds for ESAs have been studied since many decades. It is interesting to notice that until this millennium, the large majority of works were based on using a spherical wave expansion (SWE) of the EM fields radiated by the antenna to obtain bounds on bandwidth and gain [22–29]. The SWE is based on the representation of the EM field in a region with zero free-charge density using vector spherical harmonics [57–59]:

$$\left\{ \begin{array}{c} \mathbf{E} \\ -j\eta\mathbf{H} \end{array} \right\} = -\sum_n \sum_m a_{mn} \left\{ \begin{array}{c} \mathbf{M}_{mn} \\ \mathbf{N}_{mn} \end{array} \right\} + b_{mn} \left\{ \begin{array}{c} \mathbf{N}_{mn} \\ \mathbf{M}_{mn} \end{array} \right\}, \tag{5.1}$$

where

$$\mathbf{M}_{mn} = \nabla \times \mathbf{r}\psi_{mn}, \qquad \mathbf{N}_{mn} = \frac{1}{k} \nabla \times \mathbf{M}_{mn} \tag{5.2}$$

$$\psi_{mn} = \frac{1}{kr} \widehat{Z}_n(kr) P_n^m(\cos\,\theta) e^{jm\phi} \tag{5.3}$$

Here ψ_{mn} is the characteristic function that is the solution of the scalar Helmholtz differential equation in spherical coordinate system, i.e. $\widehat{Z}_n$ denotes Schelkunoff type of spherical Bessel or Hankel functions [59], $\mathbf{r}$ is the radial vector and k and η denote the wavenumber and the wave impedance of the considered medium, respectively. It is interesting to see if the use of SWE could help us to understand the loss mechanisms of implantable antennas and obtain some bounds on their performances.

Before doing this, let us recall the pioneering results obtained by Chu [22] on bounds for the quality factor and by Harrington [24] on the gain of ESAs. Using the

SWE and an equivalent circuit, Chu showed that the smallest possible quality factor of a lossless antenna, that can be circumscribed by a sphere of diameter *a*, radiating only the dominant spherical mode (either TE or TM) is given by [22]:

$$Q_{\min} = \frac{1}{ka} + \left(\frac{1}{ka}\right)^3 \tag{5.4}$$

where *k* is the wavenumber. Adding the second mode to the dominant mode (again TE or TM) would increase the quality factor to

$$Q_2 = \frac{3}{ka} + \frac{6}{(ka)^3} + \frac{18}{(ka)^5} \tag{5.5}$$

These results were corroborated using SWE only [29], and it was shown later that combining the first TE and TM modes leads to half the minimum quality factor obtained by Chu [44–46]. The main conclusion is that the lowest possible quality factor, thus the largest possible bandwidth is obtained for lossless antennas radiating only the dominant mode (or modes if both TE and TM modes are launched by the antennas). This quality factor is, for ESAs, proportional to $1/(ka)^3$. It is also important to note that this physical bound on the quality factor is absolute as it cannot be broken (except if we consider active antennas including for instance non-Foster matching networks). Moreover, it is very hard to reach. Importantly, this means that in order to achieve a low quality factor, the antenna needs to be omnidirectional, as only the dominant spherical mode(s) are radiated. Directivity requires higher order modes, thus a directive antenna will have a higher bound on its quality factor, as was suggested already in Chu's work.

In [25], Harrington studied the maximal gain of a directive lossless antenna, and obtained that this gain was bounded by:

$$G_{\max} = 2N + N^2 \tag{5.6}$$

where *N* is the number of modes radiated by the antenna. This bound is fundamentally different from the bound on the minimal quality factor, in the sense that it is not absolute. It increases with the number of modes radiated by the antenna. The next step is thus to try to understand what could limit the number of modes an antenna can radiate successfully, and if this number depends on the size of the antenna. To this aim, let us consider the wave impedance of the spherical modes launched by an electric source (TM modes), where k_0 is the free space wavenumber and *r* the radial distance:

$$Z_{mn}^{TM} = \frac{E_\theta}{H_\phi} = -\frac{E_\phi}{H_\theta} = j\eta \frac{\widehat{H}_n^{\prime(2)}(k_0 r)}{\widehat{H}_n^{(2)}(k_0 r)} \tag{5.7}$$

It is noticeable that the mode impedances Z_{mn}^{TM}, where *m* is the mode index related to the azimuth angle φ and *n* the mode index related to the polar angle θ, is independent of *m*. Figure 5.5 shows the ratio between the imaginary and real parts of the wave impedance of these modes versus $k_0 r$ for the first four modes, i.e. *n* taking the values from 1 to 4.

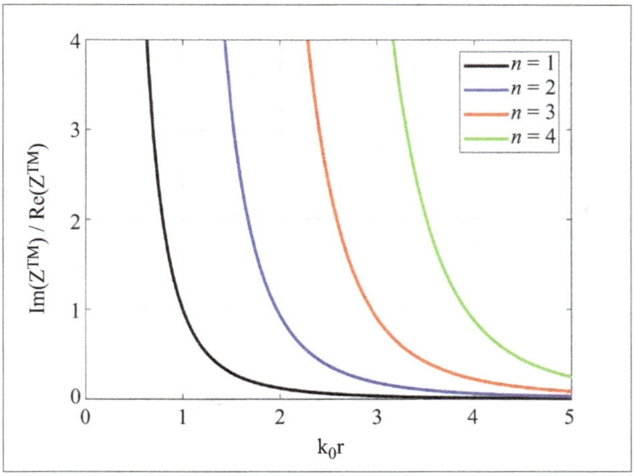

Figure 5.5 Ratio of the imaginary over the real part of the wave impedance of TM spherical modes, for n = 1–4

We notice that the wave impedance is mostly reactive for distances smaller than the mode index, thus for $k_0 r < n$. This means that at such distances, there is a predominance of stored reactive energy. Thus, in order to keep the quality factor of an antenna low and have a good bandwidth, an antenna of electrical size $k_0 a$ should radiate only modes of order $n < k_0 a$. This leads to the well-known practical limitation on the gain proposed by Harrington [24]:

$$G_{\max} = 2N_{\max} + N_{\max}^2 \tag{5.8}$$

where it is understood that $N_{\max}$ is an integer equal or larger to one, and smaller than $k_0 a$.

In other words, we have here not an absolute bound of the gain, but a practical limit of the maximum gain achievable while guaranteeing a reasonable bandwidth. We will come back to this distinction between absolute and practical bounds when we discuss bounds for implantable antennas in Section 5.3.

5.3 Loss mechanisms and efficiency in implanted antennas

The radiation efficiency of an implanted antenna is defined as the ratio between the power reaching outside the lossy host body and the power leaving the lossless encapsulation surrounding the antenna. There are three main loss mechanisms to consider. First, we have the losses due to the reactive near field close to the antenna, which can couple to the dielectric losses of the host body. Second, there are the losses incurred by the propagating field traveling through the body and finally the reflection at the interface between the host body and the free space, see Figure 5.6.

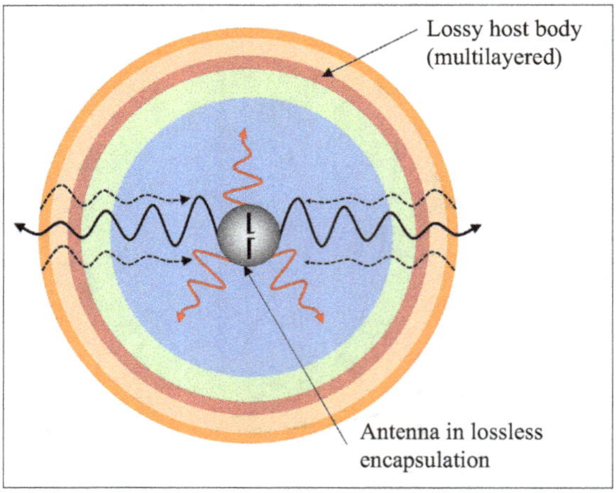

Figure 5.6 Loss mechanisms in lossy media. Red line: reactive near fields, black line: propagating fields, dashed line: reflected fields

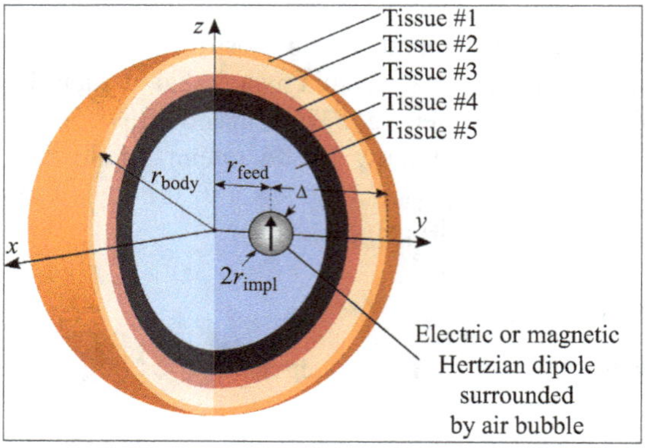

Figure 5.7 Canonical implantable antenna scenario: a multishell spherical phantom and a Hertzian dipole surrounded by a lossless spherical shell

The loss due to the propagating field through the host body is unavoidable, as are the reflection losses if no matching layer is used. Thus, the near field losses are the only ones that can be played upon to improve the overall radiation efficiency of an implantable antenna in a specific scenario. To do this efficiently, it is necessary to understand these different loss mechanisms.

5.3.1 A spherical multishell model for the lossy host medium

Let us consider the canonical implantable antenna scenario depicted in Figure 5.7. It consists of a multilayered spherical lossy host phantom in which an elementary dipole antenna is implanted. The dipole is surrounded by a lossless shell of radius

r_{impl}, the antenna is placed at a distance r_{feed} from the center of the phantom and the radius of the phantom is r_{body}.

A model based on SWE was proposed in [40] enabling the efficient study of this canonical structure. As an example, let us consider a homogenous lossy phantom of $r_{body} = 90$ mm radius and permittivity $\varepsilon_r = 43.50 - j34.75$ [47], [IEEE Head model] with the dipole located in its center. A bubble of air of radius 1mm surrounds the dipole, which is emitting a power of 1 mW. Figure 5.8 shows, for both an electric and a magnetic elementary dipole, the total power reaching out of a sphere of radius r.

The three different loss mechanisms are clearly seen in Figure 5.8: The reactive near-field absorption close to the source, much larger, as expected, for the electric dipole than for the magnetic dipole. Indeed, an electric source will produce mostly electric near fields, which will couple with dielectric losses whereas a magnetic source will produce mostly magnetic near fields. In the median part of the figure, the exponential behavior of the propagating field absorption appears, and the effect of the reflection at the interface between phantom and air is seen at a radial distance of 9 cm, corresponding to the end of the phantom. Figure 5.9 depicts

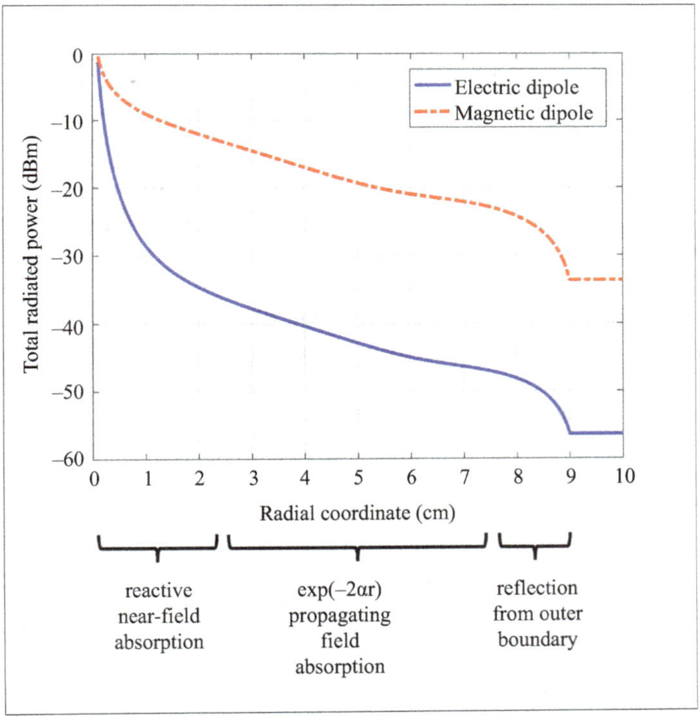

Figure 5.8 Total radiated power in the phantom reaching out of a sphere of radius r (radial coordinate) for an electric and a magnetic source placed at the center of the phantom at 403 MHz, $r_{impl} = 1mm$ (after Skrivervik et al. [39], © IEEE 2019)

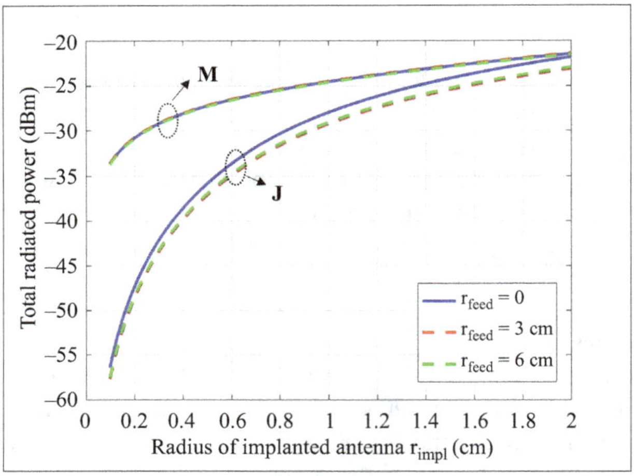

Figure 5.9 Total radiated power reaching free space as a function of the radius of the air sphere encapsulating the source (r_{impl}), for different positions of the latter (r_{feed}). Both the electric (J) and magnetic (M) sources are considered. The source is parallel to the phantom–air interface when it approaches the latter (as indicated in Figure 5.7) (after Skrivervik et al. [39] © IEEE 2019)

the total radiated power reaching free space for a similar scenario, but for three positions of the dipole inside the phantom, as a function of the radius of the lossless air sphere surrounding the implanted source. It is obvious that the radius of the air bubble has a large impact on the power able to reach free space, especially for the electric dipole. Indeed, as the bubble increases, a larger portion of the near field remains inside the lossless bubble and does not contribute to the losses. As can be seen, this effect reaches saturation when the bubble reaches the size for which most of the near field is in the lossless area.

Figure 5.10 presents the total radiated power reaching out of a sphere of radius r for the first three spherical modes, for both an electric and a magnetic dipole. It is interesting to note that the first mode has the smallest losses, and that this is due to the near field losses. Indeed the range of distance where near field absorption is noticeable seems to become larger with the order of the mode considered. Figure 5.11 shows the mode impedance of the first three spherical modes as a function of the distance to the source, and illustrates clearly this effect. The mode impedance varies rapidly close to source in the region dominated by near fields to stabilize to a nearly constant value after a certain distance, when the propagating fields are dominant. It is remarkable that the distance marking the transition from near field region to propagating field region is given by

$$|kr| = n \tag{5.9}$$

where n is the order of the mode.

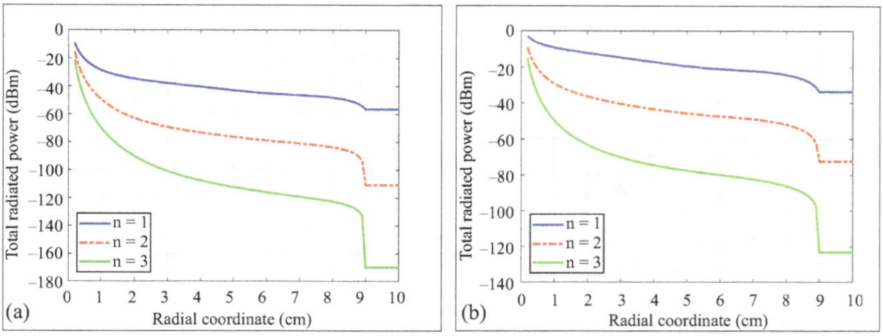

Figure 5.10 Total radiated power for specific modes along different radii away from the source: (a) electric dipole case and (b) magnetic dipole case

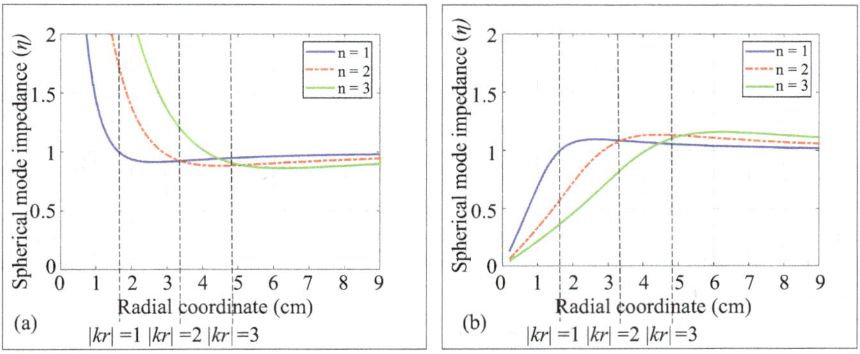

Figure 5.11 Magnitude of spherical mode impedance evaluated for specific spherical modes along different radii away from the source: (a) electric dipole case and (b) magnetic dipole case

This result is very similar to Harrington's findings when he studied the maximum gain of a practical antenna, where his results also stated that the near field of a spherical mode is confined in a sphere having an electric dimension equal to the order of the mode. If we summarize our findings so far, we have learned that:

- The losses of an implantable antenna will depend on three main contributions: the near field losses, the propagating field losses, and the reflection at the body/air interface.
- The near field losses depend on the type of the radiating source (electric or magnetic) and the electric size of the lossless encapsulation around the antenna.
- For an electrically small encapsulation, a magnetic type of antenna is more favorable than an electric one.
- In order to avoid unnecessary losses, the antenna should only excite modes with order $n < kr_{impl}$, to keep the near fields as much as possible in the lossless encapsulation of the antenna. Indeed, as seen in Figure 5.11, the distance of influence of the near fields increases with the order of the mode.

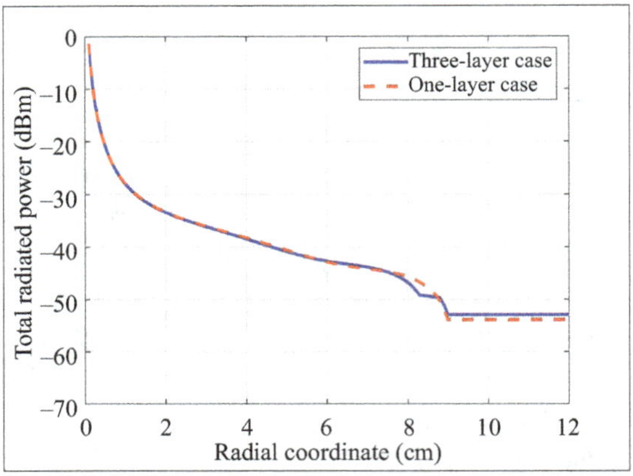

Figure 5.12 Comparison between homogeneous and multilayered phantoms: Total radiated power in the phantom reaching out of a sphere of radius r (radial coordinate)

The results in Figures 5.8–5.10 were obtained for a homogeneous phantom. To clarify how realistic this is, we compare them with the losses occurring in a multilayered phantom, comprising for instance muscle, fat and bone. This comparison is depicted in Figure 5.12 for the example of an implant located in muscle. The antenna, with r_{impl}, is located at the center of a spherical phantom having s radius of 9 cm. As before, the power emitted is 1 mW (0 dBm). The homogeneous phantom is compared to a multilayered phantom made of three concentric shells as depicted in Figure 5.7, where the inner shell is made of muscle (radius 8.3 cm) covered by a 0.5-cm thick shell of fat and a 0.2-cm shell of skin. The permittivity of different tissues was calculated using [49,50] and the frequency considered is 403 MHz. Figure 5.12 depicts the total radiated power reaching out of a sphere of radius r.

Except for the differences due to the reflections at the interfaces in the layered case, both results are very similar. This shows that for deep implants as is the case here, a homogeneous phantom is good enough to both design the antenna and obtain a good estimation of the losses due to the phantom. It can be shown that if the implant is shallow or located in a thin layer, the homogeneous approximation will still be satisfactory to perform the antenna design, however not good enough to estimate the global losses.

5.3.2 *Bounds on power density reaching the surface of the lossy phantom*

The SWE model has enabled us to understand essential loss mechanisms for implanted antennas. It also enables us to derive first bounds on the performance of implanted antennas. Indeed, we have seen that, in addition to the losses internal to the antenna, three main factors contribute to the losses that an EM signal emitted by an implanted antenna incurs, namely the near field losses, the propagating field losses, and the reflection at the interface between the host body and the free space. If we consider the SWE model described in Section 5.2, with a homogeneous host body, and assuming that only the fundamental mode is excited by the implanted antennas, the power reaching the outside surface of the lossy phantom can be expressed using the following closed form expressions [39]. These expressions are easy to evaluate (even a scientific calculator can be used for this purpose) and they depend only on basic parameters of the considered implanted scenario. Therefore, they can be useful in initial stage of the design process in which a rough estimation of the available power levels needs to be determined.

The power reaching outside the host body can be related to the power leaving the lossless encapsulation around the antenna as:

$$P_{\substack{\text{total}\\\text{radiated}\\\text{power}}} = P_{\substack{\text{entering}\\\text{the}\\\text{body}}} \cdot e_{\substack{\text{losses}\\\text{in the}\\\text{reactive}\\\text{near-field}}} \cdot e_{\substack{\text{propagating}\\\text{field}\\\text{absorption}\\\text{losses}}} \cdot e_{\substack{\text{losses}\\\text{due to}\\\text{reflections}}} \tag{5.10}$$

If η denotes the complex wave impedance in a tissue defined as $\eta = \eta_0/\sqrt{\varepsilon_r}$, k denotes the complex wave propagation coefficient $k = (2\pi/\lambda_0)\sqrt{\varepsilon_r} = \beta - j\alpha$, and $\Delta = r_{body} - r_{feed}$, then the expressions for different sources of losses and corresponding efficiencies are the following:

$$\begin{aligned} e_{\substack{\text{losses}\\\text{in the}\\\text{reactive}\\\text{near-field}}} &= \frac{f_1(\Delta)}{f_1(r_{impl})}, \\ f_1(r) = \mathrm{Re}\Bigg\{\eta \cdot \Bigg(|k|^2 &+ \frac{2\alpha}{r} + \left(1 - \frac{k^*}{k}\right)\frac{1}{r^2} - j\frac{1}{kr^3}\Bigg)\Bigg\} \end{aligned} \tag{5.11}$$

represents the losses due to the near field for an electric source and

$$e_{\substack{\text{losses}\\\text{in the}\\\text{reactive}\\\text{near-field}}} = \frac{f_2(\Delta)}{f_2(r_{impl})}, \qquad f_2(r) = |k|^2 + 2\alpha/r \tag{5.12}$$

represents the losses due to the near field for a magnetic source, while

$$e_{\substack{\text{propagating}\\\text{field}\\\text{absorption}\\\text{losses}}} = \exp(-2\alpha(\Delta - r_{impl})) \tag{5.13}$$

represents the propagating field losses and the reflection losses are given by

$$e_{\substack{\text{losses}\\\text{due to}\\\text{reflections}}} = \frac{\text{Re}\left\{|T|^2/Z_{air}\right\}}{\text{Re}\left\{1/Z_{body}\right\}}, \qquad T = \frac{2Z_{air}}{Z_{air} + Z_{body}} \tag{5.14}$$

Z is the spherical characteristic impedance depending on the spherical mode and on the complex permittivity of the medium (body or air). For the dominant TM and TE spherical modes, as we consider here, the spherical characteristic impedance is equal to

$$Z_{11}^{TM} = \eta \frac{1 - j/kr - 1/(kr)^2}{1 - j/kr}, \qquad Z_{11}^{TE} = \frac{\eta^2}{Z_{11}^{TM}} \tag{5.15}$$

The reflection losses can be approximated well using large-radius variation resulting in:

$$e_{\substack{\text{losses}\\\text{due to}\\\text{reflections}}} \approx \left| \frac{2\sqrt{\varepsilon_{body}}}{1 + \sqrt{\varepsilon_{body}}} \right|^2 / \text{Re}\left\{\sqrt{\varepsilon_{body}}\right\} \tag{5.16}$$

Note that (5.11)–(5.15) are exact for the case of implanted antennas located in a center of a spherical homogeneous phantom, and they present the upper bound of the total radiated power (i.e. of the antenna efficiency) for deep implants with $r_{feed} \neq 0$.

Similarly, we can express the power density reaching free space (i.e. the maximum value of real part of power density resulting from the tangential EM field components at the body surface) as

$$W_{\substack{\text{reaching}\\\text{free}\\\text{space}}} = W_{\substack{\text{entering}\\\text{the}\\\text{body}}} \cdot \frac{r_{impl}^2}{\Delta^2} \cdot e_{\substack{\text{losses}\\\text{in the}\\\text{reactive}\\\text{near-field}}} \cdot e_{\substack{\text{propagating}\\\text{field}\\\text{absorption}\\\text{losses}}} \cdot e_{\substack{\text{losses}\\\text{due to}\\\text{reflections}}} \tag{5.17}$$

The normalization factor r_{impl}^2/Δ^2 takes into account the spreading of the radiated EM fields (decaying of the E- and H-field components inversely proportional to the radial distance), thus the efficiency factors cover the additional losses in the electromagnetic system. In this way, an approximate expression for the available power density reaching free space is obtained allowing designers to quickly estimate the power density levels that they can expect from a certain design. Even if these expressions were developed for an antenna implanted in a sphere radiating only the dominant TE or TM mode, they represent an upper bound for the total radiated power in the same sense as the bound on minimum quality factor or maximum practical gain for electrically small antennas described in [22,24], as:

1. Any implantable antenna can be described by the spherical modes it excites.
2. Higher-order modes have more losses than the dominant mode.

3. For implants that are not too shallow, the power density reaching free space depends mostly on the depth of the implant, and not on the size nor the shape of the phantom.

5.3.3 Design rules of thumb

As an example, let us consider the case of an implant having a 1 mm radius, an antenna of the electric type, implanted in muscle at a depth of 90 mm. We can use the results of Section 5.2 in order to gain insight into the order of magnitude of the power density reaching the border of the body. To this aim, we approximate the antenna by an elementary electrical dipole encapsulated in a lossless bubble having a radius of 1 mm. This source is located at the center of a spherical homogenous phantom made of muscle (the permittivity of muscle was calculated using [49,50]) having a radius of 90 mm. The analysis of this simplified version of our design will give us an upper bound on the radiation efficiency, and a first insight into the antenna behavior. Figure 5.13 shows the power density reaching outside the phantom and the different efficiencies for this scenario, as a function of frequency [48].

At lower frequencies, the near field losses dominate the loss mechanism. This is because a lower frequency will correspond to a larger wavelength and thus a larger area of influence of the near field in the lossy tissue. At higher frequencies, the propagation losses are dominant as the losses in the tissue increase with frequency (note that for higher frequencies the body is larger in terms of the wavelength and the attenuation constant in the lossy medium is equal to $(2\pi/\lambda_0)\text{Im}(\sqrt{\varepsilon_r})$). There is thus a frequency, which is optimal for this setup. This frequency depends on the depth of the implant, the size of the lossless encapsulation and the tissue in which the antenna

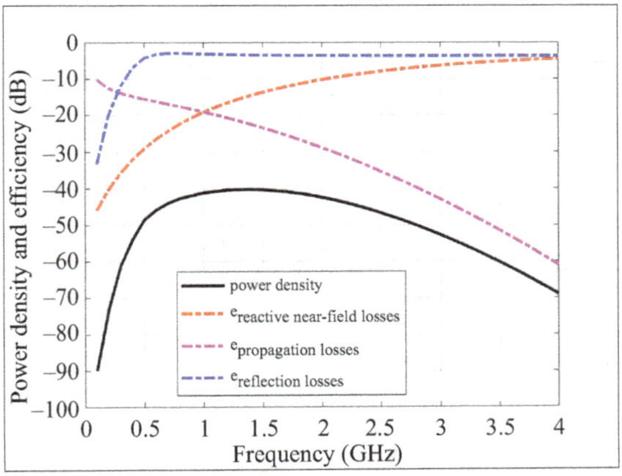

Figure 5.13 Normalized power density just outside the spherical phantom and corresponding efficiencies for an implanted antenna placed in the center of a homogeneous spherical phantom (after Sipus et al. [48] © IEEE 2021)

is implanted. Similar results can be obtained for a magnetic dipole or for multilayered phantoms [48]. This analysis gives an upper bound on the radiation efficiency, and a first insight into the antenna behavior

This type of analysis is done for canonical cases, an elementary electric (or magnetic) dipole placed in a spherical phantom radiating the dominant spherical mode. However, as the dominant mode is the mode having the least losses and as the loss mechanism of an implanted antenna depend mostly on the depth of the implant (rather than on the diameter of the phantom in which it is implanted), these results can be extrapolated to real implants. In fact, they represent an optimal case and give bounds on the achievable minimal losses between an implant and a node on or of the phantom in the same sense as Chu's formula on minimum quality factor [22]. Moreover, they give precious indication on the frequency band that should be chosen for a specific link [51].

5.4 Examples

As a first example, let us consider an existing implantable antenna and compare its performances to the canonical results presented in the previous sections. This first example is based on the antenna presented in [41] for the telemetry of implantable sensors.

The system uses two frequency bands the MedRadio band (404 MHz) for data transmission and ISM (2.45 GHz) for a wakeup signal, thus a dual band antenna is required. The antenna and the system, fully described in [41], are placed in an insulating cylinder having a diameter of 10 mm and a height of 32 mm. The implant is depicted in Figure 5.14a and the antenna in Figure 5.14b. The antenna fits in a parallelepiped of dimensions $8 \times 11 \times 4.8$ mm^3.

The implant is designed to work at a depth comprised between 1 and 3 cm, in muscle. To assess the losses of the link, the canonical model presented in Section 5.3 was used. The diameter of the sphere encapsulating the implant in the spherical wave model was taken equal to the diameter of the cylinder. Figure 5.15 shows the total radiated power reaching free space for an electric dipole and a magnetic dipole

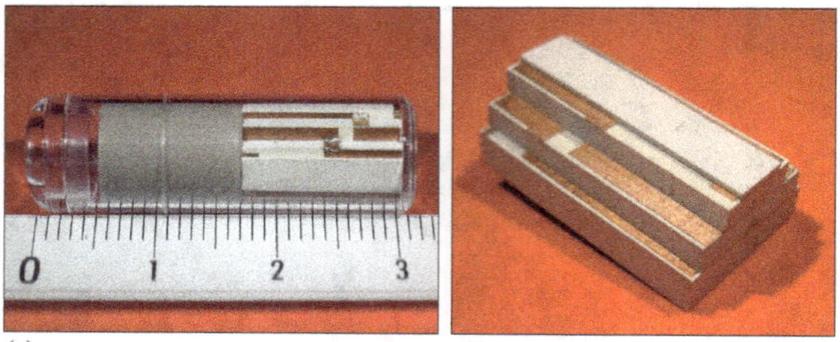

(a) (b)

Figure 5.14 Dual band antenna for telemetry of implantable sensors. (a) Implanted capsule with the antenna visible on the right. (b) Antenna in detail

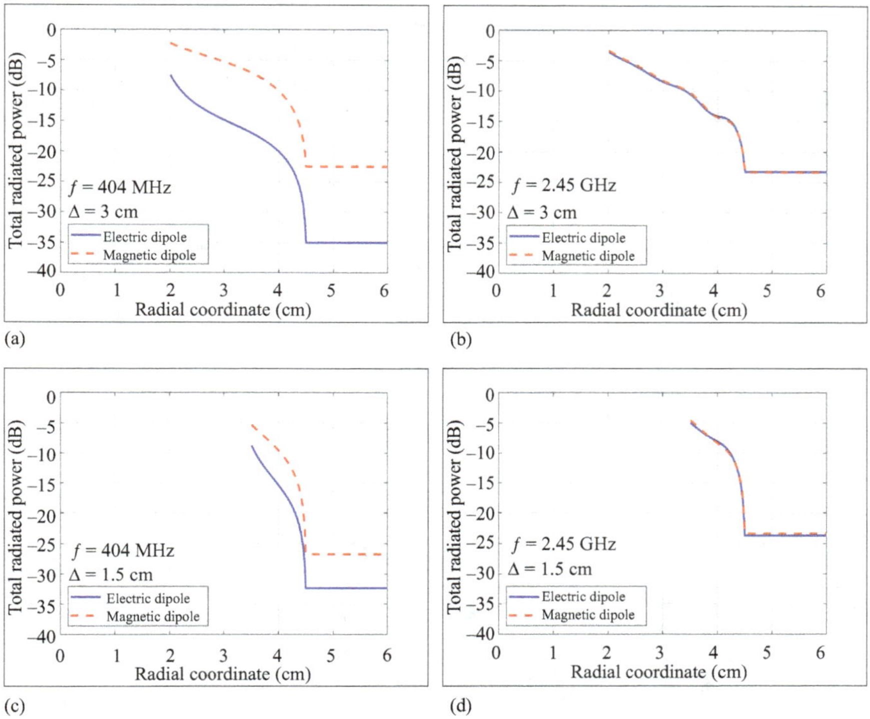

Figure 5.15 Total radiated power of an implant located in a muscle phantom having a radius of 4.5 cm. (a) 404 MHz and (b) 2.45 GHz: the implant is at a depth of $\Delta = 3$ *cm. (c) 404 MHz and (d) 2.45 GHz: the implant is at a depth of* $\Delta = 1.5$ *cm*

implanted at depths of 1.5 and 3 cm in a homogeneous muscle phantom having a radius of 4.5 cm. This power gives a good measure of the antenna efficiency as, due to the losses, the directivity of the antenna is more or less independent of the direction (the directivity of the antenna varies slowly with the angle).

At the MedRadio band of 404 MHz, there is a large difference between the results achieved by the electric dipole and the magnetic dipole, the former suffering nearly 20 dB more losses. At the ISM band of 2.45 GHz there is no noticeable difference between the two. This is due to the fact that the dimensions of the encapsulation of the antenna, a lossless sphere of 1 cm in diameter, is electrically large enough in the ISM band to contain most of the near fields, but not in the Medradio range. This effect is further illustrated in Figure 5.16, where the efficiencies due to near field losses, propagating field losses and reflection losses are depicted versus frequency for the considered case. The nearfield losses are very severe for frequencies below 0.5 GHz, and become negligible for frequencies above 2 GHz. The same figure shows that an optimal frequency for this application would be between 1 and 1.5 GHz. However, this band was not an option for the application considered.

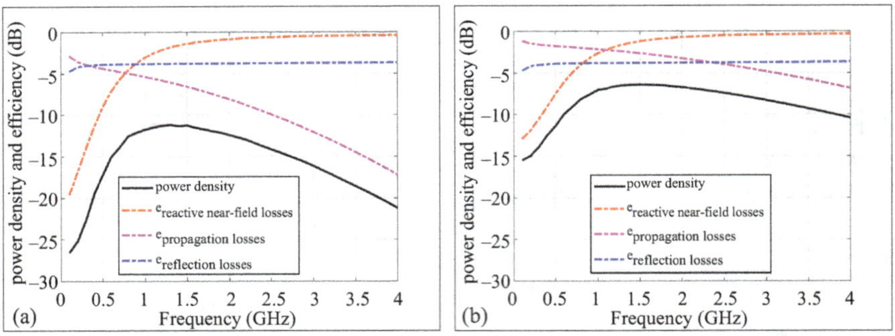

Figure 5.16 Different efficiencies versus frequency for an implant located at a depth of (a) 3 cm and (b) 1.5 cm in a muscle phantom having a radius of 4.5 cm. The radius of the antenna encapsulation is 0.5 cm

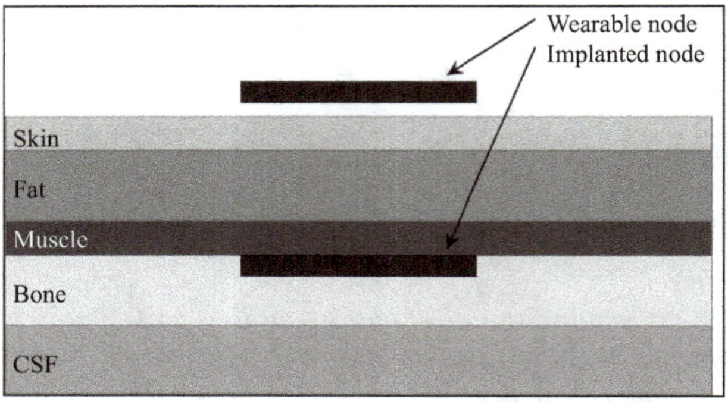

Figure 5.17 Cranial implant

These results, obtained using the canonical model, enable us to expect that the losses of the real implantable antenna could have an order of magnitude around 30 dB at 404 MHz and around 25 dB in the ISM band of 2.45 GHz.

The antenna depicted in Figure 5.14 was realized and characterized [41], both by numerical simulations using different phantoms and in vitro before being finally used in vivo. At 404 MHz, the efficiency of the antenna obtained using a commercial simulation tool was of −32.6 dB considering a homogeneous cylindrical muscle phantom and estimated at −33 dB for the in vitro measurements performed. At 2.45 GHz, the results were of −20.1 dB for the numerical simulations and estimated at −21 dB in the in vitro measurements, showing a remarkable closeness to the estimations achieved using the simple spherical model.

A second example is derived from the antenna described in [42] and consists of a device implanted in the skull behind the ear. The link consists of an implanted node and a wearable node located flush to the cranium, as depicted in Figure 5.17. The requirement for the link is a transfer function s_{21} between the antennas better than −50 dB over a bandwidth of 100 MHz. The wearable node consists of a simple microstrip patch antenna covered with a biocompatible superstrate. The maximal dimensions for the implanted antenna, including a biocompatible layer made of PEEK on top, are of 10 × 10 × 2 mm^3. The frequency of the link carrier is 5.8 GHz. The dielectric properties and thicknesses of the layers are given in Table 5.1.

The feasibility of achieving the requirements was assessed using the spherical wave expansion method. To this aim, the problem depicted in Figure 5.18 was analyzed. We consider a phantom with radius of 10 cm, and the thicknesses of bone, muscle (when relevant), fat and skin are given in Table 5.1. The source is a Hertzian dipole oriented in parallel to the tissue interfaces, and placed in an air

Table 5.1 Permittivity and thickness of cranial tissues

Tissue name	**Relative permittivity**	**Loss tangent**	**Nominal thickness (mm)**
Cerebrospinal fluid (CSF)	60.47	0.40	5
Bone cortical	9.70	0.37	5
Muscle	48.48	0.32	2.2
Fat	9.90	0.26	4.8
Skin dry	35.10	0.33	2.2

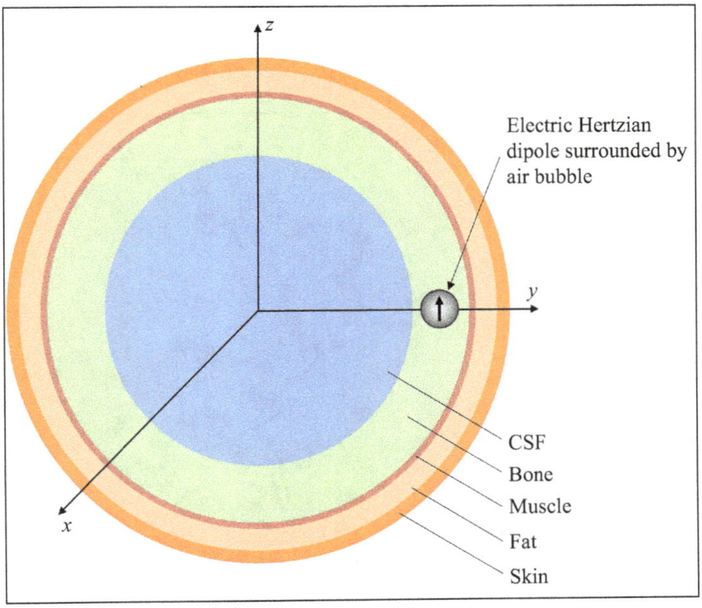

Figure 5.18 Spherical model corresponding to the problem of the cranial implant

bubble having a radius of 1 or 2 mm. These values were selected, as the thickness of the insulating biocompatible layer on top of the antenna will be comprised between these values.

Figure 5.19 shows the decay of the power density with distance from the source in the phantom. It is interesting to note that the overall losses in the media correspond in the worst case (insulation of 1 mm) to about −22 dB, which shows that is seems feasible to satisfy the requirements.

The next step is to design the implanted antenna. As the bandwidth is an important requirement, a PIFA was preferred to a patch antenna due to its better bandwidth performances. The design was performed in three steps: The antenna, including its biocompatible PEEK cover, was first designed in a homogeneous medium made of a lossless material having the relative permittivity of muscle. This was done because muscle is the layer closest to the antenna and is thus expected to have the largest impact on the performances. Discarding the dielectric losses in this step enables to assess the losses corresponding to radiation only. Once the achieved results are satisfactory, losses are added to the phantom model, still keeping it homogeneous. Finally, the model is adjusted to take into account all the layers of the phantom. The advantage of this procedure is that each step requires only small adjustments, avoiding lengthy optimizations. The antenna is depicted in Figure 5.20, and a full description can be found in [42]. The PIFA is built on a substrate having a permittivity ε_r of 9.9 and is placed in a metallic cavity (sensor) of $10 \times 10 \times 2\ \text{mm}^3$. The dimensions are given for the antenna in the multilayered structure and were very slightly adjusted in the homogeneous lossless and lossy muscle phantom.

Figure 5.21 shows the input reflection coefficient when placed in three different phantoms: homogenous muscle without and with losses and the complete multilayered phantom. It is interesting to note that there is no significant difference

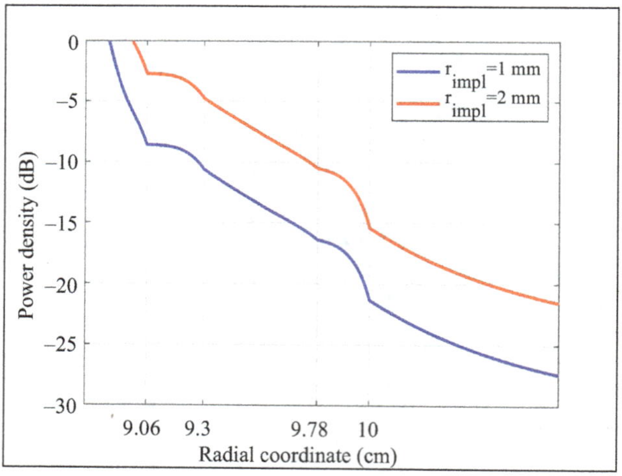

Figure 5.19 Relative available power density as a function of distance for an elementary source placed in the phantom depicted in Figure 5.18

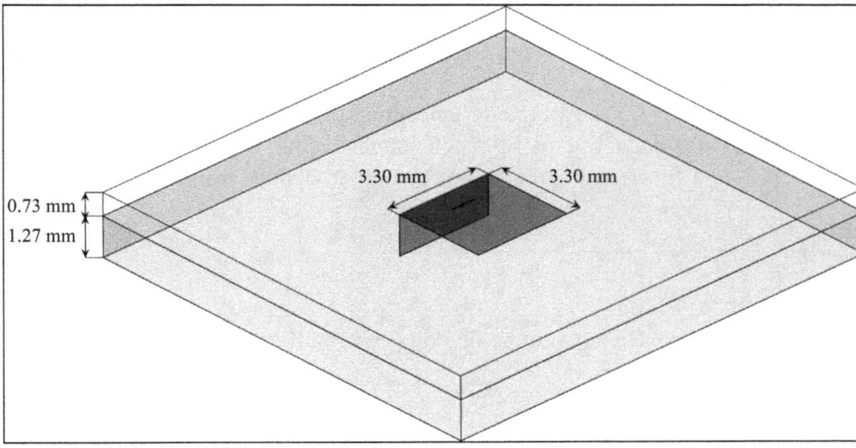

Figure 5.20 PIFA antenna for a cranial implant

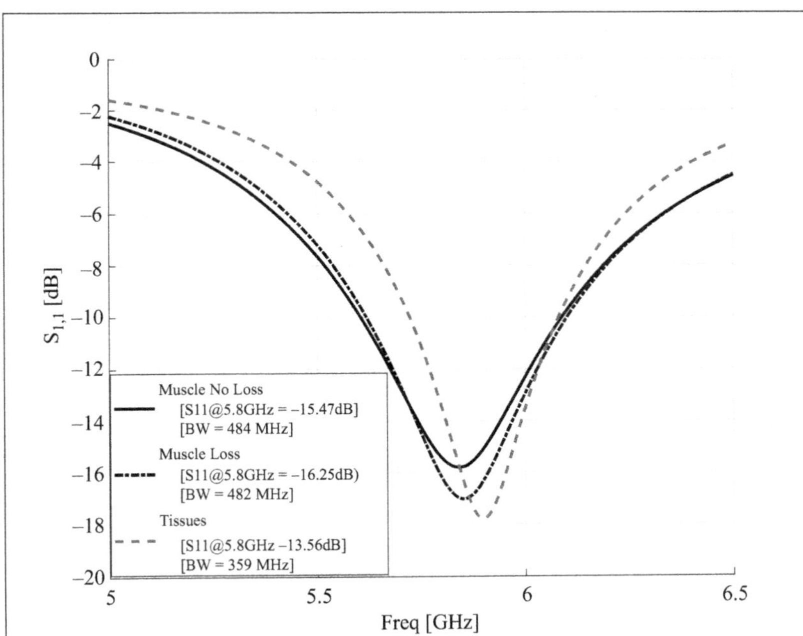

Figure 5.21 Reflection coefficient of the implanted antenna

between the three cases. This stems from two facts: first, the antenna is a PIFA, which in general has a broader bandwidth and thus less near field than patch antennas (see for instance [21]). Second, the antenna is built on a substrate having a high permittivity, which means that most of the near field is concentrated in this substrate and does not reach the lossy medium. The design procedure, where most of the design was performed using the simplest possible phantom, is thus amply justified in this case.

Finally, Figure 5.22 shows the de-embedded transmission coefficient between the implanted PIFA and a microstrip patch antenna acting as the wearable node (see Figure 5.17), as a function of frequency. The de-embedded s_{21} is given by

$$|s_{21_{DE}}|^2 = |s_{21}|^2 \frac{1}{\left(1 - |s_{11}|^2\right)} \frac{1}{\left(1 - |s_{22}|^2\right)} \tag{5.18}$$

This transmission coefficient can be de-composed in the following way [42]:

$$|s_{21}|^2 = gL_m A_e \left(\frac{1}{4\pi d^2}\right) \tag{5.19}$$

where g is the gain of the PIFA antenna in a lossless medium, L_m represents the losses in the media (bone, muscle, fat, skin), A_e is the effective aperture of the patch antenna on the wearable node, and d is the distance between the implanted PIFA and the wearable patch. In our specific case, the product of the two last terms, the effective aperture of the patch and the loss due to the spherical nature of the wave, is approximately equal to one, in a different case it could easily be computed. The

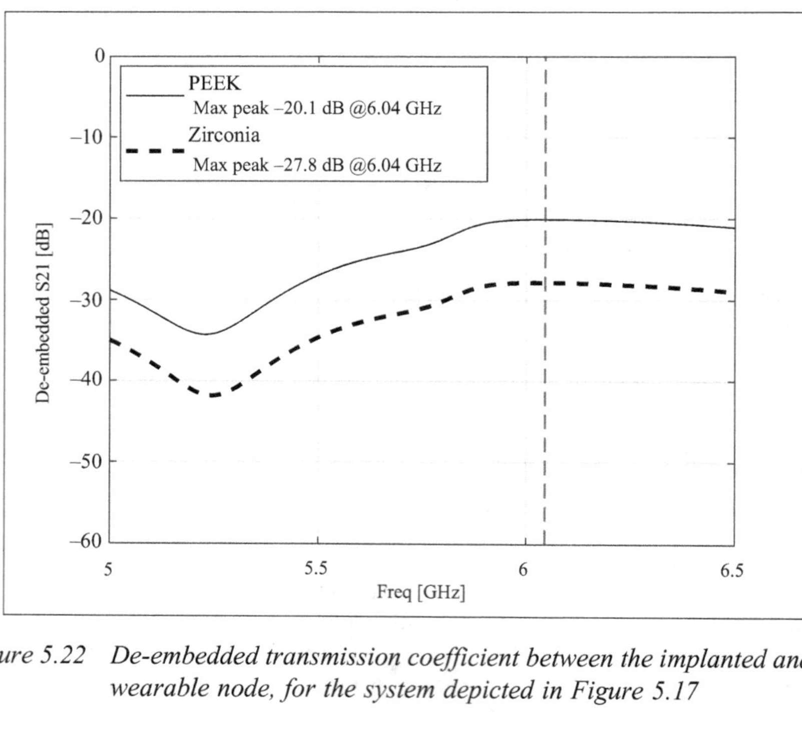

Figure 5.22 De-embedded transmission coefficient between the implanted and the wearable node, for the system depicted in Figure 5.17

product of the two first terms in (5.14) represents the power density at the level of the patch antenna. It is remarkable to note that this achieved transmission at the requested frequency of 5.8 GHz corresponds very well to what was foreseen by the SWE model in Figure 5.19, where a relative power density of −20 dB was foreseen for a superstrate thickness of 1 mm.

5.5 Measurements

The experimental characterization of implantable antennas is far from trivial, as in their final environment they are, by definition, implanted in a living host. It is of course out of the question to characterize the antenna performances in this final setup, only system performances (e.g. the success of the telemetric link) can be assessed in vivo. However, before implanting anything into a living human or animal, it is essential to be sure that all the components of the system to be implanted work, as they should. Realistic body phantoms mimicking the dielectric characteristics of the biological tissues are used to this aim. There are several levels of complexity in these phantoms used for measurements, in the same way, as there are several levels of complexity in the phantoms used for numerical simulations. Phantoms and their realization are out of the scope of this contribution, but relevant information can be found in for instance [52,53]. In this section, the difficulties of the EM characterization of implantable antennas will be addressed. Indeed, implantable antennas are usually electrically small, which are notoriously difficult to measure [21,54,55].

The problem in measuring electrically small antennas are the so-called cable currents, which are at the origin of spurious unwanted radiation. In order to understand the origin of this difficulty, we need to remember that at low frequency the interconnection between elements like resistors, capacitors, or active elements are done using connecting wires, leading to circuits modeled by Kirchhoff's model. At microwave frequencies, components are interconnected using transmission lines and other guiding structures (e.g. waveguides, microstrip lines, etc.). The difference lies in the fact that the dimension of the components and of the connecting lines are not negligible with respect to the wavelength. The model describing microwave components, the scattering parameters, or impedance matrix, define the characteristics of the components between ports, which are located on the transmission lines interconnecting the components. A port must fulfill a certain number of conditions:

- At the location of the port, only one mode of the transmission line is propagating. If several modes may propagate, as many ports as modes need to be defined.
- The port is located far enough of any discontinuity to ensure that higher order modes are attenuated.

The difference between low frequency circuits and microwave circuits is depicted in Figure 5.23.

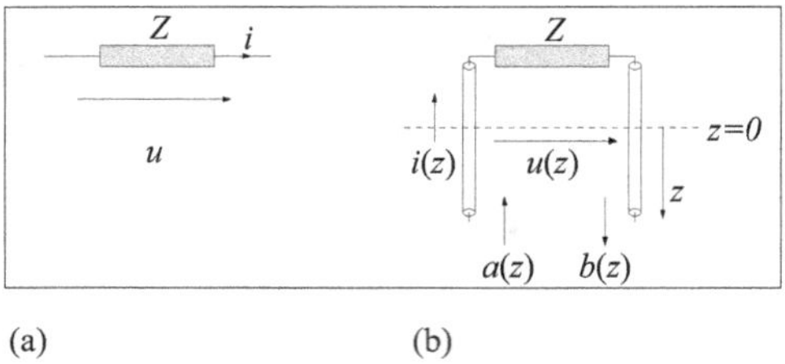

Figure 5.23 Definition of ports in circuit theory. (a) An impedance Z defined between two points; (b) an impedance defined at a port

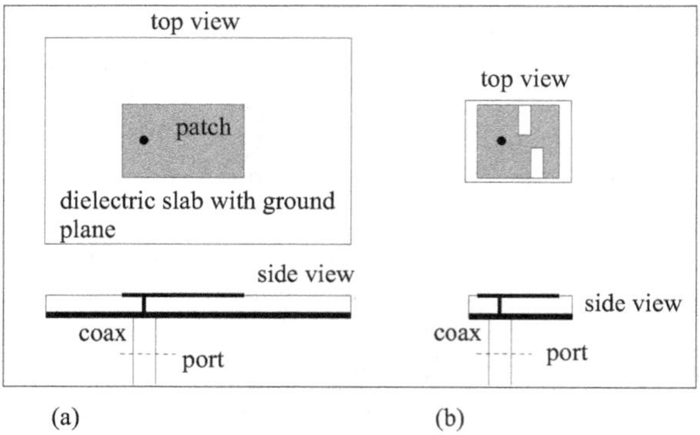

Figure 5.24 (a) Microstrip patch antenna. (b) Miniature patch antenna

In Figure 5.23a, we have a low frequency component defined by its impedance Z, and we have

$$u = Zi \tag{5.20}$$

The microwave component depicted in Figure 5.23b is defined at the port indicated by a dashed line somewhere on the transmission line (or other guiding structure) connecting the component to the rest of the world. The port is the interface between the component and the line. It is thus a single port device defined by

$$Z = \frac{u(z=0)}{i(z=0)}; \ \rho = \frac{b(z=0)}{a(z=0)} \tag{5.21}$$

where $u(z)$ and $i(z)$ are the voltage and current waves travelling on the transmission line and $a(z)$ and $b(z)$ are the forward travelling and backward travelling normalized wave amplitudes on the line [56].

Electrically small antennas often do not have a well-defined port as described above. To illustrate this let us imagine a microstrip patch antenna excited by a coaxial probe, as depicted in Figure 5.24a. The port is well defined, somewhere in the connector or on the coaxial cable, and the propagating mode defined at the port is the typical signal to ground or unbalanced mode. If we consider the miniaturized patch antenna depicted in Figure 5.24b, the situation is much less clear. On the one hand, the coaxial port and transmission line would expect the classic signal to ground mode; on the other hand, the miniaturization process has drastically reduced the size of the antenna's ground plane, which is now too small to act as a ground. Indeed, that antenna is now nearly a symmetrical structure, expecting for a symmetric or balanced mode at its port. The effect is the same as connecting a dipole directly to a coaxial cable without using an unbalanced to balanced converter (so called balun). As a result, spurious currents will circulate on the mantle of the coaxial cable, and it is clear that these currents will contribute to the radiation. In practice, the miniature patch antenna has a connection point supporting two modes, the balanced and the unbalanced, and can thus not be considered as a proper port. If used to connect the antenna to measurement equipment (e.g. a VNA), a current corresponding to the balanced mode will circulate on the mantle of the coaxial cable and perturb the measurements. This is not a problem for the use of the antenna in practice, as it will usually be connected, in the sense of Figure 5.23a, to the transceiver. The problem arises only for characterizations purposes, where a port as depicted in Figure 5.23b is expected.

For conventional electrically small antennas radiating into free space the effect of the cable current is predominant on the characterization of radiation quantities, like pattern or gain. Usually, the circuit characterization of the input impedance or reflection coefficient is usable even if not correct.

The situation is worse for implantable electrically small antennas, as eventual cable currents circulating on the mantle of a coaxial cable will be short-circuited by the lossy medium of the phantom during the characterization, creating a spurious reflected signal on the port. The effect is that in many cases, even circuit characterizations cannot be trusted if no precaution is taken.

This is illustrated by the example of the implantable antenna depicted in Figure 5.14, and described in [41]. In practice, the antenna will be connected directly to the transceiver, without using any cable, thus simulations are done using a lumped excitation placed at the point of the connection to the transceiver. However, a cable is required to measure the antenna's input reflection coefficient. Let us consider the three in vitro measurement setups depicted in Figure 5.25. The implantable capsule is in all cases immerged in a liquid having the same dielectric properties as muscle tissues.

Figure 5.26 shows the measured reflection coefficients obtained for these three setups compared to the antenna simulated using a lumped excitation port, thus without any cable. The effect of the cable current in these measurements is clearly seen in the results: for setup (a), this current is in direct contact with the lossy phantom, the later short-circuiting the current, which has a large impact on the measured reflection coefficient. In case (b), the presence of a vacuum shell around the cable provides some insulation between the cable current and the lossy medium, but there is still a large difference between the simulated and measured results showing the non-negligible effect of this cable current. Only setup (c), where the

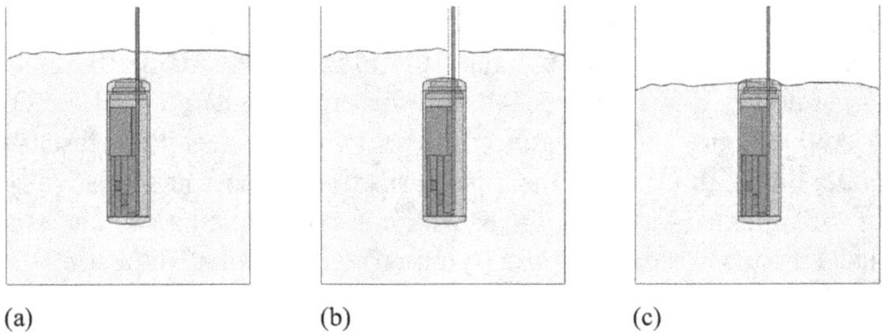

Figure 5.25 In vitro setups for the measurement of s_{11}. (a) The cable is in direct contact with the lossy medium; (b) the cable is surrounded by a lossless tube containing air or vacuum; (c) the cable is above the level of the lossy medium

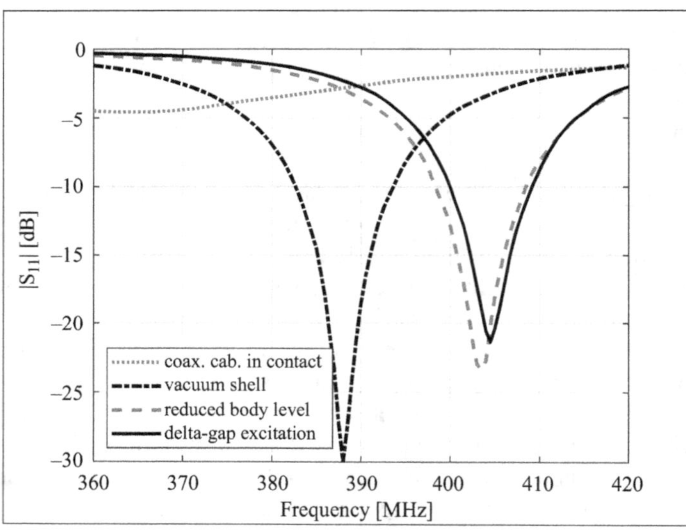

Figure 5.26 Measured s11 corresponding to the setups depicted in Figure 5.25 compared to the simulation using a lumped port

cable is entirely surrounded by free space, we obtain a measured result similar to the simulation. Indeed, in this case the current circulating on the mantle of the cable is not perturbed by any obstacle and can flow freely to the VNA, where it will be absorbed by the chassis of the latter. The measurement of the reflection coefficient is thus not perturbed by this current. However, keep in mind that this cable current would contribute to radiation, and invalidate any attempt to measure the antenna's pattern. In the case of an antenna radiating into a lossless medium, this can be alleviated by using a choke on the cable.

5.6 Conclusion

Antennas for implants are in general based on well-known designs of electrically small antennas. However, the fact that they radiate into a lossy medium greatly influences their radiation characteristics. In order to perform an efficient design of such an antenna, the influence of the lossy medium surrounding the antenna on both, the near and the propagating field terms have to be understood. This can be done efficiently using a spherical wave expansion and simplified canonical models, enabling us to gain insight and propose bounds on radiating efficiency and power density reaching the surface of the host body. The presented simple expressions for these bounds enable us to estimate the obtainable power levels in the initial stage of the design process.

The understanding of these loss mechanism and evaluation of the bounds can assist the antenna engineer to make the right choices when designing an antenna for an implant: what type of antenna to select as a starting point, what is the best volume–performance tradeoff, etc. The final result will of course depend on the actual scenario, as the depth and dimensions of the implant, and the model of the host body will greatly influence the results. However, having an initial estimate can significantly simplify and accelerate the feasibility study of a design.

Finally, the in vitro measurements of implantable antenna characteristics are even more challenging than for the case of electrically small antennas radiating into free space. As presented, great caution should be taken to eliminate the influence of the interaction of the lossy medium and the measurement setup when performing measurements and interpreting the results.

References

[1] K.S. Nikita, *Handbook of Biomedical Telemetry*, Hoboken, NJ: Wiley, 2014.

[2] T. Björninen, R. Muller, P. Ledochowitsch, *et al.*, "Design of wireless links to implanted brain-machine interface microelectronic systems," *IEEE Antennas and Wireless Propagation Letters*. vol. 11. pp. 1663–1666, 2011, 10.1109/LAWP.2013.2239252.

[3] J.M. Rabaey, M. Mark, D. Chen, *et al.*, "Powering and communicating with mm-size implants," 2011 Design, Automation & Test in Europe, 2011, pp. 1–6, doi: 10.1109/DATE.2011.5763123.

[4] A. Kiourti and K.S. Nikita, "A review of in-body biotelemetry devices: implantables, ingestibles, and injectables," in *IEEE Transactions on Biomedical Engineering*, vol. 64, no. 7, pp. 1422–1430, 2017.

[5] A. Kiourti and K.S. Nikita, "A review of implantable patch antennas for biomedical telemetry: challenges and solutions [wireless corner]," *IEEE Antennas and Propagation Magazine*, vol. 54, no. 3, pp. 210–228, 2012, doi:10.1109/MAP.2012.6293992.

[6] A. Darwish and A.E. Hassanien, "Wearable and implantable wireless sensor network solutions for healthcare monitoring," *Sensors*, vol. 11, no. 6, pp. 5561–5595, 2011.

[7] P.S. Hall and Y. Hao, *Antennas and Propagation for Body-Centric Wireless Communications*, Norwood, MA: Artech House, 2006.

[8] D. Nikolayev, M. Zhadobov, R. Sauleau, and P. Karban, "Antennas for ingestible capsule telemetry," in *Advances in Body-Centric Wireless Communication: Applications and state-of-the-art*, Q. H. Abbasi, M. Ur-Rehman, K. Qaraqe, and A. Alomainy, Eds., pp. 143–186, IET Press, 2016.

[9] S. Ma, T. Björninen, L. Sydänheimo, M.H. Voutilainen and L. Ukkonen, "Double split rings as extremely small and tuneable antennas for brain implantable wireless medical microsystems," *IEEE Transactions on Antennas and Propagation*, vol. 69, no. 2, pp. 760–768, 2021, doi: 10.1109/TAP.2020.3016459.

[10] B. Rana, J.-Y. Shim and J.-Y. Chung, "An implantable antenna with broadside radiation for a brain–machine interface," *IEEE Sensors Journal*, vol. 19, no. 20, pp. 9200–9205, 2019, doi:10.1109/JSEN.2019.2924948.

[11] X. Hu, K. Aggarwal, M.X. Yang, *et al.*, "Micrometer-scale magnetic-resonance-coupled radio-frequency identification and transceivers for wireless sensors in cells," *Physical Review Applied*, vol. 8, 014031, Published 26 July 2017.

[12] M.X. Yang, X. Hu, D. Akin, *et al.,* "Intracellular detection and communication of a wireless chip in cell," *Science Report*, vol. 11, 5967, 2021. https://doi.org/10.1038/s41598-021-85268-5.

[13] D. Nikolayev, M. Zhadobov, L. Le Coq, P. Karban and R, Sauleau, "Robust ultraminiature capsule antenna for ingestible and implantable applications," *IEEE Transactions on Antennas and Propagation*, vol. AP-65, No. 11, pp. 6107–6119, 2017.

[14] R.S. Mackay, "Radio telemetering from within the body: inside information is revealed by tiny transmitters that can be swallowed or implanted in man or animal," *Science*, vol. 134, no. 3486, pp. 1196–1202, 1961.

[15] W.H. Steinberg, F.A. Mina, P.G. Pick and G.H. Frey, "Heidelberg capsule. in vitro evaluation of a new instrument for measuring intragastric pH," *Journal of Pharmaceutical Sciences*, vol. 54, no. 5, pp. 772–776, 1965.

[16] A. Rosen, "Microwave application in cancer therapy, cardiology and measurement techniques: a short overview", *IEEE MTT Newsletters*, pp. 17–20, 1990.

[17] A. Guy, J.F. Lehman and J.B. Stonebridge, "Therapeutic applications of electromagnetic power," *Proceedings of the IEEE*, vol. 62, pp. 55–75, 1974.

[18] C.M. Furse and M.F. Iskander, "Three-dimensional electromagnetic power deposition in tumors using interstitial antenna arrays," *IEEE Transactions on Biomedical Engineering*, vol. 36, pp. 977–986, 1989.

[19] A.K. Skrivervik, "Implantable antennas: the challenge of efficiency," in *2013 7th European Conference on Antennas and Propagation (EuCAP)*, 2013, pp. 3627–3631.

[20] G.A. Conway, S.L. Cotton and W.G. Scanlon, "An antennas and propagation approach to improving physical layer performance in wireless body area networks," *IEEE Journal on Selected Areas in Communications*, vol. 27, no. 1, pp. 27–36, 2009.

[21] A.K. Skrivervik, J.-F. Zürcher, O. Staub and J.R. Mosig, "PCS antenna design: the challenge of miniaturization," *IEEE AP-Magazine*, Aug. 2001, pp. 12–27.

[22] L. J. Chu, "Physical limitations on omni-directional antennas," *Journal of Applied Physics,* vol. 19, pp. 1163–1175, 1948.

[23] H. A. Wheeler, "Fundamental limitations of small antennas," *Proceedings of the IRE*, pp. 1479–1484, 1947.

[24] R. Harrington, "Effect of antenna size on gain, bandwidth and efficiency," *Journal of Research of the National Bureau of Standards*, vol. 64D. 1960, 10.6028/jres.064D.003.

[25] R. Harrington, "On the gain and beamwidth of directional antennas," *IRE Transactions on Antennas and Propagation*, vol. 6, no. 3, pp. 219–225, 1958, doi:10.1109/TAP.1958.1144605.

[26] R.E. Collin and S. Rothschild, "Evaluation of antenna Q", *IEEE Transactions on Antennas and Propagation*, vol. AP-12, pp. 23–27, 1964.

[27] R. L. Fante, "Quality factor of general ideal antennas," *IEEE Transactions on Antennas and Propagation*, vol. AP-17, pp. 151–155, 1969.

[28] R.M. Fano, "Theoretical limitations on the broadband matching of arbitrary impedances," *Journal of the Franklin Institute*, 249, pp. 57–83, see also 139–154, 1950.

[29] J.S. McLean, "A re-examination of the fundamental limits on the radiation Q of electrically small antennas," *IEEE Transactions on Antennas and Propagation*, vol. 44, pp. 672–675, 1996.

[30] A.D. Yaghjian and S.T. Best, "Impedance, bandwidth and Q of antennas", *IEEE Transactions on Antennas and Propagation*, vol. AP-53, 2005, pp. 1298–1324.

[31] M. Gustafsson, C. Sohl and G. Kristensson, "Physical limitations on antennas of arbitrary shape", *Proceedings of The Royal Society A*, vol. 463, pp. 2589–2607, 2007.

[32] M. Gustafsson, C. Sohl and G. Kristensson, "Illustrations of new physical bounds on linearly polarized antennas", *IEEE Transactions on Antennas and Propagation*, vol. AP-57, no. 5, pp. 1319–1327, 2009.

[33] M. Gustafsson, M. Capek and K. Schab, "Tradeoff between antenna efficiency and Q-factor," *IEEE Transactions on Antennas and Propagation*, vol. 67, no. 4, pp. 2482–2493, 2019, doi:10.1109/TAP.2019.2891448.

[34] R. Moore, "Effects of a surrounding conducting medium on antenna analysis," *IEEE Transactions on Antennas and Propagation*, vol. AP-11, no. 3, pp. 216–225, 1963.

[35] S. Kim, J.S. Ho and A.S.Y. Poon, "Midfield wireless powering of subwavelength autonomous devices," *Physical Review Letters,* vol. 110, p. 203905, 2013.

[36] A. Khaleghi and I. Balasingham, "On selecting the frequency for wireless implant communications," in *2015 Loughborough Antennas & Propagation Conference (LAPC)*, Loughborough, UK, 2015, pp. 1–4, doi: 10.1109/LAPC.2015.7366023.

[37] D. Nikolayev, M. Zhadobov, P. Karban and R. Sauleau, "Electromagnetic radiation efficiency of body-implanted devices," *Physical Review Applied*, vol. 9, p. 024033, 2018.

[38] D. Nikolayev, W. Joseph, M. Zhadobov, R. Sauleau, and L. Martens, "Optimal radiation of body-implanted capsules," *Physical Review Letters*, vol. 122, p. 108101 – Published 13 March, 2019.

[39] A.K. Skrivervik, M. Bosiljevac and Z. Sipus, "Fundamental limits for implanted antennas: maximum power density reaching free space," *IEEE Transactions on Antennas and Propagation*, vol. 67, no. 8, pp. 4978–4988, 2019, doi:10.1109/TAP.2019.2891697.

[40] M. Bosiljevac, Z. Sipus and A.K. Skrivervik, "Propagation in finite lossy media: an application to WBAN", *IEEE Antennas and Wireless Propagation Letters*, vol. 14, pp. 1546–1549, 2015.

[41] F. Merli, L. Bolomey, J.-F. Zürcher, G. Corradini, E. Meurville and A.K. Skrivervik, "Design, realization and measurements of a miniature antenna for implantable wireless communication systems," *IEEE Transactions on Antennas and Propagation*, vol. AP-59, no. 10, pp. 2544–2555, 2011.

[42] A.J.M. Montes, I.V. Trivino, M. Bosiljevac, M.J. Veljovic, Z. Sipus and A. K. Skrivervik, "Antenna for a cranial implant: simulation issues and design strategies," *2020 14th European Conference on Antennas and Propagation (EuCAP)*, pp. 1–5, 2020.

[43] J. Kim and Y. Rahmat-Samii, "Implanted antennas inside a human body: simulations, designs, characterizations," *IEEE Transactions on MTT*, vol. 52, pp. 1934–1943, 2004.

[44] W. Geyi, "Physical limitations of antenna," *IEEE Transactions on Antennas and Propagation*, vol. 51, pp. 2116–2123, 2003.

[45] D.-H. Kwon, "Radiation Q and gain of TM and TE sources in phase delayed rotated configurations," *IEEE Transactions on Antennas and Propagation*, vol. 56, pp. 2783–2786, 2008.

[46] D. M. Pozar, "New results for minimum Q, maximum gain, and polarization properties of electrically small arbitrary antennas," *2009 3rd European Conference on Antennas and Propagation*, 2009, pp. 1993–1996.

[47] Evaluating Compliance With FCC Guidelines for Human Exposure to Radiofrequency Electromagnetic Fields, 97–01 ed. Washington, DC: Federal Communication Commission (FCC) Std. Supplement C, OET Bulletin 65, p. 35, 2001.

[48] Z. Sipus, A. Susnjara, A. K. Skrivervik, D. Poljak and M. Bosiljevac, "Influence of uncertainty of body permittivity on achievable radiation efficiency of implantable antennas–stochastic analysis," *IEEE Transactions on Antennas and Propagation*, vol. 69, pp. 6894–6905, 2021.

[49] S. Gabriel, R. W. Lau, and C. Gabriel, "The dielectric properties of biological tissues: II. Measurements in the frequency range 10 Hz to 20 GHz," *Physics in Medicine & Biology*, vol. 41, pp. 2251–2269, 1996.

[50] S. Gabriel, R. W. Lau, and C. Gabriel, "The dielectric properties of biological tissues: III. Parametric models for the dielectric spectrum of tissues," *Physics in Medicine & Biology*, vol. 41, pp. 2271–2293, 1996.

[51] M. Gao, D. Nikolayev, M. Bosiljevac, Z. Šipuš and A.K. Skrivervik, "Rules of thumb to assess losses of implanted antennas," *2021 15th European Conference on Antennas and Propagation (EuCAP)*, 2021, pp. 1–5, doi: 10.23919/EuCAP51087.2021.9411344.

[52] K. Ito, K. Furuya, Y. Okano and L. Hamada, "Development and characteristics of a biological tissue-equivalent phantom for microwaves," *Electronics and Communications Japan, Part I, Communications*, vol. 84, no. 4, pp. 67–77, 2001.

[53] A. Alomainy, Y. Hao, Y. Yuan and Y. Liu, "Modelling and characterisation of radio propagation from wireless implants at different frequencies," *2006 European Conference on Wireless Technology*, 2006, pp. 119–122, doi: 10.1109/ECWT.2006.280449.

[54] J-F. Zürcher, O. Staub, A.K. Skrivervik and M. Hermanjat, "Accurate measurement of the maximum gain of electrically small antennas," *Microwave and Optical Technology Letters*, vol. 23, no. 6, pp. 328–331, 1999.

[55] O. Staub, J-F. Zürcher, A. Skrivervik, "Some considerations on the correct measurement of the gain and bandwidth of electrically small antennas," *Microwave and Optical Technology Letters*, vol. 17, no. 3, pp. 156–160, 1998.

[56] E. Robert, *Collin, Foundations for Microwave Engineering*, New York, NY: Mc Graw Hill, 1966

[57] W.L. Stutzman and G.A. Thiele, *Antenna Theory and Design*, 3rd ed, New York, NY: Wiley, 2012.

[58] A. Stratton, *Electromagnetic Theory*, New York, NY: McGraw-Hill, 1941.

[59] R.F. Harrington, *Time Harmonic Electromagnetic Fields*, New York, NY: McGraw-Hill, 1961.

Chapter 6
Electromagnetic coupling in biological media

Son Chu[1], *Ekaterina Shamonina*[2] *and Christopher J. Stevens*[2]

This chapter describes and explores the impact of biological environments with low conductivity (human/animal bodies and tissues) on the well-known inductive coupling using magnetic coils. Such systems are used for wireless power transfer (WPT) to implantable medical devices and near-field communications. When using this technique in biological media or in general, non-ideal media, a time varying magnetic field induces eddy currents throughout the medium, which affect the self and mutual inductance between coils. To date, the impacts of excited eddy currents have not been fully quantified. In most current literature, scientists and researchers are using an assumption in which the attenuation of the near fields is similar to that of a plane wave travelling in a conductor. It leads to numerous problems in designing and optimizing biomedical applications. This chapter introduces the problem and presents a number of solutions for power transfer and communications through skin, muscle, and bone. A brief introduction of previous and ongoing researches in the field of magnetic coupling in non-ideal media is given in Section 6.1. Section 6.2 proposes a novel equivalent circuit model for two coupled coils inside infinite biological surroundings with the emergence of two complex Kirchhoff's coefficients. Then, the mathematical method to calculate these two new terms is presented in Section 6.3. The closed-form analytical solutions are justified by comparing them to CST EMS numerical simulations in Section 6.4. Section 6.5 performs the network analysis of a two-coil system in detail, especially impacts of the environment on the resonant frequency, Q-factor, frequency splitting, and electrical current distribution. Finally, Section 6.6 examines an immediate application: WPT using near-field magnetic induction for biomedical devices.

6.1 Introduction

Near-field electromagnetic interaction is one of the most fundamental physical phenomena in the theory of classical electromagnetism [1]. First observed by

[1]School of Neuroscience and Psychology, University of Glasgow, Glasgow, UK
[2]Department of Engineering Science, University of Oxford, Oxford, UK

Michael Faraday in 1831, the law of electromagnetic induction then becomes the principle of numerous technological applications: motors, electrical generators, transformers, inductors [2], magnetic induction tomography [3], magnetic resonance imaging [4], eddy current evaluation and testing [5], and WPT [6], to name but a few. Over the last decade, worldwide researchers and technologists have paid increasing attention to the effectiveness of electromagnetic induction in transmitting wireless signals through complex propagation environments [7–13], where transverse electromagnetic waves have a number of limitations, most notably losses. Numerous applications have been found in the field of communications, for example, underground/underwater wireless sensor networks for environmental monitoring, landslide inspection, localization, and security [7–10]. Systems exploiting near-field magnetic fields have less severe challenges associated with signal transmission, e.g., propagation delay, fading and multipath propagation [14]. In addition to this, magnetically coupled resonant circuits have been proposed for applications in in vivo medical diagnosis and treatment applications, and body area networking for health-related applications [11,12,15,16] since biological tissues and fluids are relatively conductive at microwave frequencies, typically from 0.05 to 1.5 S/m [17–19]. Near-field magnetic coupling is preferable among the various methods for wireless power and data transmission for implantable biomedical devices (ultrasonic and optical methods, capacitive coupling, intra-body communication, etc.) because of its high efficiency and lower impact on living tissues [20].

Technically, dissipative transmission environments present an extreme challenge for the magnetic channel between two coupled resonators. Unlike the lossless case, eddy current loops are induced within the medium, perpendicular to the changing magnetic field according to Faraday's law. These eddy currents have strong impacts on the entire circuits, one of which is the attenuation of the coupling strength to oppose the change in flux (Lenz's law). Intensive efforts to model the coupling between two low frequency coils in a lossy medium have been taken across many research groups in widely different fields. In most of the current literature on near-field communications, to describe the effects of eddy currents in the system modeling, path loss is the key factor to be considered [8–10,13]. In particular, the skin-depth-based-plane-wave approximation, in which the mutual coupling is assumed to have the same decay with distance as a plane wave traveling in a conductor, is often used. However, the alteration of the near-field magnetic fields due to the lossy medium appears to be more complex than this simple method. The credibility of the skin-depth model was examined in [21] against other independent methods. While the skin-depth model does not take into account the phase of the mutual inductance, it has been shown via numerical simulations and laboratory experiments that the coupling generally becomes complex as a consequence of the elliptical polarization of the magnetic field [21,22]. The skin-depth approximation has no physical basis for near fields and indeed is too simplistic to represent the effects of eddy currents on the path loss. Not only attenuating the magnetic coupling between the two coupled resonators, the induced eddy currents also modify the self-impedance of the circuits because for near-field systems, the absorption of the power at the receiver affects the load of the transmitter. This feature is the foundational principle for non-destructive eddy current testing [5,23,24], where the change in the properties of a coil is used to detect and characterize surface and sub-surface flaws in conductive samples. Nevertheless, the samples used in eddy current testing are often metal sheets so that the mathematical models are irrelevant to the biological media, whose conductivity is 6 or 7 orders of magnitude lower than that of metals.

Recent studies have paid more attention to the impacts of eddy currents on the self-impedance of the circuit as the eddy current losses and detuning effects have been taken into account in [21,25–27]. The mathematical models in these articles were all developed from the field's differential equations but each one has several considerable limitations. The configuration investigated in [21] was applicable when the two coupled coils are placed above a half-space conductive slab. It would probably be irrelevant to the case when the two coils are completely embedded inside a dissipative surrounding, e.g., implanted in the human or animal bodies as in implantable medical devices. In [25], the authors considered WPT for unmanned underwater vehicles using two coupled resonators in seawater, their electromagnetic field analysis was carried out for the region of medium only between the two coils. Furthermore, the detuning effects of the eddy current were disregarded, which may be acceptable at low frequencies (typically kHz), but can be problematic when either higher frequencies are used or the transmission medium has high permittivity. Better models in [26,27] give due consideration to the eddy current loss and the detuning effects but some important parameters in their analytical method were obtained through simulation and experimental means.

In this work, we construct a general equivalent circuit model for two magnetically coupled circuits in infinite homogeneous dissipative media as a combination of circuit theory and field theory. All electromagnetic and feedback effects due to the induced eddy currents are analytically incorporated in the model, represented by two new Kirchhoff's terms (generally complex and frequency-dependent). We further derive and elaborate the theory of WPT in a non-ideal medium. Our work will serve as a fundamental model for design and analysis of every system employing magnetically coupled circuits in RF-challenging media.

6.2 Circuit model of coupled coils in a general non-ideal environment

Figure 6.1 shows the basic unit of a coupled-coil-system in an infinite homogeneous non-ideal environment ($\sigma \neq 0$). It consists of two identical coils with the same orientation placed in lossless dielectric cavities ($\sigma = 0$ S/m). The purpose of the insulators is to prevent the direct leakage of current to the medium background and prevent direct conduction currents from the transmitter to the receiver.

It is clear that a dissipative transmission medium has non-negligible effects on the near-fields of the two coils. First and foremost, the mutual coupling is predictably attenuated due to induced eddy currents in the medium according to Lenz's law. Besides this, there is a change in the resistance and reactance of the coils as discussed above. As a result, two corresponding terms—*complex self-inductance* and *complex*

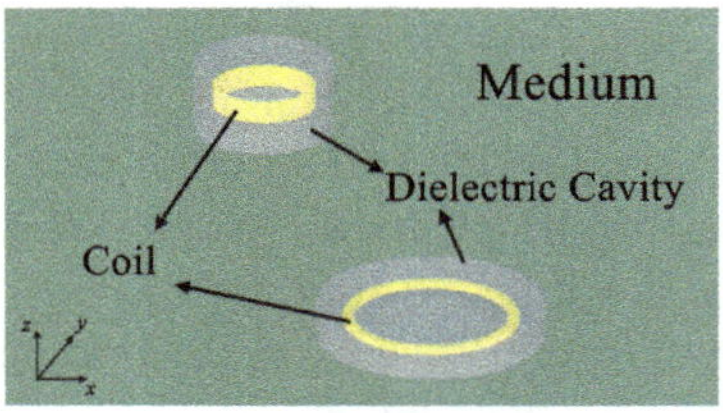

Figure 6.1 General diagram of two coupled coils in proximity with the same orientation, insulated by a small dielectric cavity, and completely immersed inside an infinite homogeneous dissipative medium. Reprinted with permission from [28]

mutual inductance—will be introduced to describe the change in the self-impedance and mutual inductance of two coils inside an attenuating medium, respectively. Applying Kirchhoff's voltage law, the general circuit model of this system is expressed by:

$$\begin{bmatrix} V_1 \\ V_2 \end{bmatrix} = \begin{bmatrix} Z_1 + j\omega\Delta\mathcal{L}_1 & j\omega\mathcal{M} \\ j\omega\mathcal{M} & Z_2 + j\omega\Delta\mathcal{L}_2 \end{bmatrix} \begin{bmatrix} I_1 \\ I_2 \end{bmatrix} \tag{6.1}$$

where V_i, I_i, and Z_i are the applied voltage, current response, and self-impedance of coil i, respectively, $\Delta\mathcal{L}_i$ is the complex self-inductances of coil i, representing the coupling of itself and the eddy currents it generates, $\mathcal{M} = M_0 + M_E$ is the complex mutual inductance between two coils inside the medium, M_0 is the mutual inductance of two coils in free space, M_E represents the coupling of a coil and the eddy currents generated by the other coil, and ω is the angular frequency. When the transmission medium is free space, i.e., $\sigma_{(medium)} = 0$ S/m, $\varepsilon_{r\,(medium)} = 1$, and $\mu_{r\,(medium)} = 1$, it has no impact on the circuits and therefore, all the additional terms—the complex self-inductance and the added mutual inductance—are zero, i.e., $\Delta\mathcal{L}_1 = \Delta\mathcal{L}_2 = M_E = 0$ and $\mathcal{M} = M_0$. Consequently, (6.1) reduces to the circuit model of two coils coupled in free space. In order to completely model the system, one need only find the values of these four terms when the medium is not vacuum.

6.3 Complex Kirchhoff coefficients: a field solution approach

In order to calculate the four complex coefficients, the differential equation for the magnetic vector potential is firstly formulated and hence, the complex self and mutual inductance of the coils can be obtained directly from the eddy-current-based fields. Figure 6.2a shows the cross-section of the two-coil-system in Figure 6.1. To simplify the configuration, the dielectric cavities made of non-magnetic insulating material are assumed to be vacuum. Also, to derive a closed-form analytical solution for the magnetic vector potential, they are then replaced by infinitely wide layers, as shown in Figure 6.2b.

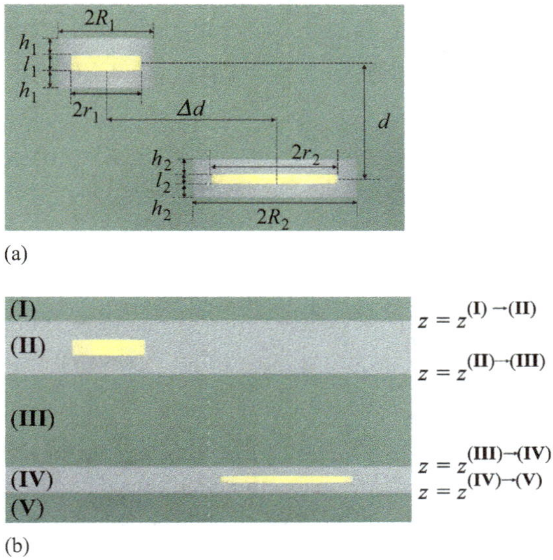

Figure 6.2 Cross-section of the configuration in Figure 6.1. (a) The finite size dielectric cavities used in the configuration are considered as infinite dielectric layers (b) in the theoretical model. Reprinted with permission from [28]

Here, the superposition theorem is exploited to reduce complexity: the field solutions for one coil are solved in the absence of the other coil. Due to the symmetry of the configuration, the magnetic vector potential has only one spatial component along the azimuthal angle φ in the cylindrical coordinates, i.e., $\mathbf{A}(0, A_\varphi, 0)$. As a result, the differential equation for the magnetic vector potential of coil 1 due to an applied voltage $\mathbf{V}_0$ is expressed by [24,29]:

$$\begin{aligned}\nabla^2 \mathbf{A}_1^{(n)} &= \left(\frac{\partial^2 A_{1\varphi}^{(n)}}{\partial z^2} + \frac{\partial^2 A_{1\varphi}^{(n)}}{\partial \rho^2} + \frac{1}{\rho}\frac{\partial A_{1\varphi}^{(n)}}{\partial \rho} - \frac{A_{1\varphi}^{(n)}}{\rho^2} \right)\hat{\varphi} \\ &= \left(\mu^{(n)}\sigma^{(n)} \frac{dA_{1\varphi}^{(n)}}{dt} + \mu^{(n)}\varepsilon^{(n)} \frac{d^2 A_{1\varphi}^{(n)}}{dt^2} \right)\hat{\varphi}\end{aligned} \tag{6.2}$$

where $A_{1\varphi}^{(n)}$ is the vector potential of coil 1 in the region n, which for symmetry reasons only depends on the cylindrical coordinates ρ and z, $\hat{\varphi}$ are the azimuthal unit vectors in cylindrical coordinates, $\mu^{(n)}$, $\varepsilon^{(n)}$, and $\sigma^{(n)}$ are the permeability, permittivity, and conductivity of region n, respectively, and t is the time variable. To further simplify the mathematical complexity, the magnetic vector potential in the time-domain is defined as the convolution of the magnetic vector potential impulse response (MVPIR) function with the transient current in the respective coils [24]:

$$A_1(\rho, z, t) \equiv \hat{A}_1(\rho, z, t) * i_1(t) \tag{6.3}$$

where

$$\hat{A}_1(\rho,z,t) = \hat{E}_1(\rho,z,t) + \hat{\Psi}_1(\rho,z,t) = \hat{E}_1(\rho,z,t) + \hat{\Psi}_1(\rho,z)\delta(t).$$

$\hat{A}_1(\rho,z,t)$ are the MVPIRs, $\hat{\Psi}_1(\rho,z,t)$ is the MVPIR generated by a unit impulse current in coil 1 in free space, $\hat{E}_1(\rho,z,t)$ is the MVPIR generated by eddy currents in the medium, and $\delta(x)$ is the Dirac delta function of variable x. By using the mathematical method in [24], the solutions for MVPIR for coil 1 problem in each region in the frequency domain and the Hankel space (α-space) are as follows:

$$\begin{aligned} &\hat{A}_1(\alpha,z,\omega) = \hat{\Psi}_1(\alpha,z,\omega) + X_1^{(\mathbf{II})}(\alpha,\omega)e^{\alpha^{(II)}z} + Y_1^{(\mathbf{II})}(\alpha,\omega)e^{-\alpha^{(II)}z}, && \text{forn} = \mathbf{II}, \\ &X_1^{(n)}(\alpha,\omega)e^{\alpha^{(n)}z} + Y_1^{(n)}(\alpha,\omega)e^{-\alpha^{(n)}z}, && \text{forn} \neq \mathbf{II}. \end{aligned} \tag{6.4}$$

The unknown coefficients $X_1^{(n)}$ and $Y_1^{(n)}$ in (6.4) can be solved by applying the boundary conditions (6.5)–(6.7) to ensure the continuity conditions for the electric and magnetic field components at the interfaces between two regions:

$$\hat{A^{(n)}}(\alpha,z,\omega)\big|_{z=z^{n\to n+1}} = \hat{A^{(n)}}(\alpha,z,\omega)\big|_{z=z^{n\to n+1}} \tag{6.5}$$

$$\left.\frac{1}{\mu^{(n)}}\frac{d\hat{A^{(n)}}(\alpha,z,\omega)}{\mathrm{d}z}\right|_{z=z^{n\to n+1}} = \left.\frac{1}{\mu^{(n+1)}}\frac{\mathrm{d}\hat{A^{(n+1)}}(\alpha,z,\omega)}{\mathrm{d}z}\right|_{z=z^{n\to n+1}} \tag{6.6}$$

$$\begin{aligned} z^{n\to n+1} &= -2h_1 - l_1, && \text{for interface I–II}, \\ &0, && \text{for interface II–III}, \\ &c, && \text{for interface III–IV}, \\ &c + 2h_2 + l_2, && \text{for interface IV–V}. \end{aligned} \tag{6.7}$$

where $z^{n\to n+1}$ is the location of the interface between consecutive zones n and $n+1$. Once all the coefficients are known, every other physically observable electromagnetic quantity can be calculated directly from the magnetic vector potential.

Complex Kirchhoff coefficients

The complex self-inductance term is defined in the same manner as the self-inductance. Whereas the self-inductance represents the interaction between the source magnetic field produced by a coil with itself to produce an EMF, the complex self-inductance can be referred to as the interaction of the secondary magnetic field generated by the eddy currents with the coil.

Meanwhile, the complex mutual inductance is a summation of the original mutual inductance in free-space and the coupling between the second coil and the eddy currents generated by the first coil, since the magnetic vector potential in the region of the second coil is an aggregate of the primary field generated by the first coil and the secondary field generated by the eddy currents, cf. (6.4). In other words, it is the mutual inductance between two coils mediated by the medium.

Consequently, the complex self-inductance of coil 1 can be calculated by integrating the eddy-current-based MVPIRs arising from currents in coil 1 in the region containing the coil itself over its volume:

$$\Delta\mathcal{L}_1 = \frac{2\pi N_1}{w_1 l_1}\int_0^{\infty}\int_{r_1-w_1/2}^{r_1+w_1/2}\int_{(-h_1-l_1)}^{(-h_1)} arJ_1(ar)\hat{E}_1^{(\mathbf{II})}(a,z,\omega)\mathrm{d}a\mathrm{d}r\mathrm{d}z \tag{6.8}$$

Then, the complex mutual inductance is determined by integrating the eddy-current-based MVPIR arising from currents in a nearby coil in the region containing the secondary coil over its volume. Applying integral transformations in [30,31] for the mutual inductance of two coils with a horizontal offset, the complex mutual inductance can be calculated by:

$$\mathcal{M} = \mathcal{M}_{21} = \frac{2\pi N_2}{w_2 l_2}\int_0^{\infty}\int_{r_2-w_2/2}^{r_2+w_2/2}\int_{c+h_2}^{c+h_2+l_2} arJ_0(a\Delta d)J_1(ar)\hat{E}_1^{(\mathbf{IV})}(a,z,\omega)\mathrm{d}a\mathrm{d}r\mathrm{d}z \tag{6.9}$$

where Δd is the horizontal displacement between two coils.

By proceeding analogously, the complex self-inductance of coil 2 and the complex mutual inductance between coil 2 and coil 1 can be calculated by using the following equations:

$$\Delta\mathcal{L}_2 = \frac{2\pi N_2}{w_2 l_2}\int_0^{\infty}\int_{r_2-w_2/2}^{r_2+w_2/2}\int_{c+h_2}^{c+h_2+l_2} arJ_1(ar)\hat{E}_2^{(\mathbf{IV})}(a,z,\omega)\mathrm{d}a\mathrm{d}r\mathrm{d}z \tag{6.10}$$

$$\mathcal{M} = \mathcal{M}_{12} = \frac{2\pi N_1}{w_1 l_1}\int_0^{\infty}\int_{r_1-w_1/2}^{r_1+w_1/2}\int_{(-h_1-l_1)}^{(-h_1)} arJ_0(a\Delta d)J_1(ar)\hat{E}_2^{(\mathbf{II})}(a,z,\omega)\mathrm{d}a\mathrm{d}r\mathrm{d}z \tag{6.11}$$

While these are complex integrals, they are susceptible to direct calculation and hence can be used to quickly and fully evaluate the impact of the conductive environment on our two coupled coils.

6.4 Simulation results of complex Kirchhoff coefficients

As one of the main assumptions in the mathematical method is that the dielectric cavities become infinitely wide, full-wave electromagnetic simulations in CST EM studio (CST EMS – statics/low-frequency package) are performed to justify the validity of the analytical method when the dielectric cavities are finite. Figure 6.3

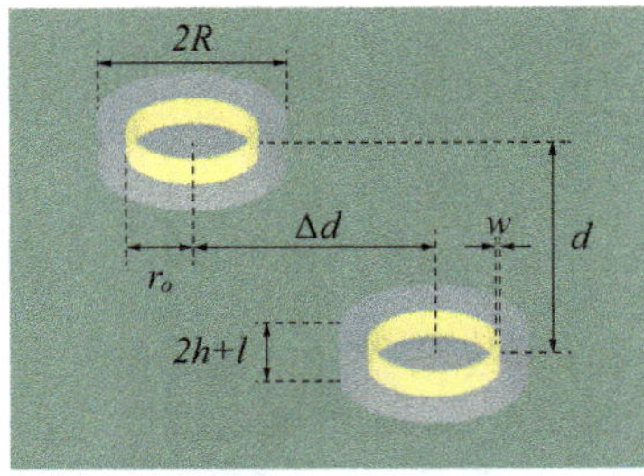

Figure 6.3 *Configuration investigated consisting of two identical circular air core coils with dielectric enclosures placed inside an infinite, linear, homogeneous, isotropic, and dissipative environment. Reprinted with permission from [28]*

Table 6.1 *Parameter list of the configuration investigated in this chapter*

Parameters	**Value**
Mean radius of coils, r_0	11 mm
Width of coils, w	1 mm
Height of coils, l	5 mm
Height of free-space cavities, $2h + l$	7 mm
Relative permittivity of medium, $\varepsilon_r^{(I)}$	78
Relative permeability of medium, $\mu_r^{(I)}$	1

Table 6.2 *Important settings in the CST EMS for all simulations in this chapter*

Settings	**Value**
Boundary condition	Open
Boundary size	A cube of at least $30r_0$
Solver type	Low frequency domain solver
Solver equation	Full wave
Solver order	3rd (high accuracy)
Mesh type	Tetrahedral
Adaptive mesh refinement	At least 10 passes

shows the geometry of the configuration investigated: two identical coils are immersed inside a conducting medium of conductivity $\sigma^{(I)}$ at a vertical distance d and a horizontal offset Δd apart from each other. The background medium is assumed to be infinite, linear, isotropic and homogeneous. Some geometric and electromagnetic parameters of the configuration are itemized in Table 6.1, while Table 6.2 lists essential settings used for all simulations.

For simplicity and without loss of generality, all results presented in the following are calculated at the frequency of 46 MHz. It is assumed that each coil is a split ring resonator made of annealed copper and its resistance is purely attributable to the skin depth effect. With the coil dimensions listed in Table 6.1, the circuit parameters are calculated by using standard formulas for loop antennas as follows: $R_1 = R_2 = 10.19$ mΩ, $L_1 = L_2 = 30.46$ nH, $C_1 = C_2 = 393$ pF, and $Q = 432$. In addition, the analytical equations for the complex coefficients can be recast as:

$$\Delta\mathcal{L} = \frac{2\pi N}{wl} \int_{0}^{\infty} \int_{r-w/2}^{r+w/2} \int_{(-h-l)}^{(-h)} \alpha r J_1(\alpha r) \hat{E}_1^{(\mathbf{II})}(\alpha, z, \omega) \mathrm{d}\alpha \mathrm{d}r \mathrm{d}z \tag{6.12}$$

$$\mathcal{M} = \frac{2\pi N}{wl} \int_{0}^{\infty} \int_{r-w/2}^{r+w/2} \int_{c+h}^{c+h+l} \alpha r J_1(\alpha r) J_0(\alpha \Delta d) \hat{E}_1^{(\mathbf{IV})}(\alpha, z, \omega) \mathrm{d}\alpha \mathrm{d}r \mathrm{d}z \tag{6.13}$$

Simulations for three different sizes of the cavities $R = 1.1r_0$, $2r_0$, and $3r_0$ (cf. Figure 6.2a) were performed to justify our assumption of infinitely wide dielectric layers. The variation of the complex coefficients across a range of common medium conductivity is the main interest since the strength of the induced eddy currents depends

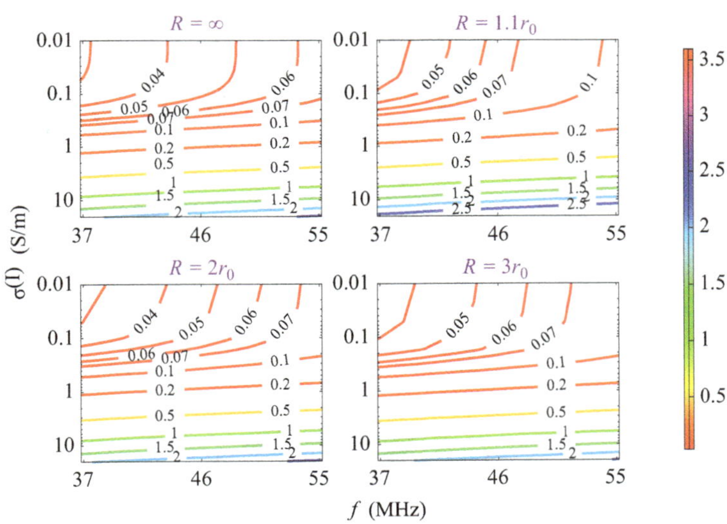

Figure 6.4 Modulus of the complex self-inductance $|\Delta\mathcal{L}|$ as a function of the working frequency and the medium conductivity. The medium conductivity varies from $\sigma^{(I)} = 0.01$ to $\sigma^{(I)} = 20$ S/m while the working frequency ranges between $f = 37$ to $f = 55$ MHz. (Top left) The analytical value obtained with (6.12) is compared to simulation results for three different sizes of the insulators: (top right) $R = 1.1r_0$, (bottom left) $R = 2.0r_0$, and (bottom right) $R = 3.0r_0$. Reprinted with permission from [28]

heavily on the electrical conduction. As a further consideration, the separation constant $\alpha^{(n)} = \sqrt{\alpha^2 + j\omega\mu^{(n)}\sigma^{(n)} - \omega^2\mu^{(n)}\varepsilon^{(n)}}$ in the Hankel transform is also a function of frequency. Accordingly, the complex Kirchhoff coefficients are frequency-dependent, unlike normal inductive terms which are always considered constant at low frequency.

Figures 6.4 and 6.5 present the modulus of the two complex terms obtained with (6.12) and (6.13) as a surface function of the conductivity of the surrounding environment and the working frequency. The frequency varies from 37 to 55 MHz while the medium conductivity is within the range between 0.01 and 20 S/m, where the conductivity of most earth and biological materials is measured. For the purpose of comparison, the simulation results are included in those figures corresponding to three different insulator radii of $R = 1.1r_0$, $2r_0$, and $3r_0$ respectively. Clearly from Figures 6.4 and 6.5, theoretical results obtained with (6.12) and (6.13) show the same trends as those obtained with CST EMS for all three different sizes of the insulators. Unlike normal inductive circuit parameters which are always considered constant at low frequency, both proposed Kirchhoff coefficients are complex and frequency-dependent.

Figures 6.6 and 6.7 further compare the real and imaginary parts of the complex self-inductance and complex mutual inductance obtained with the analytical

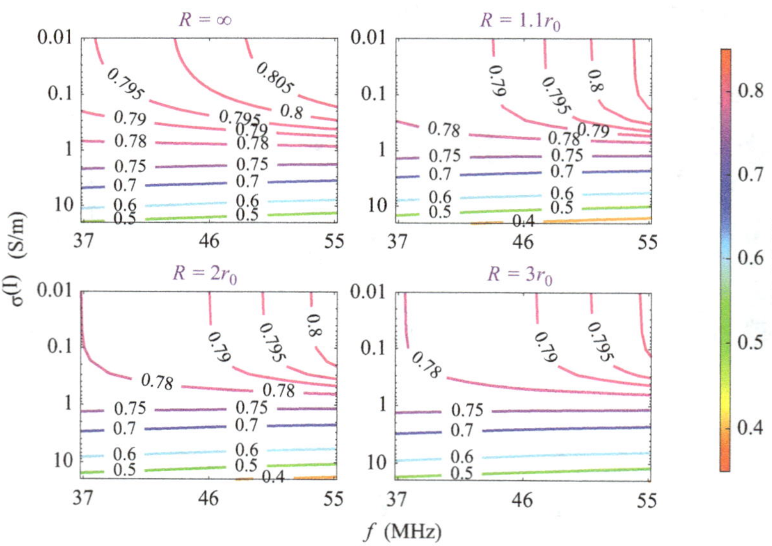

Figure 6.5 Modulus of the complex mutual inductance $|\mathcal{M}|$ as a function of the working frequency and the medium conductivity. The medium conductivity varies from $\sigma^{(I)} = 0.01$ to $\sigma^{(I)} = 20$ S/m while the working frequency ranges between $f = 37$ to $f = 55$ MHz. (Top left) The analytical value obtained with (6.13) is compared to simulation results for three different sizes of the insulators: (Top right) $R = 1.1r_0$, (bottom left) $R = 2.0r_0$, and (bottom right) $R = 3.0r_0$. Reprinted with permission from [28]

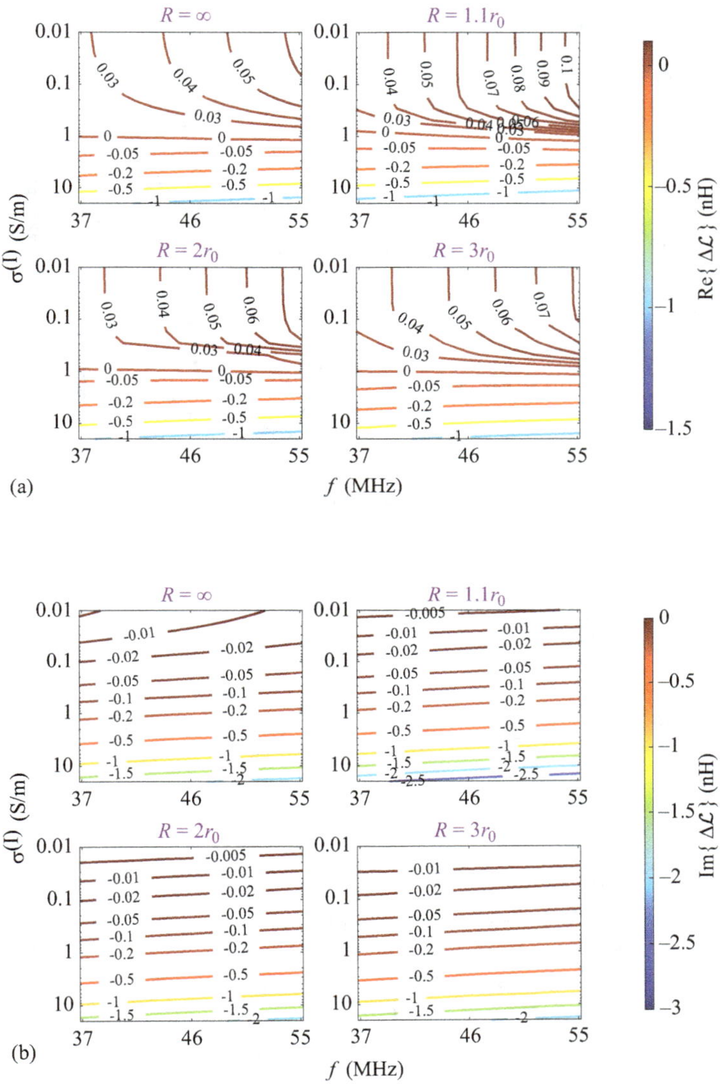

Figure 6.6 *The (a) real part and (b) imaginary part of the complex self-inductance $\Delta\mathcal{L}$ as a function of the working frequency and the medium conductivity. The medium conductivity is from $\sigma^{(\mathrm{I})} = 0.01$ to $\sigma^{(\mathrm{I})} = 20$ S/m while the working frequency ranges between $f = 37$ to $f = 55$ MHz. The (top left) analytical values ($R = \infty$) are compared to CST EMS simulation results for (top right) $R = 1.1r_0$, (bottom left) $R = 2r_0$, and (bottom right) $R = 3r_0$. Reprinted with permission from [28]*

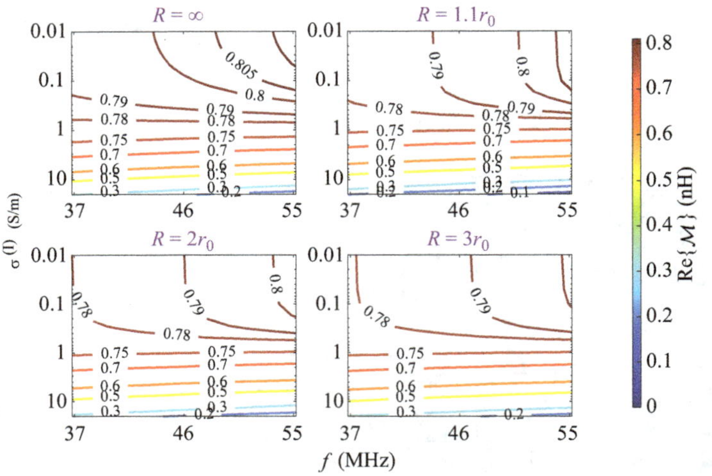

(a) Real part of complex mutual inductance

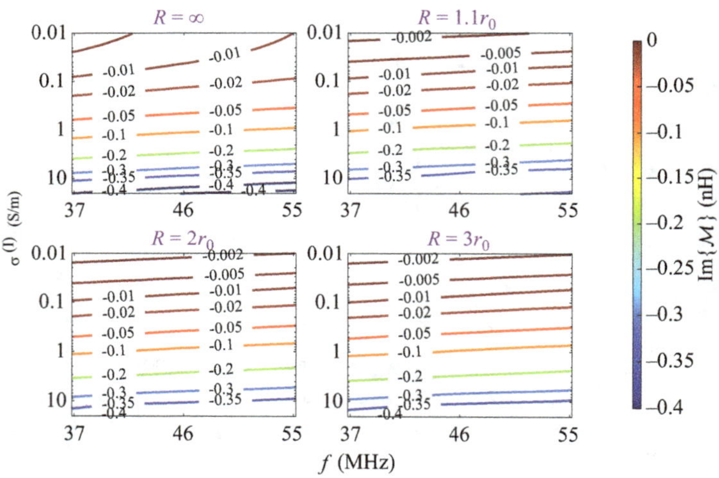

(b) Imaginary part of complex mutual inductance

Figure 6.7 The (a) real part and (b) imaginary part of the complex mutual inductance $\mathcal{M}$ as a function of the working frequency and the medium conductivity. The medium conductivity is from $\sigma^{(\mathrm{I})} = 0.01$ to $\sigma^{(\mathrm{I})} = 20$ S/m while the working frequency ranges between $f = 37$ and $f = 55$ MHz. The (top left) analytical values ($R = \infty$) are compared to CST EMS simulation results for (top right) $R = 1.1r_0$, (bottom left) $R = 2r_0$, and (bottom right) $R = 3r_0$. Reprinted with permission from [28] approach and numerical simulations, respectively. For the complex self-inductance, the magnitude of its real and imaginary parts depends proportionally on the working frequency because higher frequencies lead to stronger eddy current effects. Notably, the complex self-inductance always exhibits a negative imaginary part, representing the medium losses, as shown in Figure 6.6b. Meanwhile, depending on the electromagnetic properties of the environment, the real part of the complex self-inductance can have either sign, representing the change in self-flux of the coil. When the medium has high permittivity but low conductivity (e.g., lossy dielectric materials), the displacement eddy currents, which have an opposite sign with the conduction eddy currents, reinforce the flux into the coils. Consequently, the real part of the complex self-inductance is a positive value. As the conductivity increases, the conduction eddy current dominates their counterpart—the displacement ones. Opposite EMF induced by the conduction eddy currents reduces the self-flux of the coil as expected from Lenz's law so that the complex self-inductance's real part is negative. In any case, the complex self-inductance modifies the self-inductance of the coils, making the coils resonant at different frequencies. Else, the complex mutual inductance is indeed a complex quantity when the two coils are completely immersed in a non-ideal medium. Since the eddy currents attenuate the magnetic field generated by the transmitter, the modulus of the complex mutual inductance declines with the increasing frequency and medium conductivity, as shown in Figure 6.5.

Next, the analytical model is examined quantitatively by comparing the theoretical prediction of the complex coefficients to those obtained with numerical simulations for different cavity sizes. Based on the infinite dielectric layer approximation in the mathematical approach, it is clear that the larger the insulators are, the more accurate the model provides. A bar graph, Figure 6.8, depicts the relative absolute error (in percent) and the root-mean-square error (in arc degree) in the modulus and phase of the theoretical values for the two *Kirchhoff* terms,

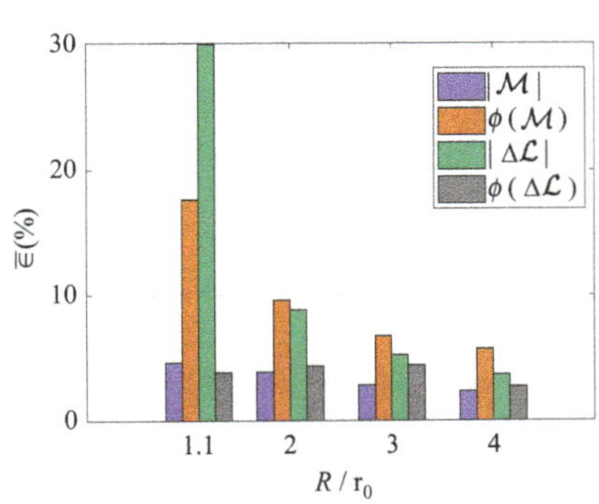

Figure 6.8 Average relative errors of the analytical solutions for the complex Kirchhoff coefficients in terms of their moduli and phases with respect to CST EMS simulation results for several different sizes of the dielectric cavities. Reprinted with permission from [28]

respectively. The errors are averaged over both the simulated ranges of conductivity and working frequency. When the radius of the cavities is at least three times larger than the coil radius $R \geq 3r_0$, (6.12) and (6.13) are found to be in excellent agreement with the finite-element simulations, with an absolute percentage error of less than 7% for the modulus and a root mean square error less than 4° for the phase. More details on the comparison for variable vertical distance and horizontal displacement can be found in [28,32].

6.5 Network analysis of a two-coil system

In the previous section, the merit of the new analytical solutions to calculate the complex Kirchhoff coefficients has been demonstrated. In comparison to the numerical simulations with CST EMS, (6.12) and (6.13) show a remarkable accuracy not exceeding 7% of the relative error for reasonable values of the parameters. In this section, the detailed network analysis is performed with only results from these expressions. It is equivalent to assuming the radius of the dielectric cavities is larger than at least three times the coil radius. Here, the notations are as follows: x is the circuit parameter of the coil in free space while $\widetilde{x}$ is that of the coil inside a dissipative medium.

6.5.1 *Resonant frequency and quality factor*

Since the induced eddy currents affect the impedance of the coils via the complex self-inductance, there are indeed some changes in resonant frequency and the quality factor. Here, the impact of the medium on the resonant frequency and the quality factor of the coils are characterized quantitatively using the equivalent circuit model in (6.1). The impedance of a resonant coil inside a non-ideal medium can be expressed by:

$$\widetilde{Z}_n = Z_n + j\omega\Delta\mathcal{L}_n = (R_n + \omega\Delta\mathcal{L}_n'') + j\omega(L_n + \Delta\mathcal{L}_n') + \frac{1}{j\omega C_n} \tag{6.14}$$

where Z_n, R_n, L_n, and C_n are the self-impedance, resistance, inductance, and capacitance of the coil n in free-space, respectively, and $\Delta\mathcal{L}_n = \Delta\mathcal{L}_n' - j\Delta\mathcal{L}_n''$ is the complex self-inductance due to the impact of the eddy currents the coil n generates (the negative imaginary part represents the medium losses).

In principle, the resonant frequency of one circuit occurs when its inductive reactance is equal to its capacitive reactance. Unlike the standard equivalent circuit theory where the electrical reactance components are ideally independent of frequency for a certain range of frequency, the complex self-inductance is frequency-dependent, as shown in the previous section. Therefore, the resonant frequency of a coil inside a conductive medium is the root of the following equation representing the zero total reactance:

$$2\pi\widetilde{f}_{0|n}\left(L_n + \Delta\mathcal{L}_{n'}(\widetilde{f}_{0|n})\right) = \frac{1}{2\pi\widetilde{f}_{0|n}} C_n \Leftrightarrow \widetilde{f}^{\,2}_{0|n}\left(L_n + \Delta\mathcal{L}_{n'}(\widetilde{f}_{0|n})\right) - \frac{1}{4\pi^2 C_n} = 0 \tag{6.15}$$

where $\widetilde{f}_{0|n}$ is the perturbed resonant frequency of the coil n inside the medium. Once the resonant frequency is determined, the coil's Q-factor inside the medium, $\widetilde{Q}_n$, is then obtained by:

$$\widetilde{Q}_n = \frac{2\pi\widetilde{f}_{0|n}\left(L_n + \Delta\mathcal{L}'_n(\widetilde{f}_{0|n})\right)}{R_n + 2\pi\widetilde{f}_{0|n}\Delta\mathcal{L}''_n(\widetilde{f}_{0|n})} \tag{6.16}$$

Using the same configuration listed in Table 6.1, the complex self-inductance can be calculated with (6.12). Figure 6.9 shows an example of the resonant frequency and the quality factor versus the medium conductivity. When the medium is a lossy dielectric material ($\sigma \ll 1$ S/m and $\varepsilon_r = 78$), the displacement eddy currents reinforce the circuit's own magnetic flux into the circuit (positive real part of the complex self-inductance). As a result of an increase in the self-inductance, the resonant frequency of the coil unsurprisingly lessens by a small amount. As the conductivity increases, the medium allows larger conduction eddy currents to take place. According to Lenz's law, the opposing fields generated by the conduction eddy currents counterbalance the self-flux of the coil, leading to a decline in the coil's self-inductance. At higher conductivities, the conduction eddy currents dominate its counterpart as the real part of the complex self-inductance becomes negative. Thus, the medium lowers the total self-inductance of the coil, and tunes the resonance to higher frequencies.

In addition to the resonant frequency, another key characteristic parameter of the coil—the Q-factor—is examined. From (6.14), the induced eddy currents not only change the self-flux of the coil but also introduce additional losses to the circuit, represented by the imaginary part of the complex self-inductance. Correspondingly, the Q-factor expectedly declines proportionally with the

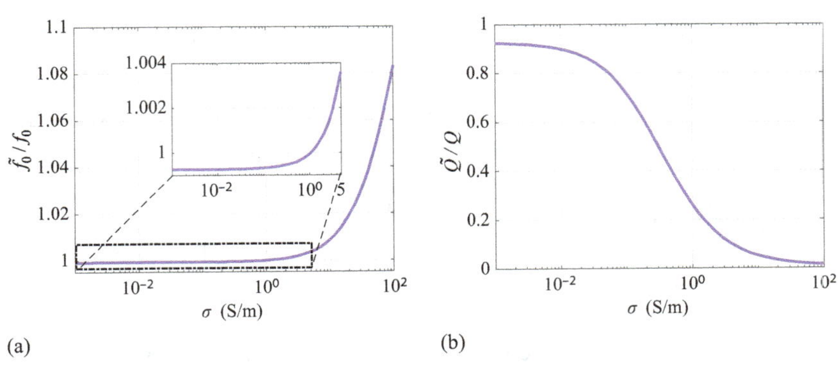

Figure 6.9 Calculated normalized (a) resonant frequency and (b) quality factor of a coil inside an attenuating medium with variable conductivity from $\sigma = 10^{-3}$ to $\sigma = 10^2$ S/m. The results obtain with (6.15) and (6.16) are normalized to those of the coil in vacuum ($f_0 = 46$ MHz and $Q = 432$). Reprinted with permission from [28]

increasing conductivity. Clearly from Figure 6.9b, the quality factor seems to be strongly affected by the medium's conductivity even for a poorly conductive medium. When the conductivity is around $\sigma = 1.059$ S/m (mean value of that of living human brains [19]), the Q-factor of the coil loses nearly 80% of its free-space value ($Q = 432$). As the conductivity reaches $\sigma = 10^2$ S/m, the losses due to the medium mostly abolish the Q-factor.

6.5.2 Frequency splitting

When the two coils are placed so that they are strongly coupled (i.e., the strength of the mutual coupling between the two resonators is larger than losses [33]), a resonance splitting exists. The current responses in the two coils then exhibit two peaks above and below the resonance frequency in the non-coupled case, namely the odd and even mode. The current responses are in phase at the even mode whilst 180° out of phase at the odd mode. To examine the effects of the dissipative medium on the frequency splitting phenomenon, the conditions for the strongly coupled regime in vacuum shall be satisfied. So the two coils above are placed coaxially at a distance of 30 mm apart from each other.

Since the Q-factor of the coil is usually high in vacuum, the internal resistance of the coil can be neglected in order to distinguish the impact of medium losses on the frequency splitting. The undamped resonant frequencies for two coupled coils can be sought by performing the eigenvalue analysis. The natural response of a two-coupled-coil-system can be written as:

$$\begin{bmatrix} \widetilde{Z} & j\omega\mathcal{M} \\ j\omega\mathcal{M} & \widetilde{Z} \end{bmatrix} \begin{bmatrix} I_1 \\ I_2 \end{bmatrix} = \begin{bmatrix} 0 \\ 0 \end{bmatrix} \tag{6.17}$$

$$j\omega(L_0 + \Delta\mathcal{L}') \begin{bmatrix} -j\dfrac{\Delta\mathcal{L}''}{L_0 + \Delta\mathcal{L}'} + 1 - \dfrac{\widetilde{\omega}_0^2}{\omega^2} & \dfrac{\mathcal{M}}{L_0 + \Delta\mathcal{L}'} \\ \dfrac{\mathcal{M}}{L_0 + \Delta\mathcal{L}'} & -j\dfrac{\Delta\mathcal{L}''}{L_0 + \Delta\mathcal{L}'} + 1 - \dfrac{\widetilde{\omega}_0^2}{\omega^2} \end{bmatrix} \begin{bmatrix} I_1 \\ I_2 \end{bmatrix} = \begin{bmatrix} 0 \\ 0 \end{bmatrix} \tag{6.18}$$

where $\widetilde{\omega}_0 = 1/\sqrt{(L_0 + \Delta\mathcal{L}')C_0}$ is the angular frequency of the coils inside the medium. Equation (6.18) can be rearranged as:

$$\begin{bmatrix} -j\dfrac{\Delta\mathcal{L}''}{L_0 + \Delta\mathcal{L}'} + 1 & \dfrac{\mathcal{M}}{L_0 + \Delta\mathcal{L}'} \\ \dfrac{\mathcal{M}}{L_0 + \Delta\mathcal{L}'} & -j\dfrac{\Delta\mathcal{L}''}{L_0 + \Delta\mathcal{L}'} + 1 \end{bmatrix} \begin{bmatrix} I_1 \\ I_2 \end{bmatrix} = \frac{\widetilde{\omega}_0^2}{\omega^2} \begin{bmatrix} I_1 \\ I_2 \end{bmatrix} \tag{6.19}$$

Therefore, $\widetilde{\omega}_0^2/\omega^2$ is the eigenvalue of the matrix on the left-hand side of (6.19). After simple mathematical operations, the splitting frequency can be determined as:

$$f_{e,o} = \frac{\widetilde{f}_0}{\sqrt{1 \pm \frac{\mathcal{M}}{L_0 + \Delta\mathcal{L}'} - j\frac{\Delta\mathcal{L}''}{L_0 + \Delta\mathcal{L}'}}} \tag{6.20}$$

where $f_{e,o}$ denotes the even and odd mode respectively, and $\widetilde{f}_0$ is the resonant frequency in the medium. As the Kirchhoff coefficients are generally complex, the eigenvalue $(\widetilde{\omega}_0^2/\omega^2)$ and thereby, $f_{e,o}$ may be complex with the imaginary part representing the losses added by the dissipative environment. In practice, the new splitting resonances of two strongly coupled coils are not much different, often within a factor of one tenth of the resonance in the non-coupled case. The complex coefficients, therefore, can be approximated to the values at the resonance frequency in the conductive medium, $\mathcal{M} = \mathcal{M}(\widetilde{\omega}_0)$ and $\Delta\mathcal{L} = \Delta\mathcal{L}(\widetilde{\omega}_0)$. Finally, because the Q-factors of the coils in free space are usually high, the even and odd modes are approximated by removing the imaginary part associated with the loss in the system as follows:

$$f_{e,o} \approx \frac{\widetilde{f}_0}{\Re\left[\sqrt{1 \pm \frac{\mathcal{M}(\widetilde{f}_0)}{L_0 + \Delta\mathcal{L}'(\widetilde{f}_0)} - j\frac{\Delta\mathcal{L}''(\widetilde{f}_0)}{L_0 + \Delta\mathcal{L}'(\widetilde{f}_0)}}\right]} \tag{6.21}$$

Figure 6.10 compares the theoretical predictions of the splitting resonances to the simulation results. The theoretical values are calculated using (6.21) while the simulation data is obtained by computing the current responses in both coils with the impedance matrix in (6.1), given that the excitation voltage of 1 V is applied to one coil. The odd and the even mode in the simulations are then detected by finding resonant peaks in the current response of either coil.

As can be seen from Figure 6.10, the theory curves and the numerical data appear to be in excellent agreement. When the medium conductivity is relatively small ($\sigma < 1$ S/m), the odd and even modes are practically equal to those in the free space case. These resonances are shifted to higher frequencies in the same fashion as that of a single isolated resonator in the medium with increasing conductivity. The mutual coupling between the two coils tends to zero when the medium is very conductive, resulting in the merger of the two modes in the theoretical curves. In the simulations,

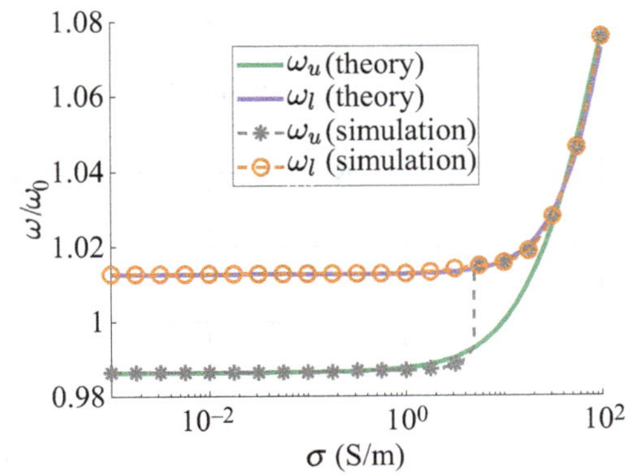

Figure 6.10 Comparison of the theoretical values and simulation results for splitting resonant frequencies of two coupled coils inside an attenuating medium. The results are normalized to $f_0 = 46$ MHz. Reprinted with permission from [28]

at $\sigma \approx 4.9$ S/m, the current response exhibits only a peak at the higher resonance instead. The phenomenon is driven by the medium losses, resulting in a very low Q-factor of the system and the spread of the current peaks in the frequency domain. The resonances may be still separated but the lower resonance is overshadowed by the slope of the upper one. The eigenvalue method above, however, does not take into account the stretch of the current response of the coils.

6.5.3 Electrical current distribution

Next, we examine the current distribution in the two resonators versus frequency and medium conductivity. Figure 6.11 presents the 3-D waterfall map of the modulus of the simulated current distribution in the two-coil configuration investigated in Figure 6.10. For the purpose of comparison, each current response of the two coils in the free space case is included as a black line. All the results are normalized to the maximum value of the corresponding currents in free space.

In free space, the current response splits into two almost perfectly symmetrical peaks as the two coils are ideally identical. In contrast, when embedding the two coils in a dissipative environment, the almost perfect symmetry in the moduli of current responses is broken. The even mode tends to experience higher losses as its decaying amplitude with the increasing medium conductivity is larger compared to the odd mode. When the medium conductivity is large enough, e.g., $\sigma > 4$ S/m in this case, the eddy currents can completely suppress the lower resonant frequency. As a result, both current responses in the two coils exhibit an under-coupled-like region with only one resonance. This phenomenon can also be observed in the phase difference between the two currents, see Figure 6.11b. Once the medium losses are introduced due to excited eddy currents, the slope of the phase difference becomes more gradual, which is often seen in coupled-resonator systems in free space with a low Q-factor. However, unlike those low Q-factor systems, the phase difference may move beyond the range between 0 and $-\pi$ rad with increasing conductivity, as shown in Figure 6.11b. Once the medium conductivity is sufficiently large, the phase difference curve is flattened out around $-\pi$, corresponding to the sole odd mode in the current response. Finally, the antisymmetry of the current responses indicates that resonances experience very different amounts of losses over the frequency range.

The main reason for these phenomena is the polarization loss, represented by the phase of the complex mutual inductance. It can be shown with circuit theory as follows. Given the two coils are identical, and the excitation is a voltage source applied at one coil, the circuit model is written as:

$$\begin{bmatrix} V_0 \\ 0 \end{bmatrix} = \begin{bmatrix} Z_0 + j\omega\Delta\mathcal{L} & j\omega\mathcal{M} \\ j\omega\mathcal{M} & Z_0 + j\omega\Delta\mathcal{L} \end{bmatrix} \begin{bmatrix} I_1 \\ I_2 \end{bmatrix} \tag{6.22}$$

where Z_0 is the self-impedance of the coil. From (6.22), the voltage–current relationship in the driven coil (coil 1) can be rewritten as:

$$V_0 = \left[(Z_0 + j\omega\Delta\mathcal{L}) + \frac{(\omega\mathcal{M})^2}{Z_0 + j\omega\Delta\mathcal{L}} \right] I_1. \tag{6.23}$$

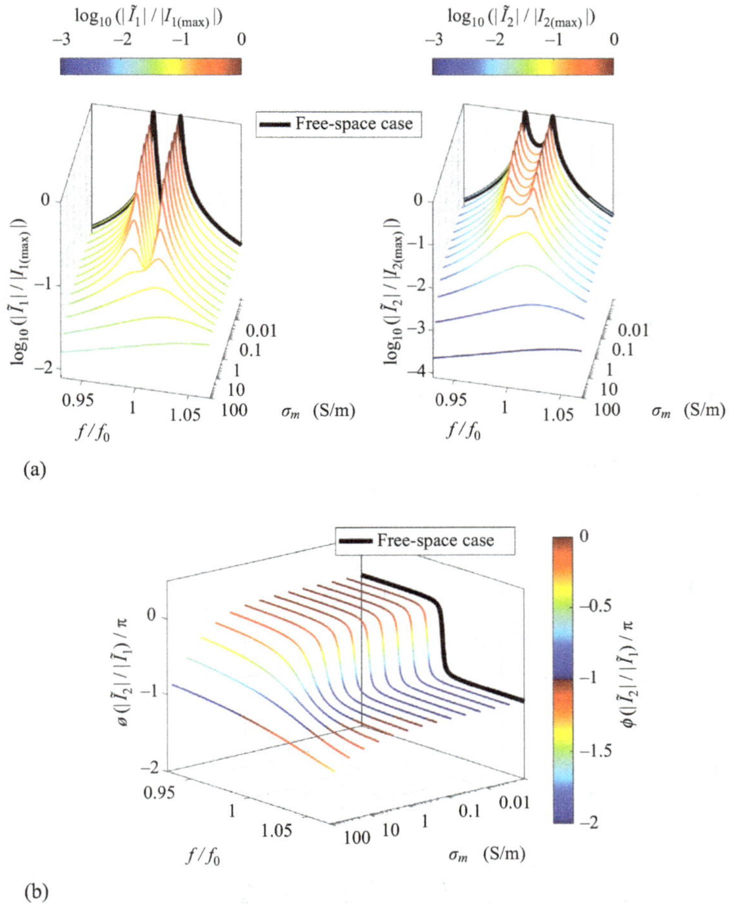

Figure 6.11 (a) Calculated 3-D waterfall maps of normalized magnitude of the current distribution versus frequency and medium conductivity in (left) the transmitter and (right) the receiver. (b) 3-D waterfall map of phase difference between the currents in the two coils versus frequency and medium conductivity. The current moduli are normalized to the maximum value of the corresponding currents in the free space while the phase difference is normalized to π. Reprinted with permission from [28]

In free space, this reduces to:

$$V_0 = \left[Z_0 + \frac{(\omega M_0)^2}{Z_0} \right] I_1 \tag{6.24}$$

Recall that $Z_0 = R_0 + jX_0$, (6.24) can be rewritten as:

$$V_0 = \left[R_0 + jX_0 + \frac{(\omega M_0)^2 R_0 - j(\omega M_0)^2 X_0}{R_0^2 + X_0^2}\right] I_1$$
$$= \left[R_0 + \frac{(\omega M_0)^2 R_0}{R_0^2 + X_0^2} + j\left(X_0 - \frac{(\omega M_0)^2 X_0}{R_0^2 + X_0^2}\right)\right] I_1 \quad (6.25)$$

As discussed above, the splitting resonances are not very separate. Therefore, it can be approximated that the additional reflected loss from the receiving coil:

$((\omega M_0)^2 R_0)/(R_0^2 + X_0^2)$ is constant within a small range of the frequency around the resonances. The well-known symmetry of the current response in coupled magnetic circuits arises as a result.

Conversely, when the environment is dissipative, the two complex Kirchhoff's coefficients arise as a result of the effects of excited eddy currents. In the particular configuration, these two coefficients can be expressed in the forms: $\mathcal{M} = \mathcal{M}' - j\mathcal{M}''$ where $0 < \mathcal{M}', \mathcal{M}'' < M_0$, and $\mathcal{L} = \mathcal{L}' - j\mathcal{L}''$ where $0 < \mathcal{L}', \mathcal{L}''$. Recall that $Z_0 + j\omega\Delta\mathcal{L} = \widetilde{R}_0 + j\widetilde{X}_0$. Using the complex forms of the complex Kirchhoff coefficients and the new coil impedance, (6.23) can be recast as:

$$V_0 = \left\{\left(\widetilde{R}_0 + j\widetilde{X}_0\right) + \frac{[\omega^2(\mathcal{M}'^2 - \mathcal{M}''2) - j2\omega^2\mathcal{M}'\mathcal{M}'']\left(\widetilde{R}_0 - j\widetilde{X}_0\right)}{\widetilde{R}_0^2 + \widetilde{X}_0^2}\right\} I_1$$
$$= \left\{\widetilde{R}_0 + \frac{\left[\omega^2(\mathcal{M}'^2 - \mathcal{M}''2)\widetilde{R}_0 - 2\omega^2\mathcal{M}'\mathcal{M}''\widetilde{X}_0\right]}{\widetilde{R}_0^2 + \widetilde{X}_0^2}\right.$$
$$\left. + j\left[\frac{\widetilde{X}_0 - 2\omega^2\mathcal{M}'\mathcal{M}''\widetilde{R}_0 - \omega^2(\mathcal{M}'^2 - \mathcal{M}''2)\widetilde{X}_0}{\widetilde{R}_0^2 + \widetilde{X}_0^2}\right]\right\} I_1. \quad (6.26)$$

where the term $\left[\omega^2(\mathcal{M}'^2 - \mathcal{M}''2)\widetilde{R}_0 - 2\omega^2\mathcal{M}'\mathcal{M}''\widetilde{X}_0\right]/(\widetilde{R}_0^2 + \widetilde{X}_0^2)$ is usually referred to as the loss from the feedback effect. Here, the numerator of this loss term is calculated as:

$$\omega^2(\mathcal{M}'^2 - \mathcal{M}''2)\widetilde{R}_0 - 2\omega^2\mathcal{M}'\mathcal{M}''\widetilde{X}_0$$
$$= \begin{cases} \omega^2(\mathcal{M}'^2 - \mathcal{M}''2)\widetilde{R}_0 + 2\omega^2\mathcal{M}'\mathcal{M}''|\widetilde{X}_0|, & \text{for } \omega < \widetilde{\omega}_0, \\ \omega^2(\mathcal{M}'^2 - \mathcal{M}''2)\widetilde{R}_0, & \text{for } \omega = \widetilde{\omega}_0, \\ \omega^2(\mathcal{M}'^2 - \mathcal{M}''2)\widetilde{R}_0 - 2\omega^2\mathcal{M}'\mathcal{M}''|\widetilde{X}_0|, & \text{for } \omega > \widetilde{\omega}_0. \end{cases} \quad (6.27)$$

where $\widetilde{\omega}_0$ is the angular resonant frequency of the coil inside the medium. With the same argument as above, split resonant peaks are relatively close in the frequency domain so that the total energy loss, $\widetilde{R}_0 = R_0 + \omega\Delta\mathcal{L}''$ is nearly constant about the resonance. However, as is evident from (6.26), the working frequency has a substantial influence on the feedback loss. As the frequency proceeds through the self-resonance of the coils in the medium ($\widetilde{\omega}_0$), the reactance of the coils ($\widetilde{X}_0$) changes from a negative value (capacitive reactance) to a positive value (inductive reactance). The increment of the working frequency diminishes the reflected loss from the secondary coil. Consequently, the even mode undergoes a much higher total loss than the odd mode, resulting in the antisymmetric current distribution as shown in Figure 6.10. Likewise, the current response in the receiver can be explained accordingly.

6.6 WPT in biological media

6.6.1 Introduction and system modeling

To wirelessly transfer power through non-ideal media (e.g., underground/underwater, human body, etc.), the simplest system comprises two magnetically coupled coils placed at a distance apart from each other. A sinusoidal voltage source is applied to one coil (primary coil) while the other coil (secondary coil) is connected to a load. The two coils are insulated by placing them in dielectric cavities to prevent current leakage to the medium background and improve current distribution. Efficiency is improved by making these coils resonant. Typically, the intrinsic capacitance of the coils is very small so that to realize WPT at lower frequencies, much larger tuning capacitors are required to form the resonant tanks on both the primary and secondary sides [34]. This tuning is essential to ensure that systems meet EM emission standards, and because the losses to the environment become a huge source of inefficiency at high frequencies. The value of tuning capacitors can be approximated from the desired resonant frequency and the self-inductance while the self-capacitance can be negligible. On the subject of capacitive tuning circuits, there are four possible simple topologies, namely series–series, series–parallel, parallel–series, and parallel–parallel, standing for the connection type of the capacitors in the primary and secondary coils, respectively [35]. Some advanced compensation networks, for example, LLC circuit, can be found in [36,37]. Using the reflected impedance theory, only the value of tuning capacitors in the series–series topology is independent of the mutual coupling [35], which depends largely on the medium's electromagnetic properties [21,22]. Therefore, series–series topology is assumed to investigate the effects of the non-ideal medium on WPT. Otherwise, no other external circuitry such as rectifiers, inverters, and regulators is considered in this chapter.

As discussed in previous sections, when embedding the coil system inside a dissipative environment, both self and mutual inductance gain lossy components due to the impacts of excited eddy current. To simplify the system, dielectric cavities providing coil insulation, which have little-to-none effects on the magnetic channel, are assumed to be free-space. The equivalent circuit model of the entire WPT system in an attenuating medium is then illustrated in Figure 6.12.

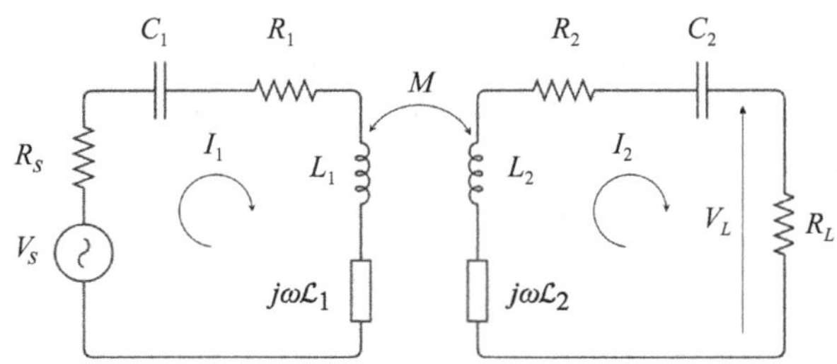

Figure 6.12 Equivalent circuit model of a WPT system consisting of two resonant coils with series–series capacitive compensation topology

Using Kirchhoff's voltage law, the circuit model can be expressed as:

$$\begin{bmatrix} R_s + Z_1 + j\omega\mathcal{L}_1 & j\omega\mathcal{M} \\ j\omega\mathcal{M} & Z_2 + j\omega\mathcal{L}_2 + R_L \end{bmatrix} \begin{bmatrix} I_1 \\ I_2 \end{bmatrix} = \begin{bmatrix} V_s \\ 0 \end{bmatrix} \tag{6.28}$$

where $\omega = 2\pi f$ is the angular frequency; V_s and R_s are the voltage and internal resistance of the source, respectively; R_L is the resistance of the load; I_i, $Z_i = R_i + j\omega L_i + 1/(j\omega C_i)$, R_i, L_i, C_i are the current response, impedance, resistance, self-inductance, and value of tuning capacitors of the coil i, respectively; $\mathcal{L}_i$ is the lossy self-inductance of coil i, representing the extra self-inductance and resistance arising due to the eddy currents it generates; $\mathcal{M} = M_0 + M_E$ is the lossy mutual inductance between two coils inside the medium; M_0 is the mutual inductance of two coils in free space (typically a real number); and M_E represents the extra mutual coupling between the coils due to the eddy currents generated in the environment. In free space, the eddy currents do not exist and hence, all the additional couplings are zero, i.e., $\mathcal{L}_i = 0$, $M_E = 0$ and $\mathcal{M} = M_0$. The lossy Kirchhoff coefficients are generally complex, i.e., $\mathcal{M} = \mathcal{M}' + j\mathcal{M}''$ and $\mathcal{L}_i = \mathcal{L}_i' - j\mathcal{L}_i''$ (where $-\mathcal{L}_i'' < 0$ because it associates with the power loss in the medium).

By solving system (6.28), the currents in each coil can be determined. Recalling that $Z_i + j\omega\mathcal{L}_i = \widetilde{R}_i + j\widetilde{X}_i$, where $\widetilde{R}_i = R_i + \omega\mathcal{L}_i''$, $X_i = \omega L_i - 1/(\omega C_i)$ and $\widetilde{X}_i = X_i + \omega\mathcal{L}_{i'}$, the output power at the load can be then calculated as:

$$\begin{aligned} P_L = |I_2|^2 R_L &= \frac{\omega^2 |\mathcal{M}|^2 V_s^2 R_L}{\left|(R_s + Z_1 + j\omega\mathcal{M}_1)(Z_2 + j\omega\mathcal{M}_2 + R_L) + \omega^2\mathcal{M}^2\right|^2} \\ &= \frac{\omega^2(\mathcal{M}'^2 + \mathcal{M}''2)V_s^2 R_L}{\begin{array}{c}\left[\left(\widetilde{R}_1 + R_s\right)\left(\widetilde{R}_2 + R_L\right) + \omega^2(\mathcal{M}'^2 - \mathcal{M}''2)\widetilde{X}_1\widetilde{X}_2\right]^2 \\ + \left[2\omega^2\mathcal{M}'\mathcal{M}'' + \left(\widetilde{R}_2 + R_L\right)\widetilde{X}_1 + \left(\widetilde{R}_1 + R_s\right)\widetilde{X}_2\right]^2\end{array}} \end{aligned} \tag{6.29}$$

Unlike the free space case, where there is no loss in the transmission medium, the non-ideal medium absorbs some of the transmitting power. The power loss in the medium therefore needs to be taken into account to calculate the power transfer efficiency:

$$\eta = \frac{P_L}{P_{in}} = \frac{|I_2|^2 R_L}{\Re\{V_s I_1^*\}} \tag{6.30}$$

6.6.2 System optimization

6.6.2.1 Optimal load

The power transferred to the load can be maximized by adjusting the load resistance. Mathematically, the optimal load can be found by solving the roots of the first-order derivative of (6.29) with respect to R_L. Here, it is noted that the expression of the optimal load is obtained at an arbitrary frequency when the resonators have reactive impedances. It is because the optimal working frequency for the WPT systems in dissipative media might not be at the resonance as assumed in most literature. It will be shown in the next section by examining the output power over a range of frequencies. Conjugate matching—a favorite matching technique when a complex input impedance is present—is not studied since this work intends to point out the limitations of the current model when applied to systems in conductive media.

Differentiating and solving $\mathrm{d}P_L/dR_L = 0$, the optimal resistive load can be obtained as follows:

$$R_L = \sqrt{\frac{\omega^4(\mathcal{M}'^2 + \mathcal{M}''2)^2 + 2\omega^2\left\{(\mathcal{M}'^2 - \mathcal{M}''2)\left(\widetilde{R}_1\widetilde{R}_2 + \widetilde{R}_2 R_s - \widetilde{X}_1\widetilde{X}_2\right) + 2\mathcal{M}'\mathcal{M}''\left[\widetilde{R}_2\widetilde{X}_1 + \left(\widetilde{R}_1 + R_s\right)\widetilde{X}_2\right]\right\} + \left[\left(\widetilde{R}_1 + R_s\right)^2 + \widetilde{X}_1^2\right]\left(\widetilde{R}_2^2 + \widetilde{X}_2^2\right)}{\left(\widetilde{R}_1 + R_s\right)^2 + \widetilde{X}_1^2}} \tag{6.31}$$

In free space, i.e., $\mathcal{M}' = M_0$ and $\mathcal{M}'' = \mathcal{L}' = \mathcal{L}'' = 0$, it becomes:

$$R_L = \sqrt{\frac{\omega^4 M_0^4 + 2\omega^2 M_0^2(R_1R_2 + R_2R_s - X_1X_2) + \left[(R_1 + R_s)^2 + X_1^2\right](R_2^2 + X_2^2)}{(R_1 + R_s)^2 + X_1^2}} \tag{6.32}$$

6.6.2.2 Optimal working frequency

Given the mutual coupling is fixed and the load is optimal, the maximum transferred power is usually achieved when both transmitting and receiving coils are tuned to the same resonance, i.e., $f_{i0} = 1/\left(2\pi\sqrt{L_iC_i}\right)$ [6]. This is easily realized in practice by adding suitable external capacitors to the windings. However, in an attenuating medium, the resonant frequency is shifted due to the effects of the eddy currents on self-inductance, as shown in Section 6.5.1. The new resonant frequency

$\widetilde{f}_{i0}$ of the coil can be calculated using (6.15). As a result, the optimal working frequency for these systems may not be the original resonance of the windings in free space (f_0).

In the literature, authors in [21] also observed this phenomenon when investigating a WPT system with a pair of magnetically coupled resonators above a half-space conductive slab. The resonant frequencies of these two coils are shifted to higher frequencies with the one closer to the slab being altered more profoundly. To address this problem, the authors suggested to retune the coil closer to the slab to the original resonant frequency by adjusting its external capacitors. Simulation results show that the retuning technique has the merit of increasing the useful power at the load in their case [21]. However, in our case, since both the two coils are submerged in a dissipative medium, the medium detunes the two coils equally. Thus, the original resonance may not be the most advantageous frequency for such WPT systems.

6.6.3 Simulation results

In this section, we reinvestigate a case study, where the WPT system consists of two identical coils placed coaxially at a distance from each other. The system operates at the lowest ISM band (6.78 MHz), following the Rezence A4WP standard [38]. The coils are assumed to have similar dimensions with the pancake transmitting coils in [39] but unlike the printed PCB design in [39], we only consider wire-wound coil designs only. To simplify the calculation, the resistance of the coils is mainly due to Ohmic loss and is approximately constant with the frequency. The resistance can then be obtained at the resonance using standard formulae for loop antenna [40]. The WPT system is then immersed inside a non-ideal environment with relative permittivity of $\varepsilon_r = 78$ (similar to that of water at room temperature), and variable conductivity up to $\sigma = 100$ S/m, which is the highest value of all earth or biological materials [17–19,41]. Table 6.3 summarizes the geometric and circuit parameters of the system. The radius of the dielectric boxes is assumed to be three times larger than the coil radius in order to accurately calculate the lossy Kirchhoff coefficients using (6.12) and (6.13).

Figure 6.13 shows the output power at the load versus frequency and load resistance for the WPT system in free space. Given the load is optimized, the output power has its maximum peak at the resonant frequency as expected. When the load resistance is small, frequency splitting occurs as the graph clearly exhibits two sharp peaks at the event and the odd mode, f_e and f_o respectively. When the load resistance is large enough, the maximum transferred power is expectedly found at resonance where both coils are tuned to. With given working frequencies, the optimal load can be predicted correctly by (6.31): $R_L = 1378.8\ \Omega$ and $6.72\ \Omega$ at f_0 and f_u, respectively.

Now, we consider the same WPT system embedded inside a non-ideal environment with variable conductivity. The primary and secondary coils are detuned and the mutual coupling between them becomes complex due to the impact of excited eddy currents. Hence, it may be questionable whether the optimal

Table 6.3 Parameter list of the system in the case study

Parameters	Value
Inner coil radius, r_i	10.5 mm
Outer coil radius, r_o	24 mm
Wire diameter, r_w	0.25 mm
Coil height, l	0.5 mm
Number of turns, N	18
Distance between coils, d	15 mm
Dielectric insulator radius, R	72 mm
Dielectric insulator height	2.5 mm
Resonance frequency, f_0	6.78 MHz
Self-inductance, L	13.248 μH
Parasitic resistance, R	0.7926 Ω
Mutual inductance (free-space), M_0	2.1098 μH
AC voltage source, V_s	10 V
Internal source resistance, R_s	5 Ω
Medium relative permittivity, ε_r	78
Medium relative permeability, μ_r	1
Medium conductivity, σ	10^{-2}–10^2 S/m

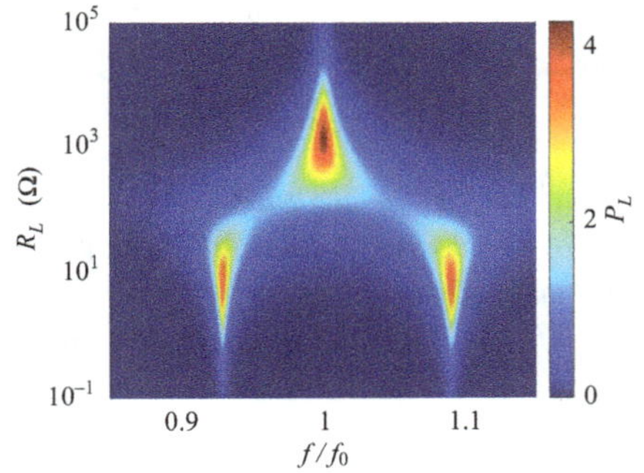

Figure 6.13 Calculated color map of the output power P_L as a surface function of the normalized working frequency f and the load resistance R_L. The frequency is normalized by the resonance of the coils in the isolated case. The WPT system operates in free space

frequency is still the coil's self-resonance in free space. Figure 6.14 presents the simulation results for the output power at the original resonance and the optimal frequency versus the increment of medium conductivity. The optimal frequency and load resistance determined by the location of the maximum power are given in Figure 6.14c and d. For the purpose of comparison, the retuning method, suggested in [21], is included.

In the retuning method [21], the resonant frequency of the resonators is fixed by adding suitable additional capacitors to compensate for the impacts of the eddy

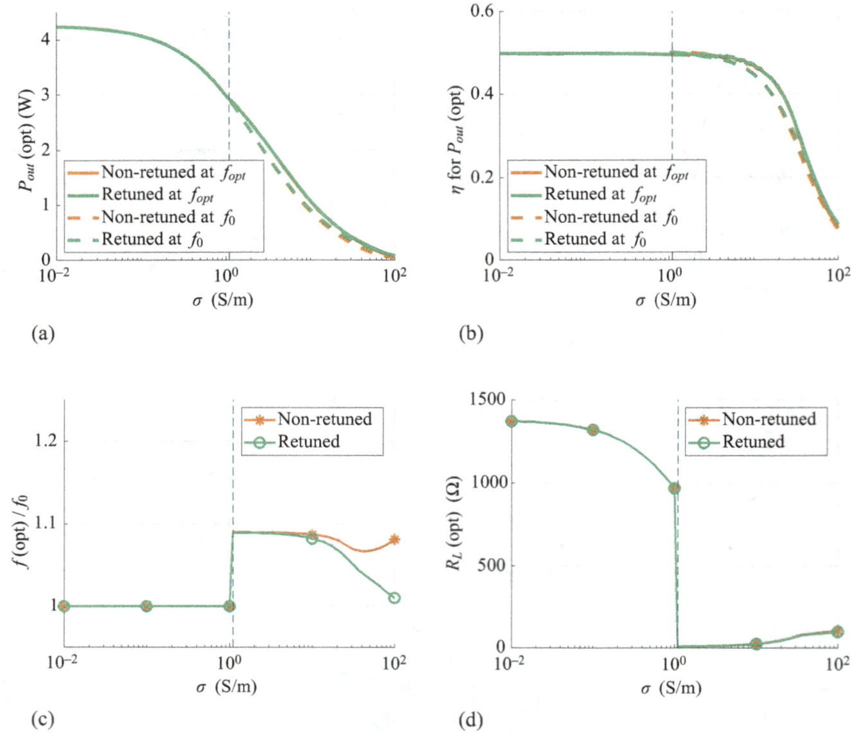

Figure 6.14 (a) Calculated active power at load at the optimal frequency and the resonance, (b) the corresponding efficiency, (c) optimal frequency, and (d) optimal load resistance. The output power in the case study is compared when the compensation capacitances are kept the same as in free space (non-retuned) and when the compensation capacitances are added according to the prediction of the detuning effect based on the novel circuit model (retuned)

currents via the lossy self-inductance. In Figure 6.14a, the retuning method appears to be not favorable compared to the non-retuned case as the two curves almost coincide. Furthermore, in reality, lumped capacitors are not ideal and often modelled as a capacitance in series with a resistance, called "Equivalent Series Resistance" (ESR) [42]. Inserting more capacitors into the circuit may cause more losses due to ESR and thus degrade the performance of the WPT system. Interestingly, the output power at the original resonance in both non-retuned and retuned cases may not be the maximum power for some particular conductivities. When the conductivity is relatively small, the medium seems to have little-to-none effects on the output power as the optimal frequency is practically the original resonance, see Figure 6.14c. The optimal frequency for the maximum output power then exhibits a discontinuity jump to a higher frequency once the medium conductivity is sufficiently large. This result is somewhat consistent with those presented in [43], where it is concluded that "the optimum operating frequency of a WPT system in seawater should be larger than the resonant frequency to achieve the maximum DC-DC efficiency."

To examine the impact of medium conductivity on the change of the optimal frequency for maximum output power, Figure 6.15 illustrates the 3D color plots of the output power versus the frequency and load resistance for four different conductivities: $\sigma = 0.1$, 1, 4, and 10 S/m. Apparently from Figure 6.15, the system undergoes different amounts of losses at variable frequencies due to the frequency-dependent feature of the eddy currents. The power peaks at the even-mode and the self-resonance suffer from larger attenuation with the increasing conductivity compared to the odd-mode one. When the losses due to the surrounding environment are high enough, the lower resonance is completely suppressed whilst the peaks at the f_0 and f_u coalesce. A slight drop in the optimal frequency in the non-retuned case in Figure 6.14c is the indication of the merging of these two peaks.

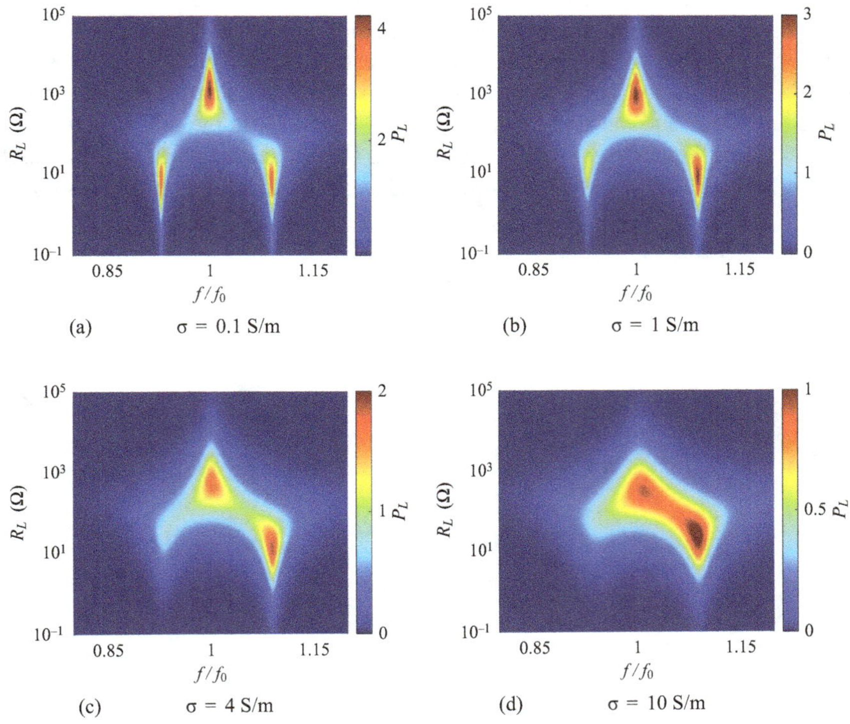

Figure 6.15 Calculated color map of the output power P_L as a surface function of the normalized working frequency f and the load resistance R_L. The frequency is normalized by the resonance of the coils in the isolated case. The WPT system operates inside a medium with the conductivity of (a) $\sigma = 0.01$, (b) $\sigma = 0.1$, (c) $\sigma = 1$, and $\sigma = 10$ S/m

These results are consistent with the analyses of the frequency splitting and current distribution of two coupled coils discussed in Section 6.5.2. The optimal frequency for larger conductivities is now the odd-mode resonance in the frequency splitting phenomenon. To allow the frequency splitting to take place, the corresponding optimal load resistance is much smaller, as shown in Figure 6.14d.

6.6.4 Criteria for the frequency switching

In the previous section, the total losses of the system, including the damping of the coils and the frequency-dependent medium losses, have been shown to be the reason underlying the discontinuity of the optimal frequency curve in Figure 6.14. Hence, it may be suggested that the quantitative relationship between the medium losses and damping in the coils can be used to determine criteria for the "switching point." To identify the influences of the damping in the windings, Figure 6.16 shows the simulation results for some lossless cases when (i) the transmitter's resistance is zero ($R_1 = 0\ \Omega$), (ii) the receiver's resistance is zero ($R_2 = 0\ \Omega$), and (iii) both transmitter's resistance and receiver's resistance are zero ($R_1 = R_2 = 0\ \Omega$).

As can be seen from Figure 6.16, while the parasitic resistance of the transmitter is an insignificant factor, the non-zero receiving coil's resistance causes the switching point in the optimal working frequency. Therefore, the position of the discontinuity can be derived as a function of the receiver's resistance by comparing the output power at the self-resonance and the odd-mode resonance. Recall P_0 is the output power at f_0 while P_u is the output power at f_u where f_u is the odd-mode resonance frequency calculated as (6.20). The output power at these two frequencies can be expressed as:

$$\begin{cases} P_0 = \dfrac{\omega_0^2|\mathcal{M}_0|^2 V_s^2 R_L}{\left|(R_s + R_1 + j\omega_0\mathcal{L}_{1,0})(R_2 + j\omega_0\mathcal{L}_{2,0} + R_L) + \omega_0^2\mathcal{M}_0^2\right|^2} \\ P_u = \dfrac{\omega_u^2|\mathcal{M}_u|^2 V_s^2 R_L}{\left|\left[R_s + R_1 + j\omega_u L_1\left(1 - \dfrac{\omega_0^2}{\omega_u^2}\right) + j\omega_u\mathcal{L}_{1,u}\right]\left[R_2 + j\omega_u L_2\left(1 - \dfrac{\omega_0^2}{\omega_u^2}\right) + j\omega\mathcal{L}_{2,u} + R_L\right] + \omega_u^2\mathcal{M}_u^2\right|^2} \end{cases} \tag{6.33}$$

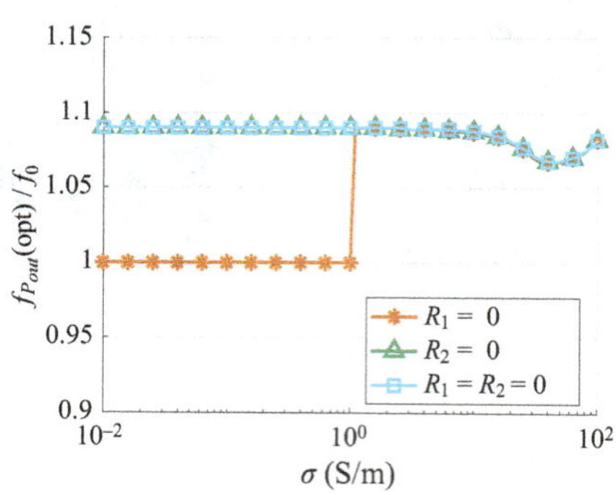

Figure 6.16 Simulated optimal frequency for the maximum output power at load for three extreme cases of the configuration: $R_1 = 0\ \Omega$, $R_2 = 0\ \Omega$, and $R_1 = R_2 = 0\ \Omega$

where $\mathcal{M}_0$ and $\mathcal{M}_u$ are the lossy mutual inductance at f_0 and f_u, respectively, and $\mathcal{L}_{i,0}$ and $\mathcal{L}_{i,u}$ are the lossy self-inductance of the coil i at f_0 and f_u, respectively. Using the load optimization (6.31) on these power equations, the maximum output at load at f_0 and f_u can be derived as:

$$\begin{cases} P_0(\max) = \dfrac{\omega_0^2|\mathcal{M}_0|^2 V_s^2}{2[A_0R_2 + B_0 + \sqrt{A_0}\sqrt{A_0R_2^2 + 2B_0R_2 + C_0}]} \\ P_u(\max) = \dfrac{\omega_0^2|\mathcal{M}_0|^2 V_s^2}{2[A_uR_2 + B_u + \sqrt{A_u}\sqrt{A_uR_2^2 + 2B_uR_2 + C_u}]} \end{cases} \tag{6.34}$$

where

$$\begin{cases} A_k = (a_k^2 + b_k^2)\left(\dfrac{\omega_0|\mathcal{M}_0|}{\omega_k|\mathcal{M}_k|}\right)^2, \\ B_k = [(a_k^2 + b_k^2)c_k + a_ke_k + b_kg_k]\left(\dfrac{\omega_0|\mathcal{M}_0|}{\omega_k|\mathcal{M}_k|}\right)^2, \\ C_k = [(a_k^2 + b_k^2)(c_k^2 + d_k^2) + e_k^2 + g_k^2 \\ \qquad + 2c_k(a_ke_k + b_kg_k) + 2d_k(a_kg_k - b_ke_k)]\left(\dfrac{\omega_0|\mathcal{M}_0|}{\omega_k|\mathcal{M}_k|}\right)^2, \\ a_k = R_s + R_1 + \omega_k\mathcal{L}''_{1,k}, \\ b_k = \omega_kL_1\left(1 - \dfrac{\omega_0^2}{\omega_k^2}\right) + \omega_k\mathcal{L}'_{1,k}, \\ c_k = \omega_k\mathcal{L}''_{2,k}, \\ d_k = \omega_kL_2\left(1 - \dfrac{\omega_0^2}{\omega_k^2}\right) + \omega_k\mathcal{L}'_{2,k}, \\ e_k = \omega_k^2(\mathcal{M}'_k2 - \mathcal{M}''_k2), \\ g_k = 2\omega_k^2\mathcal{M}'_k\mathcal{M}''_k, \\ k = \{0, u\} \end{cases} \tag{6.35}$$

Given that the load resistance is optimized, the discontinuity point in the optimal frequency curve occurs when the maximum output power at the odd-mode resonance is larger or equal than that at the self-resonance, i.e., $P_u(\max) \geq P_0(\max)$. After a few simple algebraic operations, the inequality can be rewritten as:

$$\sqrt{A_0}\sqrt{A_0R_2^2 + 2B_0R_2 + C_0} - \sqrt{A_u}\sqrt{A_uR_2^2 + 2B_uR_2 + C_u} \geq (A_u - A_0)R_2 + B_u - B_0 \tag{6.36}$$

Once the coefficients A_k, B_k and C_k are known, the inequality (6.36) is reducible to a quadratic inequality of variable R_2. Consequently, the location of the switching point can be expressed as a function of the parasitic resistance of the receiver, R_2.

Now we reinvestigate the case study in the previous section. For each conductivity, the value of the secondary coil's resistance corresponding to the switching point can be obtained by solving the inequality (6.36). Figure 6.17 presents the 2D color-map of the optimal working frequency as a function of the medium conductivity and the receiving coil resistance. The solid line refers to the value when the strict equality occurs in (6.36), i.e., the maximum output power at the original resonance and the upper resonance are equal $P_0(\max) = P_u(\max)$.

As is evident in Figure 6.17, the optimal working frequency is always the upper resonance without regard to the medium conductivity when the internal resistance of the receiver tends to zero. This result is indeed compatible with the simulation data for some extreme cases, presented in Figure 6.16. When the medium conductivity is sufficiently large, e.g., $\sigma \approx 63$ S/m for this case study, the optimal working frequency is irrevocably at the upper resonance as the equilibrated line exhibits a vertical asymptote. It clearly demonstrates the optimal working frequency in highly conductive environments is the odd-mode resonance due to the insurmountable medium losses at f_0.

Ultimately, the color-map constructed by using (6.36) is exceptionally useful for the practical design of WPT systems in common attenuating media. As operating at a single predefined frequency, preferably at the self-resonance of the coils in free-space, is desirable for most WPT systems, the upper bound of the medium conductivity, where the system can function seamlessly, can be determined based on the receiver resistance and vice versa. In addition, in the case that the system design allows the flexibility in selecting the working frequency, the switching point is remarkably applicable for the operation of WPT systems in time-varying environments, such as oceans with dynamic currents, weather-dependent soils, or human body, given that the receiver's resistance is known in advance. In other respects, if

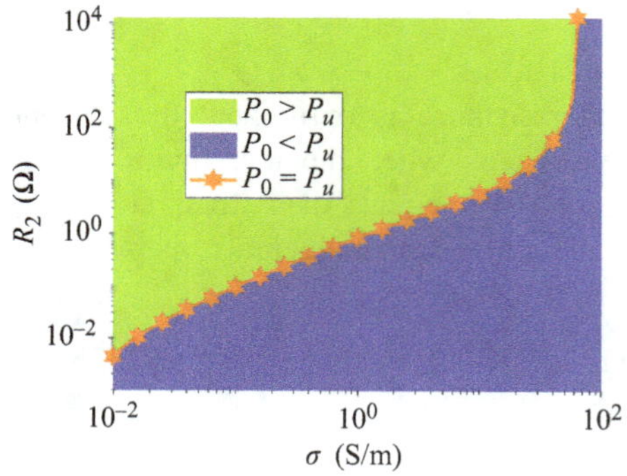

Figure 6.17 Calculated colormap guidance by using (6.36) for the optimal frequency selection as a function of the receiver resistance and the medium conductivity. The solid line refers to the values when the output powers at the original resonance and upper resonance are equal

the working frequency is predetermined, employing electronically tunable coils would enable tracking the best performance for WPT systems in variable-conductivity media.

6.7 Conclusions

The study of near-field coupling is a long-standing topic of much interest. Since the beginning of the twenty-first century, with the emergence of metamaterials, this near-field coupling has attracted more attention from scientists and researchers worldwide as a key design element in numerous practical applications such as WPT, near-field imaging, antenna and waveguide, and light manipulation. Nevertheless, there seems to be a lack of literature on near-field coupling in biological media like the skin, bone, and tissues. The aim of this chapter is to investigate the coupling of magnetic resonators in these environments.

In most bioelectromagnetic applications, small-size magnetically coupled resonators, acting as transmitting–receiving antennas, are often immersed partly or completely in the human body and tissues. Some examples include wearable medical devices, implantable sensors, and swallowable electronics capsules. We, as a result, investigate the configuration when two magnetic resonators insulated by small dielectric cavities are completely immersed in an infinite homogeneous non-ideal medium. A novel equivalent circuit model, which incorporates for the first time all electromagnetic coupling and feedback effects, has been developed with the existence of complex Kirchhoff coefficients. They are called complex self-inductance and complex mutual inductance, which represent a change in the self-inductance of the circuits and the mutual inductance between circuits mediated by the medium, respectively. These complex terms depend mainly on the working frequency, geometry of the coils and most importantly, electromagnetic characteristics of the surrounding environment. In this chapter, we developed a mathematical method to calculate the complex terms for two air-core circular coils placed in the same orientation. Whereas this analytical method provides excellent accuracy in predicting these coefficients, 3D electromagnetic simulation software may be required for more complicated coil geometries.

With the novel circuit model being postulated, the network analysis of coupled-coil systems has been presented. When immersing a magnetic resonant circuit in a biological environment, the resonant frequency indeed varies with the real part of the complex self-inductance, whereas the Q-factor is degenerated because of the medium losses represented by the imaginary part. Regarding the current distribution, the almost perfect symmetry of the splitting resonant peaks is noticeably broken with the even-mode frequency peak smaller than the odd-mode one. It can be explained with the circuit theory with the main reason being the additional phase to the circuit added via the complex mutual inductance. Additionally, the positions of the even and odd modes can be predicted by performing the eigenvalue method.

Finally, we consider an immediate application of near-field couplings – WPT for medical devices. Unlike the free-space case, the working-frequency selection requires more attention. Simulation results show that the optimal frequency is either the uncoupled resonance or the odd-mode one in the frequency splitting. Surprisingly, the criteria for choosing the best working frequency is the quantitative relationship between the damping in the receiving coil and the eddy current losses. The formula for frequency selection, which is reducible to a solvable quadratic inequality, can be used to construct a guidance map, based on the receiver's resistance and medium's electromagnetic properties.

In a broader context, we believe that the proposed equivalent circuit model will serve as a fundamental model for systems employing inductive coupling techniques in biological media. All the effects of the excited eddy currents have been fully examined. These findings above are very significant for medical and healthcare applications functioning in real-time dynamic environments like in the human body and tissues. Our analysis, therefore, can offer insights into the process of designing and optimizing these applications.

References

[1] Griffiths DJ. *Introduction to electrodynamics*. 4th ed. Cambridge: Cambridge University Press; 2017.

[2] Sadiku MN. *Elements of electromagnetics*. Oxford: Oxford University Press; 2014.

[3] Griffiths H. Magnetic induction tomography. *Meas Sci Technol*. 2001;12(8): 1126.

[4] Zheng Y, Miller GW, Tobias WA, *et al.* A method for imaging and spectroscopy using γ-rays and magnetic resonance. *Nature*. 2016;537(7622): 652.

[5] Sophian A, Tian G, Taylor D, *et al.* Electromagnetic and eddy current NDT: a review. *Insight*. 2001;43(5):302–306.

[6] Kurs A, Karalis A, Moffatt R, *et al.* Wireless power transfer via strongly coupled magnetic resonances. *Science*. 2007;317(5834):83–86.

[7] Sun Z, Akyildiz IF. Magnetic induction communications for wireless underground sensor networks. *IEEE Trans Antennas Propag*. 2010;58 (7):2426–2435.

[8] Kisseleff S, Akyildiz IF, Gerstacker WH. Survey on advances in magnetic induction-based wireless underground sensor networks. *IEEE Internet Things J*. 2018;5(6):4843–4856.

[9] Gulbahar B, Akan OB. A communication theoretical modeling and analysis of underwater magneto-inductive wireless channels. *IEEE Trans Wireless Commun*. 2012;11(9):3326–3334.

[10] Li Y, Wang S, Jin C, *et al.* A survey of underwater magnetic induction communications: fundamental *issues, recent advances, and challenges. IEEE Commun Surveys Tuts*. 2019;21(3):2466–2487.

[11] Movassaghi S, Abolhasan M, Lipman J, *et al.* Wireless body area networks: a survey. *IEEE Commun Surveys Tuts*. 2014;16(3):1658–1686.

[12] Golestani N, Moghaddam M. Theoretical modeling and analysis of magnetic induction communication in wireless body area networks (WBANs). *IEEE J Electromagn, RF, Microw Med Biol*. 2018;2(1):48–55.

[13] Sharma AK, Yadav S, Dandu SN, *et al.* Magnetic induction-based non-conventional media communications: a review. *IEEE Sensors J*. 2016;17 (4):926–940.

[14] Domingo MC. Magnetic induction for underwater wireless communication networks. *IEEE Trans Antennas Propag*. 2012;6(60):2929–2939.

[15] Marmugi L, Renzoni F. Optical magnetic induction tomography of the heart. *Sci Rep*. 2016;6(1):1–8.

[16] Ciuti G, Menciassi A, Dario P. Capsule endoscopy: from current achievements to open challenges. *IEEE Rev Biomed Eng*. 2011;4:59–72.

[17] Miklavcic D, Pavselj N, Hart FX. *Electric properties of tissues*. Washington, DC: American Cancer Society; 2006. p. 1–12.

[18] Keller DUJ, Weber FM, Seemann G, *et al.* Ranking the influence of tissue conductivities on forward-calculated ECGs. *IEEE Trans Biomed Eng*. 2010;57(7):1568–1576.

[19] McCann H, Pisano G, Beltrachini L. Variation in reported human head tissue electrical conductivity values. *Brain Topogr*. 2019;32(5):825–858.

[20] Kim HJ, Hirayama H, Kim S, *et al.* Review of near-field wireless power and communication for biomedical applications. *IEEE Access*. 2017;5: 21264–21285.

[21] Vallecchi A, Chu S, Solymar L, *et al.* Coupling between coils in the presence of conducting medium. *IET Microw Antennas Propag*. 2019;13(1):55–62.

[22] Chu S, Vallecchi A, Stevens CJ, *et al.* Fields and coupling between coils embedded in conductive environments. *EPJ Appl Metamater*. 2018;5(2): 1–8.

[23] Garca-Martn J, Gómez-Gil J, Vázquez-Sánchez E. Non-destructive techniques based on eddy current testing. *Sensors*. 2011;11(3):2525–2565.

[24] Luloff MS, Desjardins D, Morelli J, *et al.* Solution for a transmit-receive eddy current probe above a layered planar conductive structure. *NDT E Int*. 2018;96:1–8.

[25] Kim J, Kim K, Kim H, *et al.* An efficient modeling for underwater wireless power transfer using z-parameters. *IEEE Trans Electromagn Compat*. 2019;61(6):2006–2014.

[26] Zhang K, Zhang X, Zhu Z, *et al.* A new coil structure to reduce eddy current loss of WPT systems for underwater vehicles. *IEEE Trans Veh Technol*. 2019;68(1):245–253.

[27] Zhang K, Ma Y, Yan Z, *et al.* Eddy current loss and detuning effect of seawater on wireless power transfer. *IEEE Trans Emerg Sel Topics Power Electron*. 2020;8(1):909–917.

[28] Chu S, Luloff MS, Yan J, *et al.* Magnetoinductive waves in attenuating media. *Sci Rep*. 2021;11(1):1–12.

[29] Chu SC. *Near-fields in attenuating media*. Oxford: University of Oxford; 2019.

[30] Conway JT. Inductance calculations for noncoaxial coils using Bessel functions. *IEEE Trans Magn*. 2007;43(3):1023–1034.

[31] Conway JT. Mutual inductance for an explicitly finite number of turns. *Prog Electromagn Res*. 2011;28:273–287.

[32] Chu S. *Near-fields in attenuating media*. Oxford: University of Oxford; 2020.

[33] Ahn D, Hong S. A study on magnetic field repeater in wireless power transfer. *IEEE Trans Ind Electron*. 2012;60(1):360–371.

[34] Wang B, Yerazunis W, Teo KH. Wireless power transfer: metamaterials and array of coupled resonators. *Proc IEEE*. 2013;101(6):1359–1368.

[35] Zhang Z, Pang H, Georgiadis A, *et al.* Wireless power transfer: an overview. *IEEE Trans Ind Electron*. 2019;66(2):1044–1058.

[36] Zhang W, Mi CC. Compensation topologies of high-power wireless power transfer systems. *IEEE Trans Veh Technol*. 2016;65(6):4768–4778.

[37] Lu F, Zhang H, Hofmann H, *et al.* An inductive and capacitive integrated coupler and its LCL compensation circuit design for wireless power transfer. *IEEE Trans Ind Appl*. 2017;53(5):4903–4913.

[38] Tseng R, Von Novak B, Shevde S, *et al.* Introduction to the alliance for wireless power loosely-coupled wireless power transfer system specification version 1.0. In: 2013 IEEE Wireless Power Transfer (WPT). IEEE; 2013. p. 79–83.

[39] Liu Z, Zhong Z, Guo YX. In vivo high-efficiency wireless power transfer with multisine excitation. *IEEE Trans Microw Theory Techn*. 2017;65 (9):3530–3540.

[40] Balanis CA. *Antenna theory: analysis and design*. New York, NY: John Wiley & Sons; 2016.

[41] Palacky GJ. Resistivity characteristics of geologic targets. In: Electromagnetic methods in applied geophysics. Tulsa, OK: Society of Exploration Geophysicists; 2012; pp. 52–129.

[42] Pu XS, Nguyen TH, Lee DC, *et al.* Fault diagnosis of DC-link capacitors in three-phase AC/DC PWM converters by online estimation of equivalent series resistance. *IEEE Trans Ind Electron*. 2012;60(9):4118–4127.

[43] Yan Z, Zhang Y, Kan T, *et al.* Frequency optimization of a loosely coupled underwater wireless power transfer system considering eddy current loss. *IEEE Trans Ind Electron*. 2018;66(5):3468–3476.

Chapter 7

Electromagnetic resonators and metamaterials-based structures for chemical and biological sensing in body-centric wireless healthcare applications

Muhammad Usman Ejaz[1], Tayyaba Irum[1], Qammer H. Abbasi[2] and Akram Alomainy[1]

Electromagnetics (EM) and associated technology has been providing the engineering and scientific community with brilliant devices such as antennas, absorbers, lenses, and EM-based sensors having variety of applications in the fields of communication, defense and security, chemical and biomedical industry. Demand for wireless EM-based measurement and sensing systems in medicine and chemical industry has surged significantly, for concentration estimation, and monitoring of important industrial use chemicals, early diagnostics of various diseases and monitoring of several physiological parameters where antennas, resonators, and metamaterial (MTM)-based sensors are key components of such systems. EM-based measurement and sensing devices are auspicious for mentioned divergent applications due to their simplicity, cost effectiveness and non-destructive nature. A specific frequency band (Industrial, Scientific, and Medical (ISM) radio band) has also been dedicated internationally and reserved for the use of EM energy for industrial, scientific, and medical purposes.

In this chapter, novel, compact, non-invasive, and cost effective EM MTMs-based microfluidic sensors with high sensitivity are discussed to be used for concentration estimation and detection of various chemicals and biochemicals. Highly sophisticated tools are required to investigate and analyse various chemicals and biomaterials demanding highly sensitive sensing mechanisms. Comprehensive and remarkable research has been done in the field of EM resonators and MTMs for their potential in healthcare applications. Various applications of EMs in field of biomedical engineering (wireless endoscopy, cancer detection, wireless drug delivery and cardiovascular healthcare) have been discussed. Section 7.1 will discuss scope of MTMs and resonators in sensing devices. Section 7.2 will discuss

[1]Queen Mary University of London, London, UK
[2]University of Glasgow, Glasgow, UK

applications of MTMs in chemical sensing. Section 7.3 will discuss role of EMs, antennas, MTMs and resonators for biomedical applications and healthcare. Generally, investigation of EM sensors, antennas and resonators are discussed in this chapter for chemical and healthcare applications.

7.1 MTMs and resonators in sensing Devices

The word "meta" means "beyond" in Greek, and therefore, term "metamaterials" refers to "beyond conventional materials." MTMs are carefully fashioned homogenous EM structures with extraordinary properties not found in nature. It consists of cells with dimensions smaller than the guided wavelength and are usually arranged in a periodic fashion. They are mostly composed of periodic resonators providing features like negative refraction, chirality, and perfect absorption, which cannot be found in regular materials. EM MTMs attract great attention due to the possibility of engineering their EM properties to values not encountered in nature. Negative refractive index metamaterials (NIM) or left-handed materials have a variety of applications in super lenses, antennas, potential energy absorbers and sensing devices. There are many peculiarities of NIMs which can be utilized to enhance the sensors performance. EM sensing and diagnostic devices developed using different type of MTMs are getting significant importance because of their high speed, low cost, high sensitivity, and selectivity. Previously, these characteristics have played an important role in improvement of different EM devices such as antennas, absorbers, lenses, polarizers, modulators, and cloaking devices but also opened new degrees of freedom in sensor design by providing a sensitivity boost. MTMs can be utilized for sensing purposes from optical to microwave frequency range. Veselago theoretically proved that if electrical permittivity and magnetic permeability of any material are simultaneously negative at any given frequency, the material will have negative refractive index (NRI). In 1996, Pendry carried out an experimental study, and showed NRI could be obtained. Later, interest in such materials increased steadily. In the 2000s, several experimental studies have confirmed that both electrical and magnetic permeability can be simultaneously negative [1–3].

The structural parameters ε and μ form the refractive index as given where ε_r and μ_r are relative electrical permittivity and magnetic permeability, respectively [4]:

$$n = \pm\sqrt{\varepsilon_r \mu_r} \tag{7.1}$$

Resonators and MTM-based sensors operate on the working principle of change in the resonant frequency and quality factor of the given design with dimensions determined by properties like permittivity and permeability. When a sample is introduced, part of the EM boundary conditions changes which result in a change in its resonant frequency and quality factor. The sample under test is either part of the resonator structure or is in the proximity of sensing device. Most of resonator or MTM-based sensors are implemented with planar techniques. Some examples of such type of sensors are based on SRR (split ring resonators) topology, micro-strip-line resonators, and patch antennas. In general, these sensing devices have higher accuracy and sensitivity as the measurement principle is based on a shift in resonant frequency or variation in quality factor of the given design [5,6].

Melik *et al.* (2010) investigated an SRR-based sensor which was capable of wireless strain measurement [7]. Withayachumnankul *et al.* (2013) designed and fabricated a sensor capable of microfluidic measurements for concentration sensing. Resonance frequency and bandwidth are influenced by the complex dielectric values of the liquid sample. The experimental relationship between the resonance frequency of the sensor and complex dielectric values of the liquid samples was evaluated in this case [8]. Amir Ebrahimi (2015) investigated microfluidic sensor working in microwave frequency range for determination of glucose concentration in water. The motive of this research was to extend his investigation in field of biomedical engineering for evaluation and monitoring of glucose content in human blood to predict the probable case of diabetes [9]. Mehmet Bakir (2017) numerically and experimentally investigated EM-based microfluidic sensor in which square shaped split resonators were used in the design. Targeted application of this design was to evaluate and differentiate between different concentrations of solutions of ethanol and methanol [10]. Similarly, Olcay Altintas (2017) investigated MTM-based sensor consisting of circular split ring resonators coupled with microstrip line. The resonance frequency and corresponding dimensions of this design were carefully selected such that the samples to be sensed have linear dielectric characteristics at that particular frequency. Sensor was evaluated for concentration estimation of ethanol in aqueous solutions, and for strain and rotation sensing [11]. André Soffiatti (2018) used two patch antennas with MTMs to evaluate the percentage of two chemicals (sodium chloride (NaCl) and ethanoic acid ($C_2H_4O_2$)) in aqueous solutions. Transmission coefficient was measured between two antennas and a linear increase in transmission (dB) was observed with increase in percentage of chemicals in corresponding solution samples at four different frequencies (1.4 GHz, 2.4 GHz, 3.4 GHz, and 4.1 GHz) [12].

Luigi La Spada investigated how to control simultaneously all the metasurface properties like amplitude, phase, and bandwidth. Main target of his research was to emphasize on EM characteristics of materials to develop metasurface structures with high selectivity and sensibility to be used further for advance sensing and diagnostics platforms. Two main configurations were considered for realization of metasurfaces which were SRR and complementary SRR (CSRR). High EM field concentration areas were identified as most suitable platforms in both configurations for sensing purpose. Dielectric properties of different chemical solutions and biological tissue samples are measured. Frequency responses of corresponding samples are the results of the interaction between the EM radiations and ingredients of samples at molecular or cellular level. Any changes in composition of samples result in significant changes in corresponding dielectric properties which lead to a resonant frequency shift or amplitude variation in the output signal of the metasurface sensor. The shifts in resonant frequencies are more related to real part of electrical permittivity while variation in amplitude is dependent on imaginary part (losses) of electrical permittivity. Both parameters (resonant frequency and amplitude) can play a vital role in sensing depending on samples nature and applications [13].

EM sensors could also be utilized to measure the physical integrity and detect induced damage in various dielectric, composite, and porous materials in the automotive and aerospace industry. These damages were estimated by sensing the difference between EM properties of these materials when exposed to heat damage as it was found that dielectric properties change when materials are exposed to heat or high temperatures. Dielectric properties are described by complex electrical permittivity ($\varepsilon = \varepsilon' - i\varepsilon''$) and complex magnetic permeability ($\mu = \mu' - i\mu''$) where real part participates in shift in resonance frequency and imaginary part participate in resistive losses [14].

Chemical and biomedical sensors are emerged for medical, healthcare, and environmental applications, beverages, chemicals and refinery industries, explosive detection, and other applications. In the next move, researchers will put more attention to implantable sensing devices and sensing of multiple parameters using a single sensing device. These sort of sensing devices have contributed to non-invasive detection of cancerous cells at initial stage and innovation in drug delivery methods in different medical research centers.

7.2 EMs in chemical sensing

Innovative MTM-based microfluidic chemical sensors are quite prominent among other EM sensors. MTMs are artificially defined and carefully fashioned homogenous EM structures with extraordinary properties not found in nature, also exhibiting negative permittivity, and permeability simultaneously. Careful selection of design parameters and appropriate formulation lead to versatile sensors having various applications in the microwave and millimeter wave frequency domain. Their behavior can be determined and evaluated by their shape, geometry, orientation, and way of electric/magnetic field excitation. Scientists have designed and fabricated multifunctional miniaturized sensor structures with higher sensitivity as compared to conventional sensors. Most of them have carried out their research and studies on the X-band frequencies and stated that similar features and functions can be achieved in THz frequencies by adjusting the structure size and geometry. These sensors are adequate and conclusive for chemical, biological, and pressure measurements. The dielectric characteristics of this type of sensor structures are also sensitive to temperature, humidity, density, strain and pressure changes. These changes can be interpreted and translated by comparing the frequency response at initial and final states [15].

Mehmet Bakir in 2017 investigated MTM-based sensor for microfluidic applications. Sensor structure was made up of SRR coupled with microstrip line. The sensor has shown promising applications which include concentration estimation of ethanol, methanol, acetone, ammonia, and polyethylene glycol (PEG 300, PEG 400) in respective aqueous solutions along with temperature sensing. This investigation was validated after comparison between numerical and experimental results, both showing consistency for sensing applications and reveal that the investigated microfluidic sensor operating on the principle of change in resonant frequency with change in dielectric properties, can be utilized for different sensing tools being used in chemical industry [10]. Similarly, in another research investigation, Mehmet Bakir *et al.* in 2018 investigated chiral MTM sensor (CMS) designed using a split ring resonator structures fabricated on both front and back side to evaluate different concentration levels of methanol, ethanol, and isopropyl alcohol (IPA), also designed to evaluate the purity of gasoline by estimating the percentage of kerosene in selected samples of gasoline. Proposed sensor structure was mainly composed of rectangular shaped split ring resonators being placed with 180° angle difference with each other to increase mutual capacitance and mutual inductance effects on substrate. All the measurements including complex permittivity values and reflection coefficients were taken using a vector network analyzer, dielectric probe kit and waveguide setup including sample holder. An example of measured dielectric properties (electrical permittivity values of samples of Ethanol from 10% to 90% concentration in water) and measurement setup has been provided in Figures 7.1 and 7.2. Design was numerically investigated in CST Microwave Studio program and then fabricated for experimental investigation. Complete sensor structure with split ring resonators can be regarded as a series of a RLC resonance circuit for which resonance frequency can be formulized as,

$$f = \frac{1}{2\pi\sqrt{LC}} \tag{7.2}$$

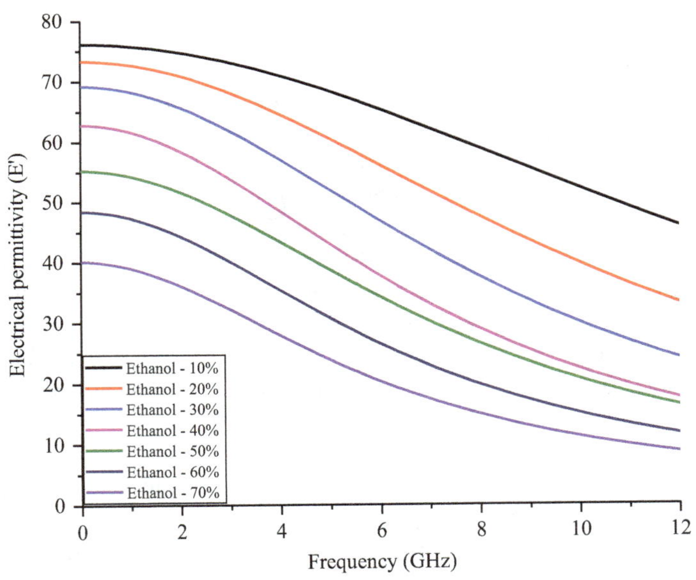

Figure 7.1 Electrical permittivity of ethanol samples from 10% to 70% concentration in water measured at room temperature

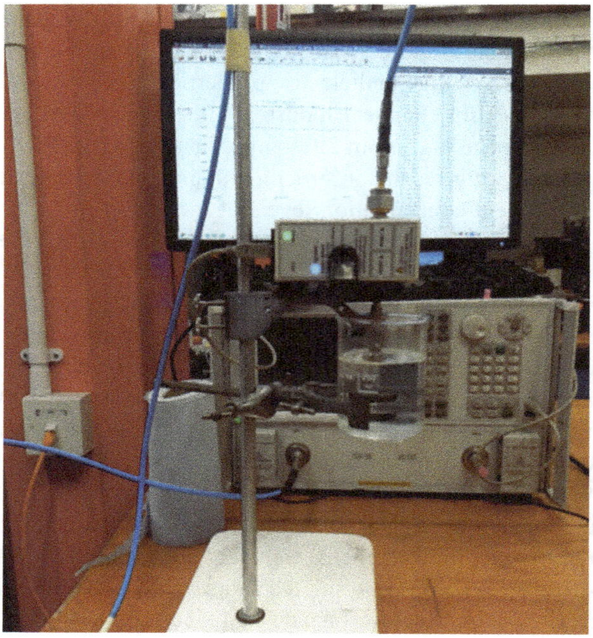

Figure 7.2 Measurement of electrical permittivity of liquid samples using dielectric probe kit & network analyzer

It was found that simulation and experimental results are quite relative which validates sensor design for proposed applications [15].

Ejaz *et al.* investigated a multifunctional MTM sensor operating in frequency range of X-band (8–12 GHz) for chemical, biomedical, and strain sensing. In the proposed design, sensing layer is structured between two resonator structures as shown in Figure 7.3. Methanol and edible fat solution samples were investigated for concentration sensing. The reflection coefficient (S_{22}) was measured where sensor structure was resonating at 9.0 GHz when simulated with vacuum in sample layer and resonant frequency showed consistent decreasing behavior with increase in concentration of methanol and edible fat, and also, with increase in thickness of sample layer for the case of strain sensing. This investigation is a steppingstone leading to novel and efficient sensor designs for chemical and biomedical sensing after experimental validations [16].

In another study, a highly sensitive MTM-based sensor was designed and investigated to differentiate between diesel samples of different quality (different level of impurities). Sensor structure consisted of a sensor layer along with the radiating edge of patch antenna printed on FR4 substrate. Since fuel quality is a common problem in Asian countries, there are different methods to investigate this issue like gas chromatography and other density measurement methods, and fiber grating sensors. These methods have their own pros and cons like they provide

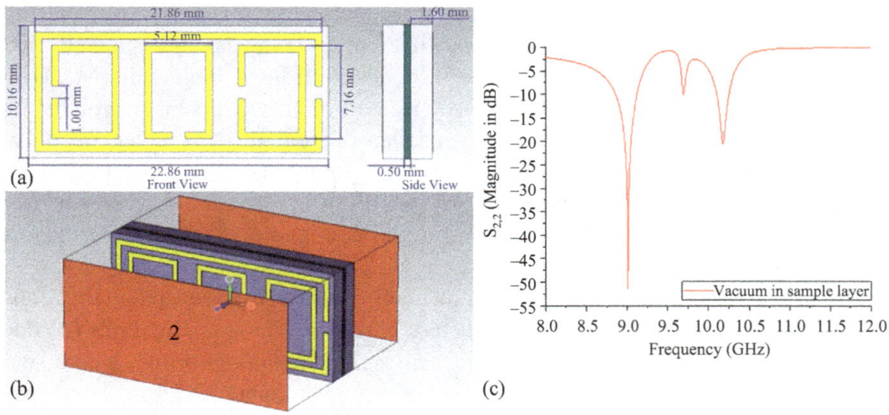

Figure 7.3 *(a) Structure dimensions, (b) profile view, and (c) scattering parameters with vacuum as sample in sample layer. © [2020] IEEE. Reprinted, with permission, from [16]*

higher precision and accuracy, but they require costly laboratory equipment and also take sufficient amount of time to show correct results. So, the proposed sensor design provides an alternative and highly cost effective solution against these methods which can differentiate between high and low quality diesel [17].

Olcay Altintas *et al.* (2017) investigated an MTM-based sensor for concentration measurements of ethanol, and similarly, for strain and rotation sensing measurements. The sensor structure consisted of circular split ring resonators coupled with microstrip line with two ports. The resonance frequency and corresponding dimensions of this sensor design were carefully selected such that the samples to be sensed have linear dielectric characteristics in selected frequency range [1].

Similarly, Şekip Dalgac *et al.* in (2020, 2021) investigated two different MTM-based sensors to evaluate the water content in grease oil and the condition of motor lubricant. Humidity in grease oil is not desirable as it deteriorates the operation life of different components in machines, but it cannot be eliminated. Therefore, a MTM-based sensor was proposed which was investigated numerically and experimentally with humid grease oil samples with different content of water. The proposed sensor was successful in determination of the water content in these oil samples. Significant shift in resonant frequency was observed with increase in humidity in those oil samples [18].

Second, a chiral MTM-based sensor was investigated to evaluate the condition of motor lubrication oil. Sensor structure was designed and simulated using Computer Simulation Technology (CST) Microwave Studio program. To achieve high sensitivity and maximum efficiency, higher frequency shifts are desired. Sensor structure was mainly composed of copper (conductivity is 5.8×10^6 S/m and thickness of 0.035 mm) based resonators whereas the chiral mediums were obtained by asymmetrical placement of the resonator to the front and back side of the substrate layer [19].

Hamzah Hayder investigated split ring resonator-based EM sensor operating at 2.5 GHz to differentiate between different pH levels and showed that sensor was highly sensitive to change in concentration of table salt (NaCl) in aqueous solutions [20]. Zarifi *et al.* investigated microwave sensor which was basically designed using microwave resonators coated with layer of polydimethylsiloxane (PDMS). PDMS layer was acting as a sorbent layer and changed its permittivity and permeability resulting in variation in resonant frequency when acetone vapor concentration was changed and respectively absorbed by PDMS layer [21].

Microwave sensing is highly recommended for diverse fields and applications as they are non-invasive and non-destructive in nature. These applications include dielectric characterization, microfluidic sensing, health monitoring, biomedical sensing, and healthcare. An economical, compact, and contactless resonator-based microwave sensor for microfluidic applications was proposed for dielectric characterization of liquids. Sensor design structure is based on complementary circular split-ring resonators fabricated on cost effective FR-4 substrate, operating at a resonant frequency of 2.45 GHz without sample and various resonant frequency peaks corresponding to different samples placed in a capillary tube placed in high electric field region [22].

The proposed design structure has promising results and provided a convenient and cost-effective solution for the investigation of chemicals in their aqueous solutions. The quality of organic solvents (like ethanol, methanol, acetone, and other industrial standard alcohols) is often compromised by various contaminants (like water or other chemicals) which is reason of investigation of such resonator based EM sensors that can be an economical and effective selection for concentration estimation in these liquids. A contactless, simple, and convenient technique is carried out to detect variations in concentration in different liquid samples. This technique employs a small glass tube adjusted normal to the ground plane. Design of proposed microwave sensor is introduced with its equivalent circuit model is shown in Figures 7.4 and 7.5 [22].

The capacitive area in this sensing structure was the area where maximum electric fields were concentrated. The sample was placed such that it is closer to high electric field territory of the sensor. The capacitance of sensing structure would have been changed due to the change in dielectric characteristics of material where electric field would have been distorted by the sample under test. The variation in capacitance would depend upon the dielectric properties of sample and the resonant frequency would change accordingly. Hence, the change in dielectric characteristics of material can be detected through variations and shift in resonant frequency due to change in capacitance. Similarly, the dielectric losses by the same material will be quantified by change in Q-factor. Schematic of measurement setup, an example of measurement results and formulated complex permittivity has been provided in Figures 7.6–7.8. It can be concluded that the proposed resonator-based microwave sensor was a good selection for accurate dielectric characterization and microfluidic measurements while offering a convenient and low-cost operation [22].

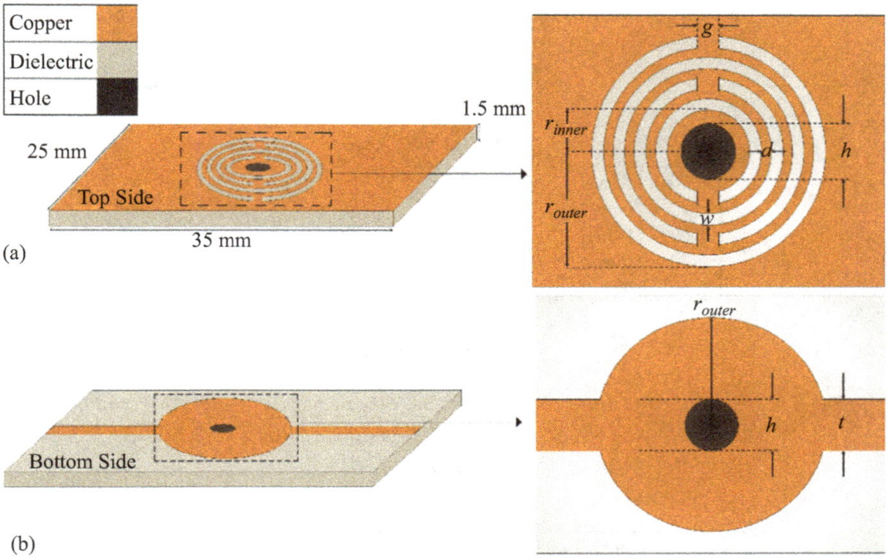

Figure 7.4 *Proposed microwave sensor with dimensions. (a) Top view with dimensions: $r_{outer} = 4.65$ mm, $r_{inner} = 2.35$ mm, $t = 2.65$ mm, $g = 0.7$ mm, $d = 0.3$ mm, $w = 0.35$ mm, $h = 1.5$ mm. (b) Bottom view. © [2020] IEEE. Reprinted with permission from [22].*

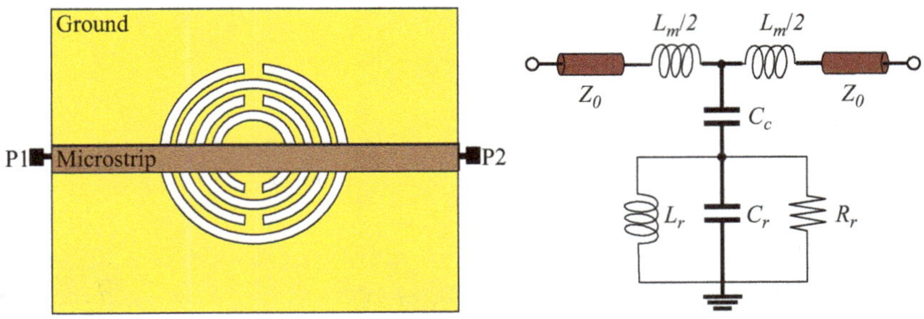

Figure 7.5 *(a) Schematic diagram of proposed sensor and (b) proposed equivalent circuit model associated with resonator-based EM sensor design. © [2020] IEEE. Reprinted with permission from [22]*

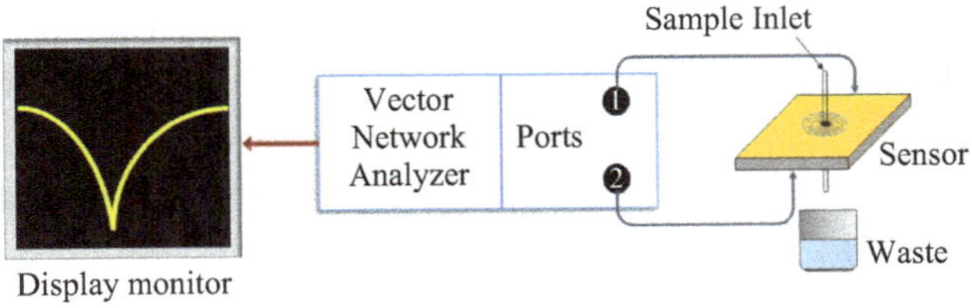

Figure 7.6 *Measurement setup used for dielectric characterization. © [2020] IEEE. Reprinted with permission from [22]*

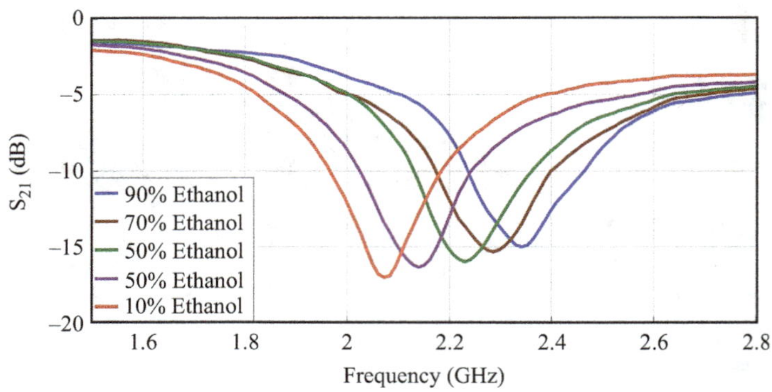

Figure 7.7 Measured results for samples of ethanol, where the concentration is varied from 10% to 90% with step size of 20%. © [2020] IEEE. Reprinted with permission from [22]

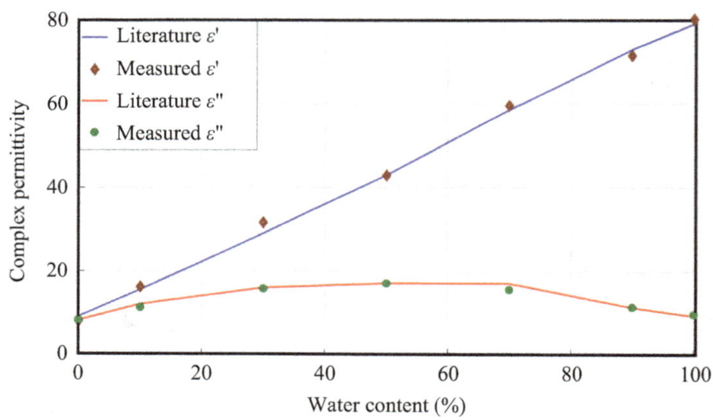

Figure 7.8 Measured values of complex permittivity of samples of Ethanol solutions and comparison with reference values mentioned in literature. © [2020] IEEE. Reprinted with permission from [22]

7.3 EMs in biomedical and healthcare applications

EM-based microwave sensors and resonators constitute a valuable share in the area of biomedical sensing and healthcare. These sensors have huge advantage over their counter parts because of the fact that they offer an increased precision and resolution in comparison to high-frequency ultrasound devices. Operating principle of these sensing devices is based on nonionizing radiation and also offers better potential and ability to penetrate through, skin, body tissue, and other biomaterials in contrast to other optical techniques. Different techniques and approaches are outlined in this topic which are suitable for numerous biomedical applications and solutions for novel healthcare diagnostic devices such as selectively connecting metabolic transitions to variation in the dielectric properties of a sensing substrate or material or even the human tissue itself. There has been a huge research interest in the modeling and realization of the dielectric measurements at microwave and millimeter-wave frequencies of different biological materials and tissues, such as skin, fat, blood, cholesterol, cerebral tissue, and intestine tissue. More effort is required for development of dielectric models of various biomaterials and tissues to step up the ladder in this field. Progression in this field will facilitate implementations of more reliable biosensors in future, hence, providing an appealing solution for remote or localized non-invasive and contactless biosensing applications. Microwave sensing is a modern technology developing for a vast range of applications including but not limited to chemical industry applications including estimation of moisture content in different organic liquids, construction industry for non-invasive evaluation of structures and healthcare for monitoring of glucose and various other important nutrients or biochemicals. A huge benefit of using microwave sensors over its counter parts is that they are capable enough to measure without any contact and affiliated destruction, from a short distance using penetrating waves. Non-ionizing nature of this technology also removes severe health hazards which can be introduced by other counter parts, particularly in biomedical sensing and healthcare [23–27].

Bioelectronics is one of the influential and driving aspect of electronic devices which are designed to operate in the vicinity of human body. Extensive research work has been done to make flexible, small, cost effective and biocompatible electronic sensing devices for variety of applications for healthcare and medical diagnostics. These bioelectronic devices are further categorized into implantable and epidermal devices. An implantable device with huge potential, has opened new paths for real-time monitoring of various parameters and bio fluids from different sections of human body including skin, heart, and various other organs. In contrary, the epidermal devices are mostly investigated to monitor skin temperature and cardiac electrotherapy. A classic bioelectronic device having a versatile range of features and functionalities incorporate a power source (battery or similar), an antenna system for communication with external environment, control circuits with ultra-low power circuitry, and tiny sources or containers to carry medication or drug formulas. These devices collect different signals like electroencephalogram (EEG) and electrocardiogram (ECG), temperature and measurements of bio fluids emerged from surrounding organs or tissues, which then transmitted to the external environment for processing using an antenna and connected systems. Therefore, a wireless electronic device can provide the opportunity of real time monitoring and processing of various bodily parameters and signals associated with different measurements. Every corresponding application requires an efficient and cost effective antenna design for wireless communication between bioelectronic device and external receiver. The emerging interest in flexible materials led to development of soft, conformal, and biocompatible antennas systems, facilitating the impeccable integration of these systems in bioelectronics. The propagation of EM waves through skin and body tissues is more effective at low frequency which lead to larger antenna dimensions. So, strenuous miniaturization techniques are investigated for integration of such antenna in compact implant or bioelectronic devices. In literature, various techniques and methods are explicated to design an efficient and cost effective antenna for such applications, which can operate at industrial scientific and medical (ISM) band, medical implant communication service (MICS) band (402–405 MHz), and MedRadio (401–406 MHz) band. Although low-frequency systems were utilized for implants and other healthcare applications because of less attenuation of signal by skin and other tissues, such devices are restricted as they have low data transmission rate and lower range. For devices operating at higher frequencies, the EM signal is highly distorted by body tissues but have higher data rates and longer range [27–43].

The emergence of smartphones has surged the expansion of applications in the field of health monitoring and maintaining one's wellbeing which is evolving the overall model of healthcare. Health-care contributors and end users, both are specifically curious about new smartphone updates and technologies. This evolution in changing is the conventional healthcare environment by providing convenient, effective, and personal method of communication between physicians, patients, and other healthcare professionals. Although this modernization is just starting and still require a lot of development in different sectors. Majority of available sensing equipment is dependent on manual data entry, which is vexatious to users mainly physicians and other managing health professionals [40–42]. It is anticipated that the leading edge of smartphone technology should contemplate integration of various sensing devices in a single unit to record health monitoring parameters and clinical data at a single target as shown in Figure 7.9. This step will assist early diagnoses and accurate treatment especially in the case of deadly and severe

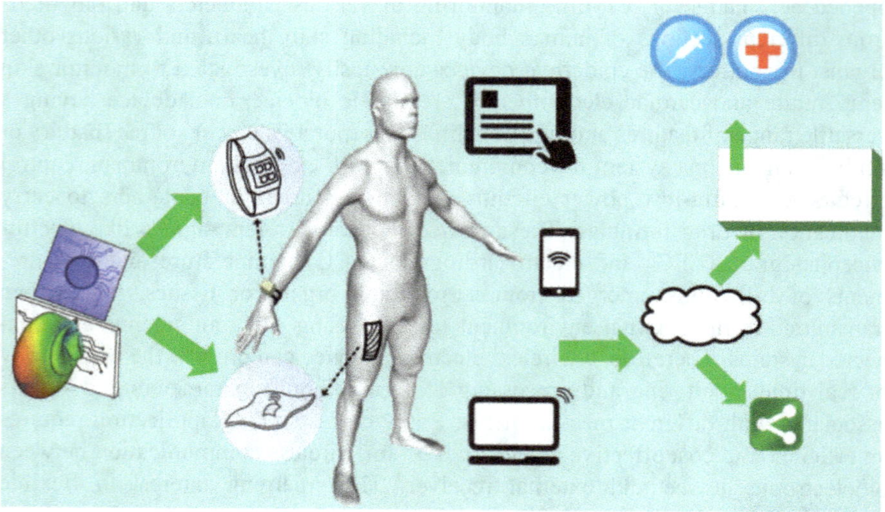

Figure 7.9 A description of a mobile health and wellness monitoring system.

diseases. There should be three characteristics in every future generation sensing environment which are portability, connectivity, and usability. Connectivity is the capability of the sensor to share and record information while connected with other sensors via body area network to provide pertinent information and data to physicians. Portability refers to the requirements associated with a wearable sensor, for instance, like convenience, accessible, low power consumption, and small form factor. Usability is defined as the potential to create an environment for sensors to continuously compute various human body parameters in absence of an external stimulus and without causing any restriction in routine tasks. This refers to monitoring of various parameters, for instance, breathing, heart rate, blood pressure, body weight, body movements, and numerous chemical variables like hydration, glucose, and various other essential nutrients [41–44].

Chemical sensors to be used for health monitoring should be given huge attention in research and development so that smartphone technology can flourish in biomedical sensing and healthcare. Only few such sensors are under development and most of which have further limitations like short lifespan and also depend on optical or other type of sensors which use a probe for measurements making it impractical for monitoring applications. Therefore, EM-based microwave biosensors are genuine alternative to compute various biochemical parameters due to their interactive properties. The behavior of these sensors is different upon interaction with different materials highly depending upon their dielectric properties which in return dependent on their molecular structure. Therefore, it is practical to evaluate or estimate the composition of body tissues and various fluids using their dielectric properties. EM waves can penetrate through skin and human tissue to certain extent, facilitating the non-invasive measurement of water content and various other biochemicals in skin, organs, and other body tissues but the measurements are highly dependent on the operating frequency. The usage of these sensors has various other advantages over their counter parts like higher sensitivity, low-energy consumption, small form factor, and EM waves normally used in these sensors are nonionizing unlike X-rays [44].

The crucial part behind the sensing ability of EM-based microwave biosensors in detecting biomaterials and biochemicals is that all biological samples containing different cells, proteins, or DNA (deoxyribonucleic acid) have different dielectric properties like electrical permittivity when compared with vacuum, air and water. These sensors are designed and configured to translate changes into a computable signal. Although the transducer alone cannot translate and evaluate the changes in the dielectric properties, as these changes can be due to the variation in concentration of blood glucose or changes in protein content. Therefore, models are required to define the electrical permittivity of any material in particular frequency range. Debye relaxation equation provides the dielectric constant of polar molecules such as water [43,44]. Debye equation is given as:

$$\varepsilon_r = \varepsilon_\infty + \frac{\varepsilon_s - \varepsilon_\infty}{1 + j\omega\tau} - \frac{j\sigma}{\omega\varepsilon_0} \tag{7.3}$$

where ε_s is the low-frequency permittivity, ε_∞ is the high-frequency permittivity, τ is the relaxation time, ω is the radian frequency, ε_o is the permittivity of free space, and σ is the ionic conductivity. Biological samples usually contain huge amounts of water like skin and other body tissues contain 70–80% water which is the reason dielectric properties of skin, fat, tissues, blood, and various body proteins are significantly governed by the dielectric constant of water at microwave frequencies [44–47]. However, dielectric data from biological tissues commonly exhibit relaxation behavior that is far broader than the simple Debye type. This could arise from the superposition of several relaxation processes. Data from biological samples can be fitted nicely by a modified form known as the Cole–Cole function:

$$\varepsilon_r = \varepsilon_\infty + \frac{\varepsilon_s - \varepsilon_\infty}{1 + (j\omega\tau)^\alpha} - \frac{j\sigma}{\omega\varepsilon_0} \tag{7.4}$$

The Cole–Cole model has same parameters as Debye model except one exponential factor α. Using this relation, complex electrical permittivity can be computed over a wide frequency range providing reasonable information regarding tissue structure and dynamics for healthcare diagnostics and monitoring of body tissues. In short, huge potential lies in favor of EM-based microwave sensors for monitoring of numerous biological parameters and early diagnostics of various fatal diseases at an early stage. Spada *et al.* [48] investigated an MTM-based biosensor to detect malignant cancer and tumors. In this study, they proposed that small change in water content inside biological tissue will result in change in dielectric properties. As tumors and cancerous cells contain significantly higher concentration of water as compared to normal tissues, so, electrical permittivity will be higher in cancerous cells which will increase the capacitive effect, as capacitance is directly related to electrical permittivity as given where ε_0 is absolute electrical permittivity, ε_r is relative electrical permittivity, A is area and d is distance of dielectric region [48]

$$C = \varepsilon_0\varepsilon_r\frac{\mathrm{A}}{\mathrm{d}} \tag{7.5}$$

Therefore, proposed MTM structure has easily detected the difference between healthy and malignant tissue and theoretically, it is possible to distinguish between healthy and cancerous cells. This part of chapter articulates the recent advances in EM devices and systems specifically designed for biomedical and healthcare applications. Different biocompatible materials, fabrication methods, suitable frequencies, and many other aspects of this emerging research field in terms of healthcare applications are discussed in detail [44–48].

7.3.1 Drug delivery

The modern development in the field of biomedical engineering has enabled curative treatment using state of the art drug delivery devices for efficient drug potency. Drug delivery devices can regulate and control curative treatment by targeting malignant cancer locations with high drug potency which is exceptionally desired in the case of cancer patients. The dispensing phenomenon of drug delivery device can be amended according to the clinical requirements proposed by the physician. Drug delivery devices have transformed the course of curative treatment and provided wide range of strategies to overcome the constraints associated with supply of high drug potency at the site with infected tissues. Current research and developments in the pharmaceutical industry helped in exploring the promising actuating mechanisms that can precisely release drugs to master the curative effects allowing minimal toxicity. The wireless actuation of drug delivery is contemplated to be a promising non-invasive approach to dispense encapsulated drugs at disease affected site [33].

Drug delivery devices are categorized into active and passive devices. The size of an active device is large as it contains batteries to energize the associated electronic and control circuitry, which is the reason, microfabrication techniques are essential to minimize the size of an implantable active device for efficient drug delivery. On contrary, passive devices majorly depend on the pathophysiological characteristics and stimulus responses of the disease site, such as temperature, pH, and photo-responsive materials. The release mechanisms are also dependent on the chemical and physical properties of different tissues and organs in human body. The active systems are chosen for commercialization over their counter parts which result in small actuation force and complex packaging. Majority of active systems are wirelessly controlled and therefore facilitate patient in drug quantity and dispensing in the body. Various radiofrequency (RF) controlled drug delivery mechanisms are promising for treatment of broad range of diseases like malignant cancers, illness associated with hormonal imbalance, and many others. The recent revolution in micro-fabrication technologies enabled large variety of electronic controllers, RF and associated circuits packed in a compact drug delivery device [33,49,50].

Compact and conformal antennas are designs for wireless drug delivery devices, sparing sufficient space for integration of micro chambers. After flexible biocompatible materials are unfolded, they have facilitated fabrication of flexible antennas which can adjust without occupying significant space. Selection of antenna is highly based on efficient bidirectional communication between the implant and external receiver. External receiver contains information received from implants mostly related to drug dispense rate and timing. This information helps physicians, patients, and attendants to monitor drug delivery and its effectiveness [33].

The drug delivery devices can also be categorized into implantable and transdermal devices. The implantable devices are mostly capsules which contain electronic and control circuitry with power source (normally batteries), actuators, micro-reservoirs, and antennas for communication with external environment. This capsule, once injected or consumed wanders through the body until it reaches its destination (area of affected site). Information is transmitted wirelessly to the device which triggers the actuating process to dispense the drug, which diffuses and accumulates around the affected site more than anywhere else, causing less side effects on healthy part. This process is also utilized to focus on the site which need a certain amount of drug concentration. The concentration and triggering are controlled by a control unit and predefined drug concentration is maintained as proposed by the physician. Therefore, physiochemical and clinical pharmacokinetics behavior of drugs can be managed and mastered externally for treatment of diseased clinical sites. For example, to destroy cancer cells, high drug concentration is required for early recovery. To adapt to the requirements and condition of patient's body, the device should be miniaturized, bio friendly, and convenient to be implanted at the pathological site. In addition, to protect the drug from contamination before the dispensing from the device, a tremendous potential lies in the utilization of novel biocompatible materials for utilization in drug delivery owing to their biomaterial and biodegradable properties [33,51–55].

An example of drug delivery is through transdermal devices which is highly effective and ensure remarkably improved patient convenience. These devices are non-invasive in nature and can be self-managed with minimum variations in drug concentration. Wearable transdermal patch contains flexible circuit board, a temperature sensor, reservoirs, and NFC module. This patch is placed and pressed on the skin to transport drug into the systemic circulation at defined rate. Skin temperature is transferred to smart phone through the near field communication (NFC) antenna, which is helpful to control the quantity of drug dispense. Another advantage of NFC-enabled transdermal patch is that it can exclude the requirement of battery. The device is powered by inductive coupling between smart phone and NFC. Patch can operate under various bending conditions in frequency range of 14.2–14.6 MHz and can be energized from the smartphone. For smooth operation, the voltage required is obtained by keeping the distance between patch and smartphone to under 10 mm [33,53,54].

RF waves were also proposed as the extensive approach for non-invasive control of the drug delivery devices. Previously, the application of RF and EMs was considered rare for medical treatment due to the challenges associated with lossy nature of human body tissues in terms of dielectric properties. However, significant progress has been made in research to understand and comprehend the interaction of EM waves with the diverse nature of the human body, which has validated the use of electromagnetic waves and RF for curative treatment in various diseases. Figure 7.10 shows the dispensing mechanism of drug delivery devices at targeted disease affected

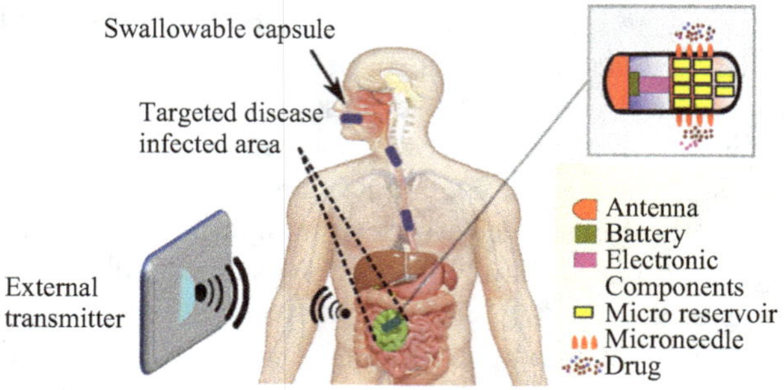

Figure 7.10 Drug delivery using capsule where drug is dispensed through microneedles upon external actuation. Reprinted from [33] with permission from AIP Publishing

sites in human body. RF wireless technologies have become significantly important for data telemetry and power transfer to implantable devices [33].

7.3.2 Cancer detection and treatment using hyperthermia

Cancer is the development of abnormal contagious cells and their uncontrolled spread in severe case. Recent development and advancement in EM-based microwave sensing devices and resonator structures inspired by MTMs can lead to a highly effective sensors in terms of precision, accuracy, sensitivity, and cost to detect presence of malignant cells in the human body. The basic operating principle of such device would be to detect changes in concentration of water inside biological tissues as contagious and affected cells have significantly higher concentration of water than normal ones. Variation in water content will bring changes in conductivity (σ) and electrical permittivity (ε) of respective tissues and higher water content means higher electrical permittivity and conductivity for the case of malignant tissues at microwave range of frequencies. The difference in electrical permittivity due to presence of affected tissues will cause variation in capacitance which will lead to shift in resonant frequency. Spada *et al.* investigated an MTM-based biosensor to detect the difference between healthy and malignant tissue and stated that, theoretically, it is possible to distinguish between healthy and cancerous cells [48].

Medical diagnostics and therapeutics are very important domains where EM-based MTM lenses are quite advantageous and have shown remarkable improvement in comparison to their typical counterparts. Hyperthermia is a treatment in which affected and contagious tissue is treated by exposure to high temperatures for specific time and is considered to be a very effective treatment for malignant cancers and various other diseases. Microwaves are employed to cure by ablating the cancerous tissue by elevating the temperature to the point that it will cause death of affected cells. This elevation of temperature will damage cell membrane of malignant tissues resulting in their destruction. In hyperthermia treatment, delicately controlled and restricted amount of heat dose is directed on cancerous part in such way that surrounding healthy tissues are not affected. The left hand MTM lens has a unique property of NRI that helps in creating a focus point where electric field is converged at the site of affected biological tissue as required in hyperthermia treatment using microwaves. This device conventionally comprises an array of split ring resonators excited by an antenna and located between a couple of parallel metallic plates. Additionally, David *et al.* [56] investigated MTM-based multi element array acting as hyperthermia applicators for their heat performance and considered appropriate for homogenous heating of larger target reaching up to 45 °C as compared their counterparts (waveguide antennas). These arrays are effective enough to transfer heat down up to 4 cm in depth using a water bolus temperature of 30 °C with moderate or high-level blood perfusion [56].

7.3.3 Wireless capsule endoscopy

Wireless capsule endoscopy (WCE) is a technique used to capture visuals of the digestive tract for healthcare applications [57]. In this approach, a small, embedded camera incorporated in a capsule is swallowed by patient under examination. The capsule passes through the digestive tract and record images. These images and visuals are transferred to receiver located outside the patient's body for examination by the physician. In a WCE system, the antenna is one of the crucial components for transmission of high quality images. The transmission at a higher frame rate is restricted to the signal-to-noise ratio (SNR) during communication between the capsule inside the body and the receiver. Therefore, an effective WCE system requires antennas which can perceive lower path losses. These path losses are normally lower at lower frequencies inside human tissue and the optimal frequency for communication between transmitter contained in a capsule inside the body and external receiver is around 450 MHz [58]. Transmitting antenna should be accommodated in a same small capsule containing the camera, and the associated frequency bandwidth should be broad enough to overcome the detuning effects caused by variations in dielectric properties of body tissues throughout the digestive tract. Embedded antennas are contained in the capsule while conformal antennas make use of capsule surface allowing inner space for other components. Receiving antennas should be compact and planar in geometry to be smoothly attached to abdomen surface and should show wideband impedance matching to counter any detuning effects caused by various parts of the patient's body [59].

Extensive research has been carried out on a communication link of a WCE system, with receiver antennas in contact and without contact with human body and it has been concluded that keeping the receiver antenna close to the human body can reduce path loss. Various operating frequencies like 402, 868, and 2400 MHz are considered for this application. A novel and compact antenna system has been investigated in the recent past for Wireless Capsule Endoscopy operating at frequency of 433 MHz in ISM (industrial, scientific, and medical) band, where a conformal antenna attached on the capsule is used inside the body for transmission of signal and for on body reception, a printed monopole antenna was designed and implemented. Equivalent model of colon as tissue phantom and Gustav voxel human body model shown in figure, built in CST were employed for simulation studies and numerical investigation with the designed antennas. Figure 7.11 shows Gustav voxel human body model with three different locations highlighted in human body for implantation of capsule for wireless capsule endoscopy (WCE). Detuning effects caused by electronic components and different tissues in vicinity of digestive tract were conquered by ultra-wideband feature of designed antennas. The path losses in communication between transmitting conformal antenna and receiver monopole antenna were restricted to less than 50 dB for any position of capsule, ensuring adequate signal level at receiver end ameliorating capsule endoscope [60].

7.3.4 Cardiovascular healthcare and other applications

In recent developments, bioelectronic devices are widespread for accurate sensing and continuous monitoring of various physiological parameters. Abundance of physiological data like measurements of heart rate, pulse, stress, sweat, and blood pressure is inspected for medical diagnostics and curative treatment. Wireless

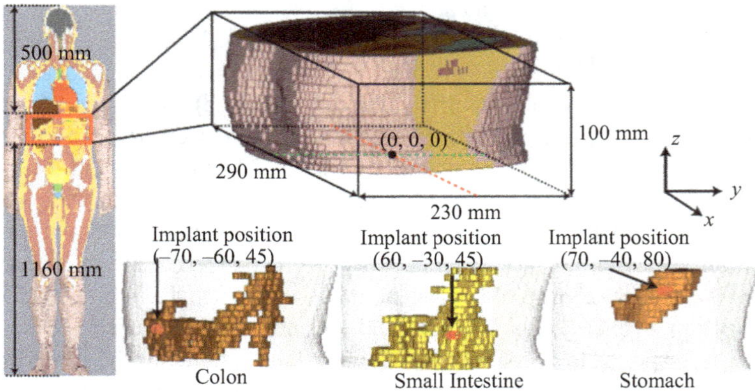

Figure 7.11 Gustav voxel human body model with three different locations of implant of the capsule. © [2019] IEEE. Reprinted with permission from [60]

technology has not only reinvigorated physiological sensing with immense support and flexibility in data acquisition but also contributed to the development of an interface for communication between sensor and broadcasting device like laptop, computers, or smartphone for graphical manifestation. Soft and conformal materials are already being investigated for fabrication of electronic components employed for cardiac monitoring, cardiac pacing, and electronic stents. Extensive range of antennas like monopole, patch and circumferential antennas are investigated for wireless pacemakers. Flexible antenna for soft cardiac wearable sensor is investigated which has capacity to transfer data associated with monitoring parameters to smart phone applications. Antenna performances were evaluated with different kind of tissue phantoms to ameliorate prospects for implantable devices like stents and pacemakers. Implantable devices have battery constraints and to conquer this challenge, a RF-powered devices like implantable rectennas are investigated which operate at 954 MHz [63–69].

Cardiovascular diseases (CVDs) are prime contributors to affect life expectancy, elevating mortality rate and augmenting health complications worldwide. Among various cardiovascular conditions, coronary heart disease and stroke or cardiac arrest are two main causes of mortality. Factors responsible for significant growth in cardiovascular risks include but not limited to chronic stress, high blood pressure, high low-density lipoprotein (LDL) cholesterol, obesity, physical inactivity, and diabetes. Extensive research is being conducted to investigate monitoring of various functions of heart for early diagnostics of cardiovascular diseases. Role of chronic stress in cardiovascular diseases is attracting significant attention due to its adverse enduring effects on both mental and physical health. Chronic stresses and negative emotional states like anxiety, depression, and childhood trauma can trigger cardiovascular effects independent of classical risk factors responsible for cardiovascular diseases. Furthermore, chronic stress negatively impacts classical risk factors (hypertension, adiposity, diabetes, and physical inactivity) as well [61,62].

Hydrocortisone is the chemical name for the cortisol hormone which has a vital role of a biomarker of chronic stress in cardiovascular diseases. Cardiovascular effects of chronic cortisol exposure can be estimated by cortisol measurements in body fluids (saliva, blood, and urine) and human hair. The hypothalamic pituitary adrenal (HPA) axis is a crucial stress response system in humans whose prime function is to facilitate successful adaptation and maintaining equilibrium with external environment. This function is performed through a chain of physiological and hormonal reactions engaging brain, organs, and glands. HPA-axis activation has an adaptive function of handling external stressful events by triggering a number of physiological reactions which include immune activation, insulin resistance, energy mobilization, and enhanced cognition. The activation of HPA-axis in response to chronic stress stimulate the production of adrenocorticotropic hormone (ACTH) from the pituitary gland which prompts the adrenal gland in kidneys to release glucocorticoids including cortisol which is the end product of initial cycle. Elevated level of cortisol contributes to various health conditions like hypertension, insulin resistance, hyperglycemia, and obesity. Repeated cortisol exposure is maladaptive and lead to long lasting alterations in physiological functions, jeopardizing the functions of the cardiovascular, nervous, and immune systems, and metabolism as well. Patients suffering from Cushing's syndrome (health condition indicating chronic cortisol excess) have a fourfold higher risk of CVD complications and mortality, and have higher cardio metabolic risk factors (e.g., obesity, hypertension, insulin resistance) compared to healthy controls. Also, patients treated with high doses of hydrocortisone have significantly high risk of cardiovascular diseases and adverse cardio metabolic markers [62]. Figure 7.12 shows the production mechanism of cortisol hormone with increase in chronic stress and its adverse cardiovascular effects.

Various methods to measure cortisol levels are discussed in literature including measurements in body fluids (saliva, blood, and urine) and hair [62]. These methods distinctly belong to controlled environment of laboratory to monitor HPA-axis response. Measurement of cortisol in hair offers several advantages over traditional specimens as it offers non-invasive, low burden and single-sample measurement of HPA-axis activity. On an average, 150 hair strands are taken close to scalp from posterior side of head. Methanol is added to hair samples to extract cortisol during an overnight incubation (16 h) at 52 °C. The methanol is then transferred into a clean glass container and is evaporated until completely dry. The samples are then dissolved in phosphate buffer saline (PBS) with pH value 8.0 and mixed thoroughly. Cortisol levels in the hair extracts are then measured using a commercial kit also used for measurement of salivary cortisol [62].

In contrast to this method of cortisol measurement, an EM-based resonator structure or MTM-based sensor can be proposed for concentration measurement of cortisol in human hair by investigating the variation in dielectric properties of cortisol or hydrocortisone solutions in response to change in concentration of respective biochemicals. A significant variation in dielectric properties can be translated into shift in resonant frequency using an EM-based sensor. Operating principle is detailed in previous topics. This proposal will give rise to new research

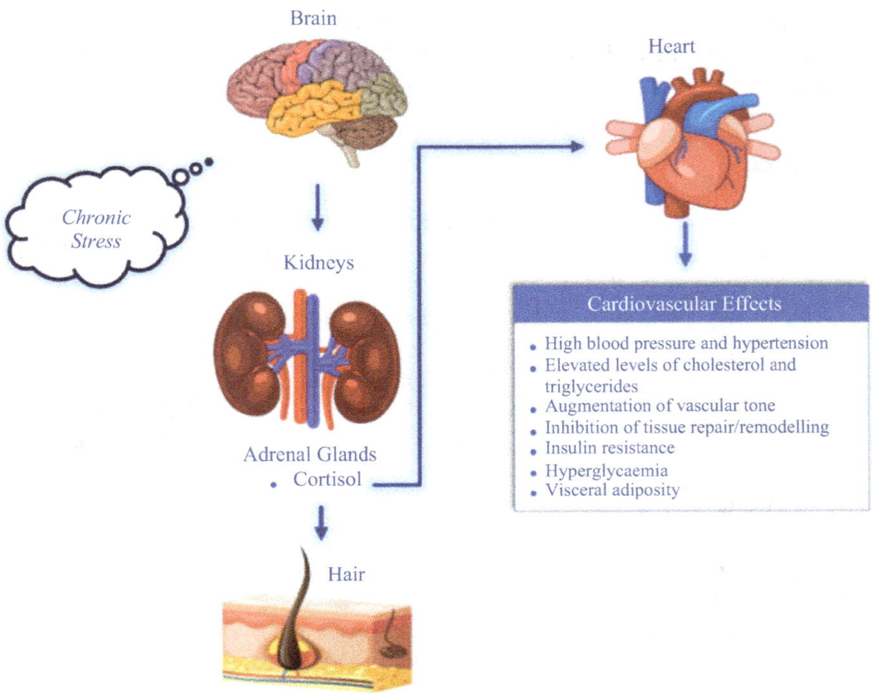

Figure 7.12 Mechanism of rise in hair cortisol with increase in chronic stress and its cardiovascular effects

area leading to simple, compact, and cost effective method to non-invasively measure concentration of cortisol in human hair.

7.4 Conclusion

This chapter has illustrated the potential and importance of EM antennas, resonators, and MTM structures in the field of chemical sensing and healthcare applications. EM resonators and MTM structures provide an appealing solution for microfluidic applications with sensors having high sensitivity for concentration estimation and detection of various chemicals and biochemicals. Comprehensive review has been provided on applications of EMs in field of healthcare applications like wireless endoscopy, cancer detection and curative treatment using hyperthermia, Wireless drug delivery, glucose sensing and cardiovascular healthcare. Remarkable amount of research has been done in field of wireless wearable sensors and implantable devices for early diagnostics and curative treatment of chronic diseases, and for real-time monitoring of crucial parameters for instance, breathing, heart rate, blood pressure, body weight, body movements, and numerous chemical variables like hydration, glucose, and various other essential nutrients in case of patients susceptible to serious health conditions. Progression in this field will facilitate implementations of more reliable sensors in future, hence, providing an extensive solution for non-invasive and contactless measurements in chemical and healthcare applications.

References

[1] V. G. Veselago, "The electrodynamics of substances with simultaneously negative values of ε and μ," *Soviet Physics Uspekhi*, vol. 10, pp. 517–526, 1968.

[2] J. B. Pendry, A. J. Holden, W. J. Stewart, and I. Young, "Extremely low frequency plasmons in metallic mesostructures," *Physical Review Letters*, vol. 76, pp. 4773–4776, 1996.

[3] J. B. Pendry, A. J. Holden, D. J. Robbins and W. J. Stewart, "Magnetism from conductors and enhanced nonlinear phenomena," *IEEE Transactions on Microwave Theory and Techniques*, vol. 47, no. 11, pp. 2075–2084, 1999, doi: 10.1109/22.798002.

[4] A. Boltasseva and V. M. Shalaev, "Fabrication of optical negative-index metamaterials: recent advances and outlook," *Metamaterials*, vol. 2, no. 1, pp. 1–17, 2008.

[5] K. Aydin, Z. Li, L. Sahin, and E. Ozbay, "Negative phase advance in polarization independent, multi-layer negative-index metamaterials," *Optics Express*, vol. 16, no. 12, pp. 8835–8844, 2008.

[6] D. Li, Z. Szabo, X. Qing, E. P. Li, and Z. N. Chen, "A high gain antenna with an optimized metamaterial inspired superstrate," *IEEE Transactions on Antennas and Propagation*, vol. 60, no. 12, pp. 6018–6023, 2012.

[7] R. Melik, E. Unal, N. K. Perkgoz, et al., "Nested metamaterials for wireless strain sensing," *IEEE Journal of Selected Topics in Quantum Electronics*, vol. 16, no. 2, pp. 450–458, 2010.

[8] W. Withayachumnankul, K. Jaruwongrungsee, A. Tuantranont, C. Fumeaux, and D. Abbott, "Metamaterial-based microfluidic sensor for dielectric characterization," *Sensors and Actuators A: Physical*, vol. 189, pp. 233–237, 2013.

[9] A. Ebrahimi, and W. Withayachumnankul, "Microwave microfluidic sensor for determination of glucose concentration in water," *2015 IEEE 15th Mediterranean Microwave Symposium*, vol. 2, pp. 1–3, 2015.

[10] M. Bakir, "Electromagnetic-based microfluidic sensor applications," *Journal of the Electrochemical Society*, vol. 164, no. 9, pp. B488–B494, 2017.

[11] O. Altintas, M. Aksoy, O. Akgol, E. Unal, M. Karaaslan, and C. Sabah, "Fluid, strain and rotation sensing applications by using metamaterial based sensor," *Journal of the Electrochemical Society*, vol. 164, no. 12, pp. B567–B573, 2017.

[12] A. Soffiatti, Y. Max, S. G. Silva, and L. M. de Mendonça, "Microwave metamaterial-based sensor for dielectric characterization of liquids," *Sensors*, vol. 18, no. 5, p. 1513, 2018.

[13] L. La Spada, "Metasurfaces for advanced sensing and diagnostics," *Sensors (Switzerland)*, vol. 19, no. 2, p. 355, 2019.

[14] E. K. Rahani, T. Kundu, Z. Wu, and H. Xin, "Heat induced damage detection by terahertz (THz) radiation," *Journal of Infrared, Millimeter, and Terahertz Waves*, vol. 32, no. 6, pp. 848–856, 2011.

[15] M. Bakır, M. Karaaslan, E. Unal, *et al.*, "Microfluidic and fuel adulteration sensing by using chiral metamaterial sensor," *Journal of the Electrochemical Society*, vol. 165, no. 11, pp. B475–B483, 2018.

[16] M. U. Ejaz and A. Alomainy, "Double-layered metamaterial structure for chemical concentration and strain sensing," in: *2020 IEEE International Symposium on Antennas and Propagation and North American Radio Science Meeting*, 2020, pp. 1403–1404, doi: 10.1109/IEEECONF35879.2020.9330089.

[17] A. Tamer, F. Karadağ, E. Ünal, *et al.*, "Metamaterial based sensor integrating transmission line for detection of branded and unbranded diesel fuel", *Chemical Physics Letters*, vol. 742, p. 137169, 2020, https://doi.org/10.1016/j.cplett.2020.137169.

[18] Ş. Dalgaç, M. Furat, M. Karaaslan, *et al.*, "Grease oil humidity sensor by using metamaterial," *Journal of Electromagnetic Waves and Applications*, vol. 34, no. 18, pp. 2488–2498, 2020.

[19] Ş. Dalgaç, F. Karadağ, M. Bakır, O. Akgöl, E. Ünal, and M. Karaaslan, "Chiral metamaterial-based sensor applications to determine quality of car lubrication oil," *Transactions of the Institute of Measurement and Control*, vol. 43, no. 7, pp. 1640–1649, 2021.

[20] H. Hamzah, J. Lees, and A. Porch, "Split ring resonator with optimised sensitivity for microfluidic sensing," *Sensors Actuators, A Physics*, vol. 276, pp. 1–10, 2018.

[21] M. H. Zarifi, A. Sohrabi, P. M. Shaibani, M. Daneshmand, and T. Thundat, "Detection of volatile organic compounds using microwave sensors," *IEEE Sensors Journal*, vol. 15, no. 1, pp. 248–254, 2015.

[22] A. Javed, A. Arif, M. Zubair, M. Q. Mehmood, and K. Riaz, "A low-cost multiple complementary split-ring resonator-based microwave sensor for contactless dielectric characterization of liquids," *IEEE Sensors Journal*, vol. 20, no. 19, pp. 11326–11334, 2020, doi: 10.1109/JSEN.2020.2998004.

[23] M. Mahmood, S. Kwon, G. K. Berkmen, *et al.*, "Soft nanomembrane sensors and flexible hybrid bioelectronics for wireless quantification of blepharospasm," *IEEE Transactions on Biomedical Engineering,* vol. 67, no. 11, pp. 3094–3100, 2020. doi:10.1109/TBME.2020.2975773.

[24] Y.-T. Kwon, Y.-S. Kim, S. Kwon, *et al.*, "All-printed nanomembrane wireless bioelectronics using a biocompatible solderable graphene for multimodal human-machine interfaces," *Nature Communications,* vol. 11, no. 1, pp. 1–11, 2020.

[25] H. Lee, Y. Lee, C. Song, *et al.*, "An endoscope with integrated transparent bioelectronics and theranostic nanoparticles for colon cancer treatment," *Nature Communications,* vol. 6, no. 1, pp. 1–10, 2015.

[26] Y. Liu, T. Yang, Y. Zhang, *et al.*, "Ultrastretchable and wireless bioelectronics based on all-hydrogel microfluidics," *Advanced Materials,* vol. 31, no. 39, p. 1902783, 2019.

[27] C. Shi, T. Costa, J. Elloian, Y. Zhang, and K. L. Shepard, "A 0.065-mm^3 monolithically-integrated ultrasonic wireless sensing mote for real-time physiological temperature monitoring," in *IEEE Transactions on Biomedical Circuits and Systems*, vol. 14, no. 3, pp. 412–424, 2020, doi: 10.1109/TBCAS.2020.2971066.

[28] H.-R. Lim, H. S. Kim, R. Qazi, Y.-T. Kwon, J.-W. Jeong, and W.-H. Yeo, "Advanced soft materials, sensor integrations, and applications of wearable flexible hybrid electronics in healthcare, energy, and environment," *Advanced Materials*, vol. 32, no. 15, pp. 1901924, 2020.

[29] S. Choi, H. Lee, R. Ghaffari, T. Hyeon, and D.-H. Kim, "Recent advances in flexible and stretchable bio-electronic devices integrated with nanomaterials," *Advanced Materials*, vol. 28, no. 22, pp. 4203–4218, 2016. doi:https://doi.org/10.1002/adma.201504150.

[30] L. Xu, S. R. Gutbrod, A. P. Bonifas, *et al.*, "3d multifunctional integumentary membranes for spatiotemporal cardiac measurements and stimulation across the entire epicardium," *Nature Communications*, vol. 5, p. 3329, 2014.

[31] F. Rangriz, A. Khaleghi, and I. Balasingham, "Wireless link for micro-scale biomedical implants using magnetoelectric antennas," in: 14th European Conference on Antennas and Propagation (EuCAP), 2020, pp. 1–4. doi:10.23919/EuCAP48036.2020.9136028.

[32] R. V. Taalla, M. S. Arefin, A. Kaynak, and A. Z. Kouzani, "A review on miniaturized ultrasonic wireless power transfer to implantable medical devices," *IEEE Access*, vol. 7, pp. 2092–2106, 2019. doi:10.1109/ACCESS.2018.2886780.

[33] A. N. Khan, A. Ermakov, G. Sukhorukov, and Y. Hao, "Radio frequency controlled wireless drug delivery devices," *Applied Physics Reviews,* vol. 6, no. 4, p. 041301, 2019. doi:10.1063/1.5099128.

[34] A. J. Bandodkar, W. Jia, C. Yardımcı, X. Wang, J. Ramirez, and J. Wang, "Tattoo-based noninvasive glucose monitoring: a proof-of-concept study," *Analytical Chemistry,* vol. 87, no. 1, pp. 394–398, 2015.

[35] X. Fan, L. Shangguan, R. Howard, *et al.*, "Towards flexible wireless charging for medical implants using distributed antenna system," *MobiCom '20*, Association for Computing Machinery, New York, NY, 2020.

[36] Y. Zhang, Z. Huo, X. Wang, *et al.*, "High precision epidermal radio frequency antenna via nanofiber network for wireless stretchable multifunction electronics," *Nature Communications*, vol. 11, no. 1, pp. 1–10, 2020.

[37] G. Yao, C. Yin, Q. Wang, *et al.*, "Flexible bioelectronics for physiological signals sensing and disease treatment," *Journal of Materiomics*, vol. 6, pp. 397–413, 2020. doi:https: //doi.org/10.1016/j.jmat.2019.12.005.

[38] C. Steiger, A. Abramson, P. Nadeau, A. P. Chandrakasan, R. Langer, and G. Traverso, "Ingestible electronics for diagnostics and therapy," *Nature Reviews Materials,* vol. 4, no. 2, pp. 83–98, 2019.

[39] M. Suzan Miah, A. N. Khan, C. Icheln, K. Haneda, and K. Takizawa, "Antenna system design for improved wireless capsule endoscope links at

433 MHz," *IEEE Transactions on Antennas and Propagation*, vol. 67, no. 4, pp. 2687–2699, 2019. doi:10.1109/TAP.2019.2900389.

[40] S. Bhattacharjee, S. Maity, S. K. Metya, and C. T. Bhunia, "Performance enhancement of implantable medical antenna using differential feed technique," *Engineering Science and Technology: An International Journal,* vol. 19, no. 1, pp. 642–650, 2016.

[41] M. Machnoor, J. Paknahad, J. Stang, and G. Lazzi, "Wireless telemetry system with independent power and data frequency resonance," *IEEE Antennas and Wireless Propagation Letters*, vol. 19, no. 4, pp. 690–694, 2020. doi: 10.1109/LAWP.2020.2976964.

[42] M. N. Islam and M. R. Yuce, "Review of medical implant communication system (MICS) band and network," *ICT Express*, vol. 2, no. 4, pp. 188–194, 2016, special Issue on Emerging Technologies for Medical Diagnostics. doi: https://doi.org/10.1016/j.icte.2016.08.010.

[43] A. Sani, M. Rajab, R. Foster, and Y. Hao, "Antennas and propagation of implanted RFIDs for pervasive healthcare applications," in: *Proceedings of the IEEE*, vol. 98, no. 9, pp. 1648–1655, 2010, doi: 10.1109/JPROC.2010.2051010.

[44] G. Guarin, M. Hofmann, J. Nehring, R. Weigel, G. Fischer, and D. Kissinger, "Miniature microwave biosensors: noninvasive applications," *IEEE Microwave Magazine*, vol. 16, no. 4, pp. 71–86, 2015, doi: 10.1109/MMM.2015.2394024.

[45] K. S. Cole and R. H. Cole, "Dispersion and absorption in dielectrics I. Alternating current characteristics," *Journal of Chemical Physics*, vol. 9, pp. 341–351, 1941.

[46] D. K. Ghodgaonkar and A. B. Daud, "Calculation of Debye parameter of single Debye relaxation equation for human skin in vivo," in: *Proceedings of the National Conference Telecommunications Technologies*, Shah Alam, Malaysia, 2003, pp. 71–74.

[47] A. P. Gregory and R. N. Clarke, "A review of RF and microwave techniques for dielectric measurements on polar liquids," *IEEE Transactions on Dielectrics and Electrical Insulation*, vol. 13, no. 4, pp. 727–743, 2006.

[48] L. La Spada, F. Bilotti, and L. Vegni, "Metamaterial biosensor for cancer detection," in: *SENSORS, 2011 IEEE*, pp. 627–630, 2011, doi: 10.1109/ICSENS.2011.6127103.

[49] M. Ochoa, C. Mousoulis, and B. Ziaie, "Polymeric microdevices for transdermal and subcutaneous drug delivery," *Advanced Drug Delivery Reviews*, vol. 64, pp. 1603–1616, 2012.

[50] G. Dogangil, O. Ergeneman, J. J. Abbott, *et al.*, "Toward targeted retinal drug delivery with wireless magnetic microrobots," *IEEE/RSJ International Conference on Intelligent Robots and Systems* (IEEE, 2008), pp. 1921–1926.

[51] M. Beccani, C. Di Natali, G. Aiello, C. Benjamin, E. Susilo, and P. Valdastri, "A magnetic drug delivery capsule based on a coil actuation mechanism," *Procedia Engineering*, vol. 120, pp. 53–56, 2015.

[52] F. Stewart, B. Cox, J. Vorstius, A. Verbeni, Y. Qiu, and S. Cochran, "Capsule based ultrasound-mediated targeted gastrointestinal drug delivery," *IEEE International Ultrasonics Symposium (IUS)* (2015), pp. 1–4.

[53] M. Vinodini Ramesh, K. S. Mohan, and D. Nadarajan, "An in-body wireless communication system for targeted drug delivery: design and simulation," *International Symposium on Technology Management and Emerging Technologies*, 2014, pp. 56–61.

[54] A. Ahmed, C. Bonner, and T. A. Desai, "Bioadhesive microdevices for drug delivery: a feasibility study," *Biomed. Microdevices,* vol. 3, pp. 89–96, 2001.

[55] G. Voskerician, M. S. Shive, R. S. Shawgo, *et al.*, "Biocompatibility and biofouling of MEMS drug delivery devices," *Biomaterials*, vol. 24, pp. 1959–1967, 2003.

[56] S. Roy Choudhury, V. Rawat, A. H. Jalal, S. N. Kale, and S. Bhansali, "Recent advances in metamaterial split-ring-resonator circuits as biosensors and therapeutic agents," *Biosensors and Bioelectronics*, vol. 86, pp. 595–608, 2016, https://doi.org/10.1016/j.bios.2016.07.020.

[57] G. Iddan, G. Meron, A. Glukhovsky, and P. Swain, "Wireless capsule endoscopy," *Nature*, vol. 405, p. 417, 2000.

[58] A. Kiourti, K. A. Psathas, and K. S. Nikita, "Implantable and ingestible medical devices with wireless telemetry functionalities: a review of current status and challenges," *Bioelectromagnetics*, vol. 35, no. 1, pp. 1–15, 2015.

[59] M. R. Basar, M. F. B. A. Malek, K. M. Juni, *et al.*, "The use of a human body model to determine the variation of path losses in the human body channel in wireless capsule endoscopy," *Progress in Electromagnetics Research*, vol. 133, pp. 495–513, 2013.

[60] M. Suzan Miah, A. N. Khan, C. Icheln, K. Haneda, and K. Takizawa, "Antenna system design for improved wireless capsule endoscope links at 433 MHz," *IEEE Transactions on Antennas and Propagation*, vol. 67, no. 4, pp. 2687–2699, 2019, doi: 10.1109/TAP.2019.2900389.

[61] A. Hafiane and J. Genest, "High density lipoproteins: measurement techniques and potential biomarkers of cardiovascular risk," *BBA Clinical*, vol. 3, pp. 175–188, 2015.

[62] E. Iob and A. Steptoe, "Cardiovascular disease and hair cortisol: a novel biomarker of chronic stress," *Current Cardiology Reports,* vol. 21, p. 116, 2019. https://doi.org/10.1007/s11886-019-1208-7.

[63] H. Kim, S. Kwon, Y.-T. Kwon, and W.-H. Yeo, "Soft wireless bioelectronics and differential electrodermal activity for home sleep monitoring," *Sensors*, vol. 21, no. 2, p. 354, 2021.

[64] Y. Gao, L. Yu, J. C. Yeo, and C. T. Lim, "Flexible hybrid sensors for health monitoring: materials and mechanisms to render wearability," *Advanced Materials,* vol. 32, no. 15, pp. 1902133, 2020.

[65] Q. Zheng, H. Zhang, B. Shi, *et al.*, "In vivo self-powered wireless cardiac monitoring via implantable triboelectric nanogenerator," *ACS Nano,* vol. 10, no. 7, pp. 6510–6518, 2016.

[66] H. Ouyang, J. Tian, G. Sun, *et al.*, "Self-powered pulse sensor for antidiastole of cardiovascular disease," *Advanced Materials*, vol. 29, no. 40, p. 1703456, 2017. doi:https://doi.org/10.1002/ adma.201703456.

[67] H. Kim, Y.-S. Kim, M. Mahmood, *et al.*, "Fully integrated, stretchable, wireless skin-conformal bioelectronics for continuous stress monitoring in daily life," *Advanced Science,* vol. 7, no. 15, pp. 2000810, 2020. doi:https://doi. org/10.1002/advs.202000810.

[68] S. Kim, M. Mahmood, Y. Lee, *et al.*, "All-in-one, wireless, stretchable hybrid electronics for smart, connected, and ambulatory physiological monitoring," *Advanced Science*, vol. 6, no. 17, pp. 1900939, 2019.

[69] L. Y. Chen, B. C.-K. Tee, A. L. Chortos, *et al.*, "Continuous wireless pressure monitoring and mapping with ultra-small passive sensors for health monitoring and critical care," *Nature Communications,* vol. 5, no. 1, pp. 1–10, 2014.

Chapter 8
In vitro test-bed design for bone fracture monitoring using implanted antennas

Symeon Symeonidis[1], *William G. Whittow*[1] *and Chinthana J. Panagamuwa*[1]

In this chapter, the development and testing of two geometrical human body phantoms based on the recipes that were investigated in Chapter 4 is presented. The first is a two-material phantom consisting of bone cortical and muscle layers simulated in Section 8.1 and measured in Section 8.2. The second is a three-material geometrical phantom consisting of bone marrow, bone cortical and muscle layers simulated and measured in Sections 8.3 and 8.4 accordingly. The specific absorption rate (SAR) evaluation of the test-bed according to the USA [1], Canada [2], and European Union [3] standards is conducted in Section 8.5. Two radio-frequency (RF) monopoles were implanted in a multi material phantom, and the condition of a bone fracture representative was replicated. The bone fracture was modelled as a cylinder residing in the mid distance between the implanted monopoles for the simulations. The technique was tested in the measurement section using three multi-material geometric phantoms. The results between simulation and measurements showed good agreement. In this section, the two-monopole system that is used for the simulations and measurements of this chapter is presented.

In Section 8.6.1, two types of biocompatible 3D-printed coatings used for the insulation of the conductive parts of the monopoles are presented. In Section 8.6.2, the blood distribution inside the fracture of a simple two material phantom is investigated in order to estimate an equivalent approximation of its geometric characteristics for the simulation design of the bone fractures. Furthermore, it was found that a cylinder consisting of the properties of blood was a sufficient design for the approximation of the blood distribution. The cylinder was used for the modeling of the bone fracture inside a voxel model of a 26-year-old female of the virtual family using CST Microwave Studio in Section 8.6.3. The results are compared with measurements inside three novel semi-solid heterogeneous bone phantoms representing the radius (arm), tibia (shin) and phalange (finger) bones in Sections 8.6.5–8.6.7 followed by an ex vivo measurement of the experiment in a lamb joint in Section 8.6.8. In

[1]Wolfson School of Mechanical, Electrical and Manufacturing Engineering, Loughborough University, UK

Section 8.6.9, the simulation and measurement results are compared and discussed. Finally, the concluding remarks of this chapter are presented in Section 8.7.

8.1 Simulation of a two-layer anatomical phantom

The geometrical characteristics of the bone layer presented in the following sections of this chapter can be adjusted to fit the geometry of any desired bone in the human body. The bone marrow and bone cortical materials of the presented phantoms have cylindrical shapes and their diameters are the same with the average diameter of a real-life femur bone found in [4]. Multi-material phantoms, such as the phantoms presented in this chapter, are used when the performance of an implanted radiator with respect to the surrounding tissues is investigated [5]. These types of phantoms usually consist of bone, fat, and muscle layers and their development requires the combination of multiple recipes and can offer a detailed representation of the investigated body part in terms of geometry and electromagnetic characteristics [6–8].

The suitability of the phantom has been tested by comparing the results of S_{11} and S_{21} of simulations and measurements of an implanted two-monopole test-bed. The CST Microwave Studio® software [9] was used for the simulations of Section 8.3. The permittivity and conductivity of the simulated phantom materials were frequency dependent and the recipes presented in Chapter 4 that mimic Gabriel's measurements [10] over the frequency range of 0.5–4 GHz were used to develop them. The dimensions of the muscle layer were 8.5 × 17 × 23 cm ($H \times W \times L$). The bone cortical cylinder centered in the muscle layer was 12 cm long with 4 cm diameter.

An antenna test-bed consisting of two monopoles was simulated for the investigation of the electromagnetic wave propagation inside the phantom. For the bone fracture monitoring that is presented in this chapter, the monopoles were implanted into the bone and muscle emulating layers of the simulated two-layer phantom. The length of each monopole was 4 cm. The monopoles' groundplanes sat at the bottom of the phantom and were outside the phantom. The dimensions of the groundplanes were chosen to be 8 × 9.5 cm (see Figure 8.1) in order to achieve good support of the

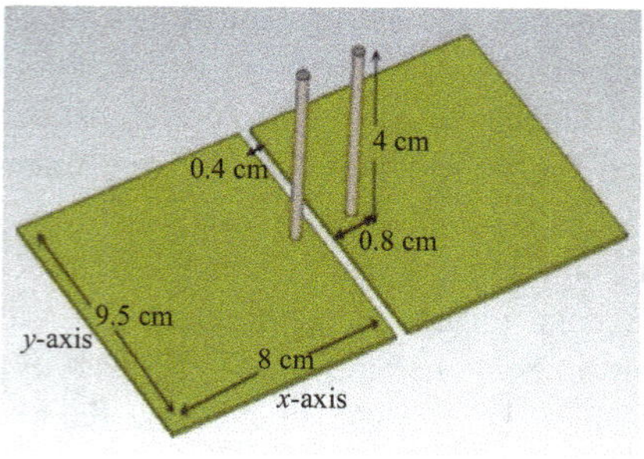

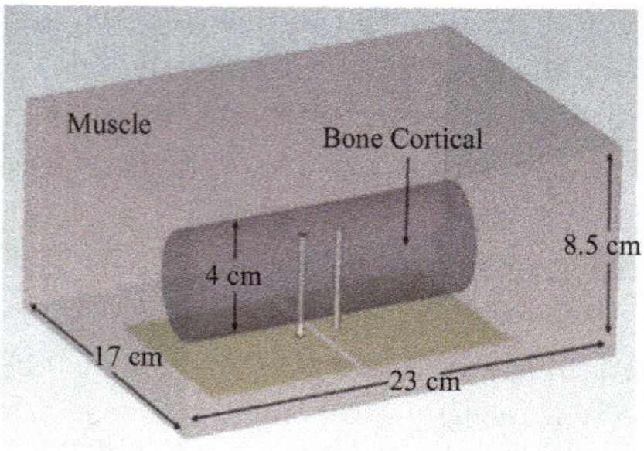

Figure 8.1 The two-monopole test bed (left); the monopoles inside the simulated two material phantom (right) taken from [15]

monopoles perpendicularly along the bone axis. The groundplanes were attached to the parts of the monopoles that were exposed on the surface of the phantom (see Figure 8.1, right). The two groundplanes were covered in plastic tape and were separated from each other by 0.4 cm. The monopoles, also covered in plastic tape, were separated from each other by 2 cm as it was found to be within the margin of screw separation distances used in real-life applications of metal plates [11–14]. Achieving a 2-cm distance between the conductors required the monopoles to be placed 0.8 cm from the edge of the groundplane in the x-axis (see Figure 8.1).

8.2 Measurement results using a two-material phantom

Using the recipes presented in Chapter 4, the phantom layers of bone cortical and muscle were developed (see Table 8.1). Bone cortical was designed to be solid, while muscle had a semi-solid form. The diameter of bone cortical was 4 cm and had a length of 12 cm. The dimensions of the muscle layer were 17 × 23 cm^2 and 8.5 cm high. An area on the top surface at the center of the muscle layer with dimensions of 4 × 4 × 12 cm ($H \times W \times L$) was removed in order to create space for the placement of the bone layer. After the bone layer was in position, the remaining air gaps in the area were carefully filled with parts of the previously removed muscle layer.

Two identical monopoles were implanted on top of the phantom into the bone layer (see Figures 8.2 and 8.3). Two stainless steel rods were used as wire parts of the monopoles coated with a 0.05-cm layer of plastic tape to avoid direct contact with the high losses of the muscle layer, and two copper sheets 0.2 cm thick were chosen as groundplanes. Figures 8.4 and 8.5 show the S_{11} and S_{21} comparison of simulated and measured results accordingly. Some disagreement can be observed mainly due to each developed recipe's dielectric properties' minor deviation from the simulated body tissue properties provided by the CST Microwave Studio® above and below the 3 GHz frequency point.

One more aspect of consideration is the difficulty in the execution of each of these measurements due to the fragile connection that is formed between the conducting part of each monopole and its groundplane, being a result of the thinness of the connector pin of the SMA launcher. In the occasion that the monopoles are forced into the bone layer, there is a high chance that the connection will break.

Table 8.1 Tissue-mimicking recipes used for the development of the two-material phantom

Ingredient (g)	Bone cortical	Muscle
Flour	80	-
Oil	30	31
Deionized water	10.5	200
Food coloring	–	–
Sugar	7.2	–
Detergent	–	31

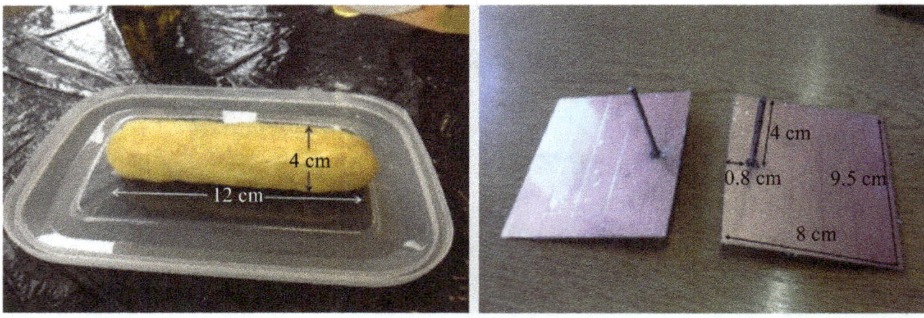

Figure 8.2 Bone cortical layer (left) and monopole implants (right)

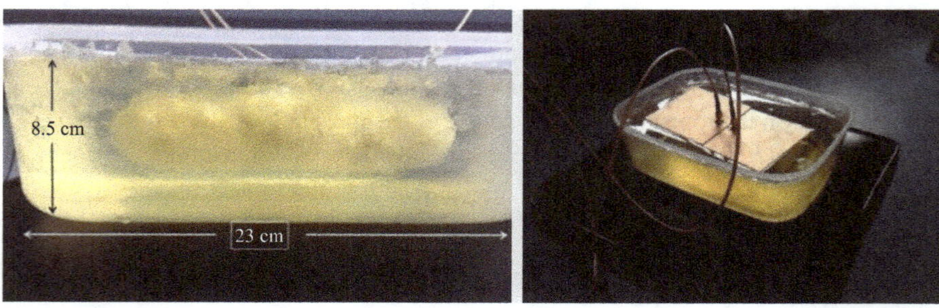

Figure 8.3 Muscle and bone cortical layers (left) and monopoles implanted into the two-material phantom (right)

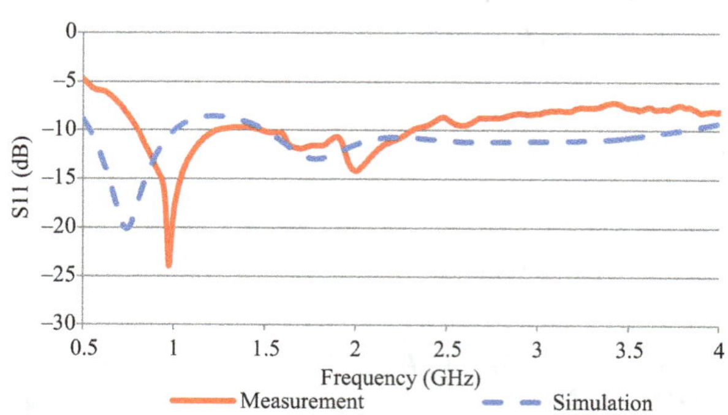

Figure 8.4 Simulated and measured S_{11} of the implanted monopoles inside the two-material phantom

To further validate the measurement results of Section 8.2, the experiment was repeated in a three-material phantom consisting of bone marrow, bone cortical and muscle layers in Sections 8.3 and 8.4.

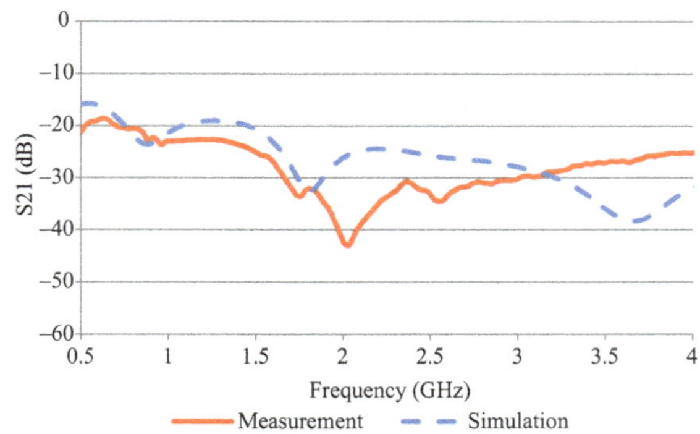

Figure 8.5 Simulated and measured S_{21} of the implanted monopoles inside the two-material phantom

8.3 Simulation of a three-layer anatomical phantom

The dimensions selected for the simulations of the muscle layer were the same as the muscle layer used in Sections 8.1 and 8.2. The bone cortical and bone marrow cylinders were 12 cm long and had diameters of 4 cm and 2 cm, respectively. The same monopole setup presented in Section 8.1 was used for the simulations of the S_{11} and S_{21} of the three-layer anatomical phantom. The complete simulation model is given in Figure 8.6.

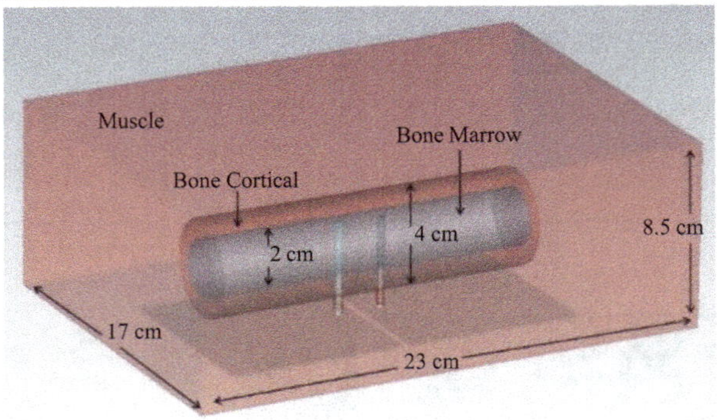

Figure 8.6 The monopoles in the simulated three-layer phantom taken from [15]

8.4 Measurement setup using a three-material phantom

For the measurements section, the bone consisted of two layers, bone marrow and bone cortical. The bone cortical layer was flattened to a thickness of 1 cm and wrapped around the bone marrow layer (see Figure 8.7). The diameter of the final bone was 4 cm and had a length of 12 cm. The dimensions of the muscle layer were 17×23 cm and 8.5 cm high and the same process of inserting the bone into the muscle phantom as in Section 8.2 was followed. The amount of ingredients used for the development of each phantom layer are shown in Table 8.2.

The two monopoles were inserted through the top of the muscle into the bone layer. Their geometry and separation distance were the same as in Section 8.2. A 0.05-cm layer of plastic tape was wrapped once again around the monopoles and stuck to the groundplanes in order to avoid short circuiting with the highly conductive muscle layer.

Figures 8.8 and 8.9 show that there is good agreement in the S_{11} and S_{21} between the simulated and the measured phantoms. This is due to the precise replication of the simulated geometry into the measured test-bed that offered a good match between the simulated and the measured power transfer of the two monopoles. A small deviation of the measured scattering parameters is expected due to the unwanted mixing of the ingredients in the contact area among the layers of the phantom. Yet, due to the high amount of gelatine used for the development of the muscle recipe, these effects are minimal. However, due to the different dielectric properties of each of the materials of the phantom, it is possible for an amount of the propagated wave to be diffracted or

Table 8.2 Tissue-mimicking recipes used for the development of the three-material phantom

Ingredient (g)	Bone marrow	Bone cortical	Muscle
Flour	80	80	–
Oil	33	30	31
Deionized water	3	10.5	200
Food coloring	1	–	–
Sugar	–	7.2	–
Detergent	–	–	31

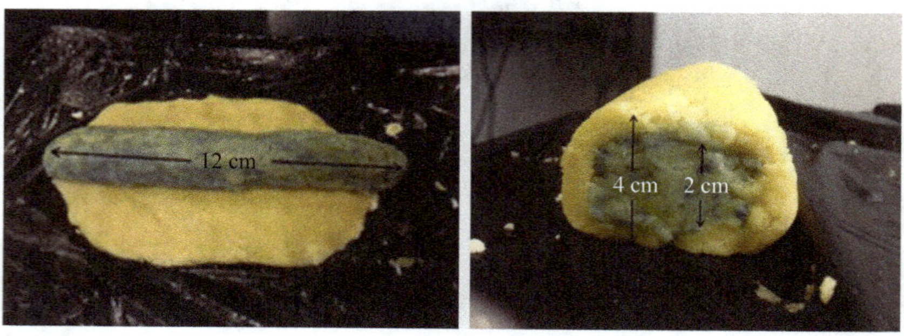

Figure 8.7 Bone marrow and bone cortical (flattened) (left) and bone cortical wrapped around bone marrow (right) taken from [15]

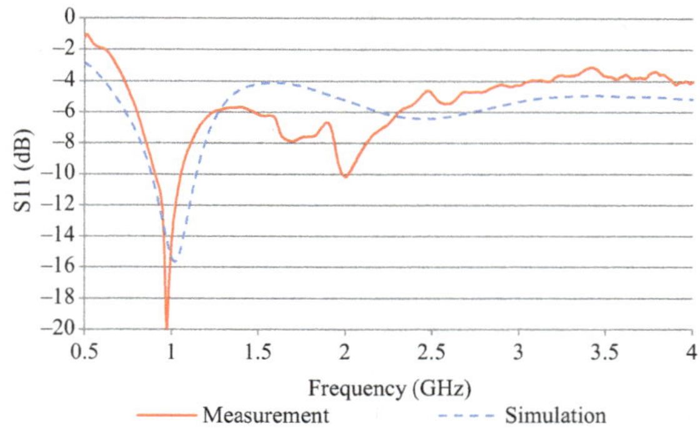

Figure 8.8 Simulated and measured S_{11} of the implanted monopoles inside the three-material phantom taken from [15]

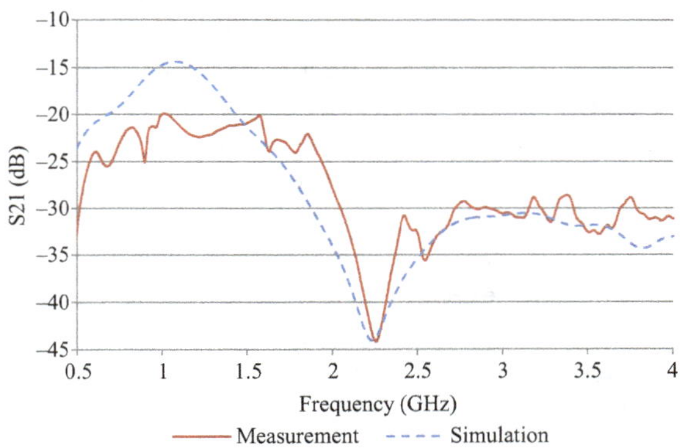

Figure 8.9 Simulated and measured S_{21} of the implanted monopoles inside the three-material phantom taken from [15]

reflected at the point of interface between the tissues of the phantom; this effect can cause minor disagreement between the simulated and measured S_{11} and S_{21}. An important aspect that will be addressed in Section 8.6.1 is the necessity for an insulation improvement for the monopoles that will ensure adequate isolation of their wire parts in occasions where liquids are introduced inside the phantom while achieving better support of the antenna and the SMA launcher.

8.5 SAR evaluation of the implanted monopoles inside a three-material phantom

According to [16], the power absorbed in the body must be limited to minimize possible adverse health effects. This is directly correlated with SAR at the frequency spectrum of 30 MHz–6 GHz [17]. SAR is defined as the rate of EM energy absorption by a unit mass, with the mass normally referring to biological tissue:

$$\mathrm{SAR} = \frac{\sigma |E|^2}{p} (\mathrm{W/kg}) \tag{8.1}$$

where σ is the conductivity of the tissue, p is the mass density of the tissue and E is the RMS value of the induced electric field. According to (8.1), the rate of EM energy absorption is proportional to the conductivity of the tissue. Thus, as the conductivity increases throughout the frequency spectrum, the absorption of the EM wave is expected to increase. Accordingly, tissues such as the muscle, that have high losses, are expected to have a higher SAR value compared to bone that has lower losses in any frequency [18].

The electromagnetic exposure of human limbs, and the human body in general, with microwave fields leads to an uneven distribution of electromagnetic power. This is due to the multitude of tissue layers in the investigated areas that have different complex permittivity values [19,20]. Therefore, it is possible that the RF power of the implanted antennas can be unevenly distributed in locations with high electric or magnetic field strengths [21]. According to the International Commission on Non-Ionizing Radiation Protection (ICNIRP), the SAR values are averaged for any period of exposure that is over 6 minutes [21]. With respect to the health and safety regulations for SAR limits, even though the exposure of a patient to EM field radiation cannot exceed this 2-minute period, the standard SAR limit (6-minute period interval) will be considered. The average localized SAR value for limb measurements for general public exposure is 4 W/kg per 10 g in the USA, Canada and the European Union [1–3,22]. The 10 g average localized SAR has been investigated for two selected frequencies, the 0.5 GHz that is the lowest frequency of operation of the antenna test bed where the EM field has the highest penetration potential in the tissues; and at 4 GHz that is the highest frequency and the overall conductivity of the phantom has the highest value, thus SAR is expected to be maximum.

Figures 8.10 and 8.11 show the average localized SAR at 0.5 GHz and 4 GHz, respectively, for the feeding power of 0.3 W. The phantom that was used for the SAR simulation was the two-material phantom that was simulated in Section 8.1. As expected, it can be observed that the EM field has more penetration into the phantom at 0.5 GHz compared to at 4 GHz. This is due to the lower conductivity of the tissues at 0.5 GHz and the larger wavelength and hence larger electrical skin depth.

The measurements with the implanted monopoles showed that simulation results can be recreated in measurements with an acceptable level of agreement. Therefore, the presented multi-material phantoms could provide a low cost, repeatable, easy to handle, and hygienic alternative to electromagnetic in vitro measurements. Furthermore, the flexibility in the design and construction means that the phantom can be adjusted to match the geometry of many bones in the

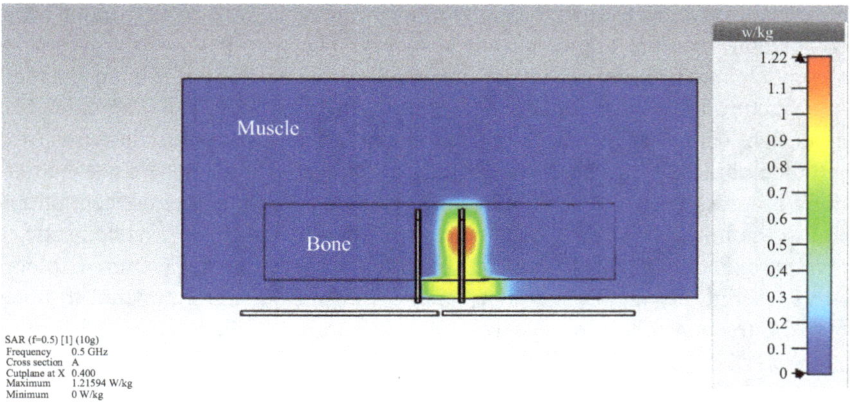

Figure 8.10 10 g SAR simulation at the frequency of 0.5 GHz for 0.3 W of feeding power

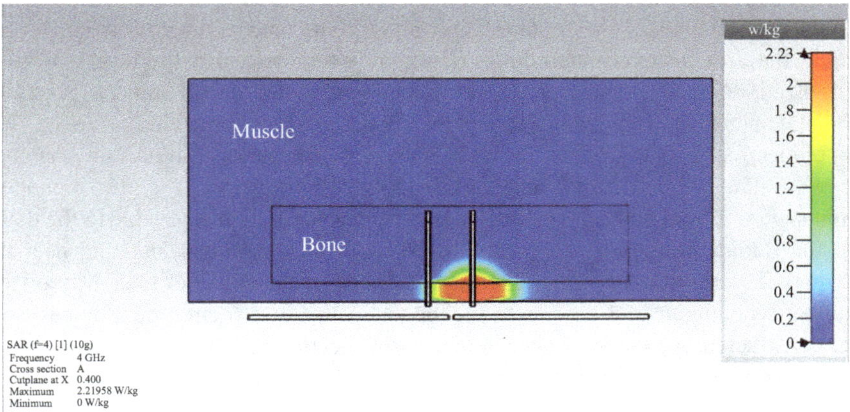

Figure 8.11 10 g SAR simulation at the frequency of 4 GHz for 0.3 W of feeding power

human body. Moreover, the obtained results have shown that the feeding power of 0.3 W for the monopole test-bed complies with the SAR specifications for the general public exposure. In the following sections, an example for monitoring the healing of severe bone fractures using a pair of RF monopoles implanted in the fractured bone will be presented.

8.6 An analysis of an antenna monitoring system for bone fractures inside multilayer phantoms

The proposed antenna monitoring system consists of two monopoles and it is based on the simulation and measurement comparisons of Sections 8.1 and 8.2. The monopole setup is envisioned to be part of a commercially available external fixator used for bone fracture alignment and recovery. The wire parts of the monopoles represent the medical screws that are implanted perpendicular to the bone either side of the fracture to provide support of the bone fragments [23–26]. Throughout this chapter, the RF power transferred from monopole 1 to monopole 2 in terms of the magnitude of the S_{21} will be investigated as an indicator of the bone's unification—healing progress as its condition will shift gradually from a higher concentration of blood in the initial fractured state, to a lower fraction towards the healed state. As shown throughout the measurements of phantom recipes of Chapter 4, blood's relative permittivity and conductivity are significantly higher compared to bone's across the frequency spectrum of 1–4 GHz, i.e. $\varepsilon_{rblood} = 59$, $\sigma_{blood} = 2.2$ S/m and $\varepsilon_{rbone} = 11.6$, $\sigma_{bone} = 0.31$ S/m at 2 GHz. According to Kerns [27], the wavenumber k for lossy media is given by: $k = \omega\sqrt{\mu_0\mu_r\varepsilon_0\left(\varepsilon_r - j\frac{\sigma}{\omega\varepsilon_0}\right)}$, where ω is the angular frequency, μ_0, ε_0 the permeability and permittivity of vacuum respectively, μ_r, ε_r the relative permeability and relative permittivity of the medium respectively, and σ the conductivity of the medium. Therefore, as the amount of blood inside and around the fracture decreases, the total relative permittivity and conductivity inside and around the area decrease accordingly. Thus, it is expected that the losses of k are minimized, resulting in more power being received by the second antenna. A pair of monopole antennas were simulated and then constructed as shown in Figure 8.2. The monopoles had a 0.25 cm diameter. For the measurements of tibia and phalange bones located in the shin and fingers of the human body respectively, the monopole lengths were 2.5 cm and the groundplanes were 6×4 cm. However, due to the depth that the Radius bone is found inside the human arm, ranging from 2 cm near the elbow to 0.5 cm near the wrist, as measured in the voxel model of CST Microwave Studio a second pair of 4 cm length monopoles with 6×5.6 cm groundplanes were also constructed. As shown in the simulations of Section 8.1, 0.5 cm of the length of each monopole was in free space connected to each groundplane and the rest of their length was inside the body placed through the muscle and perpendicularly into the bone. The gap of 0.5 cm of free space between the muscle tissue and the groundplanes was used to avoid the groundplane contact with the highly conductive muscle layer. A gap of 0.4 cm between the groundplanes was selected as adequate space to apply the needle injections of blood emulating liquid without disturbing the position of the monopoles. The use of two gap-separated groundplanes, one corresponding to each monopole wire, is also necessary. This is due to the separation distance between the monopoles (2 cm in this investigation) that is dictated by the size of the fracture. It is necessary to use larger monopole separation distances for larger fractures. Therefore, using two groundplanes and not one large groundplane that will host both monopole wires is envisioned to allow easier adjustment of the metal plate by the doctors. To maintain the 2 cm distance of the monopoles and the 0.4 cm gap between the groundplanes, the monopoles were attached to their groundplanes at a 0.8 cm distance from the right edge of the x-axis and at the middle of the y-axis of the groundplanes (see Figure 8.2).

8.6.1 The antenna coatings

Blood-emulating liquid will be injected into the phantom to mimic a bone fracture. In order to achieve total isolation of the monopoles from the liquid, two pairs of coatings were 3D printed. The coating material was biocompatible plastic polymer (PLA) [28] and it was printed using an Ultimaker 2 3D printer. These coatings were used in all of the following measurements of this chapter (see Figure 8.12). The coatings had 4 cm and 2.5 cm lengths and they were created in order to match the size of each monopole pair accordingly.

8.6.2 Visualization of the blood distribution inside the fracture

In order to understand the distribution of the blood within a fracture at the stages of inflammation and hematoma where maximum blood is inside and around the fracture, to the healed stage where the blood is minimum and the bone is restored, a measurement was conducted using a transparent gelatine phantom. The phantom was made of tap water and gelatine using the exact water to gelatine ratio (1 g of gelatine for 20 g of water) as the developed recipes presented in Section 2.3. A layer made of flour and oil was given the dimensions of the phalange bone and was embedded into the gelatine layer (see Figure 8.13). 10 ml of water with blue food dye was added in the middle of the bone phantom using a syringe gradually with a 2-ml injection step.

Figure 8.13 provides an indication on the blood fluid distribution inside a phalange (finger) bone fracture. The geometry of a bone fracture and therefore, the geometry of the hematoma, can depend on many factors, such as the angle of impact, the force of impact, and the age of the patient. Thus, the blood liquid distribution investigation of the phantom of Figure 8.13 serves as an approximation of the hematoma that will be used for the simulations of Section 8.6.3. Further on, it

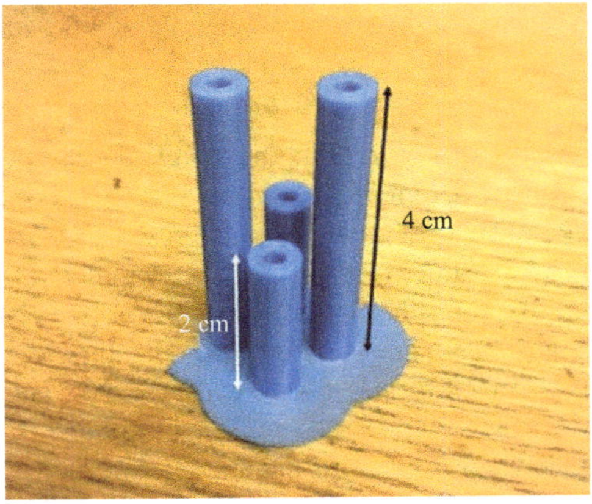

Figure 8.12 Biocompatible plastic polymer 3D printed monopole coatings, corresponding to the 2 cm and 4 cm monopoles

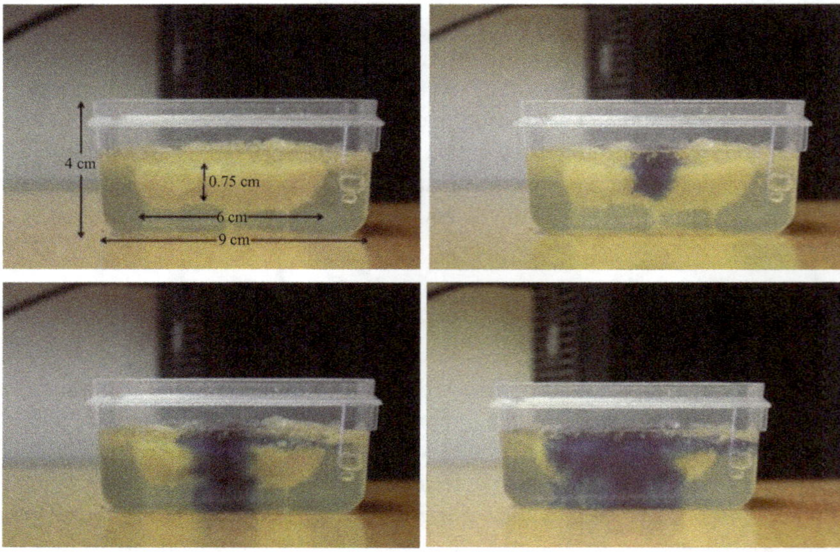

Figure 8.13 Photos (including dimensions) of the experiment to understand distribution of blood-emulating liquid in the phantom; 2 ml food coloring injection (upper right); 6 ml food coloring injection (bottom left); 10 ml food coloring injection (bottom right)

was observed that as the amount of food coloring increases in the middle of the phantom, it surrounds the bone layer approximately in a cylindrical formation. The radius of the cylinder that is composed around the bone by the food coloring is a result of the combination of the radiuses of the existing bone layer cylinder and the food coloring liquid cylinder that surrounds the bone layer. This cylinder combination has a length of 1.5 cm and radius of 1.4 cm for 2 ml of food coloring, 3.4 cm length and 1.5 cm radius for 6 ml and 5 cm length and 1.55 cm radius for 10 ml.

8.6.3 Simulations of the two monopoles inside a voxel model

For the simulations of the antenna system, the voxel model of a 26 year old female of 58 kg weight and 1.60 m height of the Virtual Family imported into CST Studio was used [9]. The permittivity and the conductivity of the model's tissues were frequency dependent, their values changed for each simulated frequency according to Gabriel's measurement results found in [10] for the range of 1–4 GHz. According to Figure 8.14, the representation of the fracture was envisioned by two coaxial cylinders. The inner cylinder of the fracture was placed within the 2 cm distance between the two monopoles and replaced the bone in that area. The diameter of the inner cylinder was equal to the diameter of the investigated bone and its length was set at 1.6 cm. For the replication of bone healing, the dielectric properties of the inner cylinder shifted from a healed state, where no blood was considered at the inner cylinder, therefore the dielectric properties of bone were

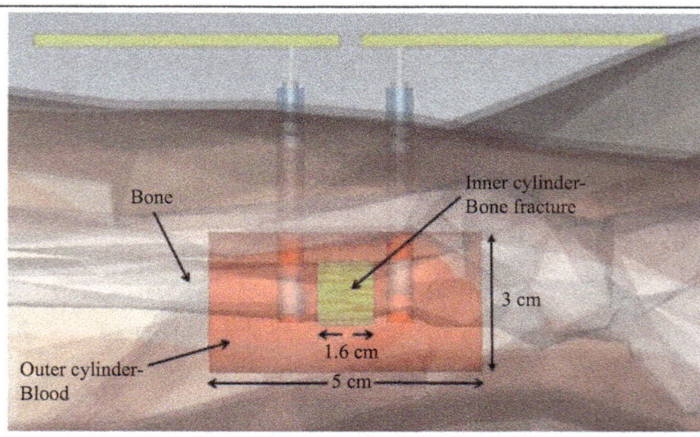

Figure 8.14 Simulation of the bone fracture using two coaxial cylinders inside a voxel model, the topology of the inner and outer cylinders is similar for arm, shin and finger bone fracture simulations taken from [26]

Table 8.3 The simulated values of the dielectric properties of the inner cylinder for 2, 6, and 10 ml of blood emulating liquid for the phalange phantom at 2.5 GHz

Injected blood emulating liquid (ml)	0.5–1.5 GHz	0.5–4 GHz
10	29.83	1.25
6	20.62	0.81
2	16.01	0.59
0	11.41	0.38

assigned, to a severed fracture state, where the maximum amount of blood was injected (10 ml for the phalange, 30 ml for the tibia and radius bone phantom simulations). The total amount of simulation stages was four, 0 ml, 2 ml, 6 ml, and 10 ml, for the phalange bone (see Table 8.3) and 0 ml, 10 ml, 20 ml, and 30 ml for the tibia and radius bones. The relative permittivity and conductivity values used for the simulations and measurements of this chapter were frequency dependent at 1–4 GHz frequency spectrum.

The outer cylinder that surrounded the fractured bone, was designated the dielectric properties of blood as it represented the hematoma inflammation [29] as reported in Chapter 4. The length and radius of the outer cylinder were given according to the measurements and calculations of blood distribution that were conducted in Section 8.6.2 (see Figure 8.14). At the initial simulation, the maximum amount of hematoma inflammation was simulated, corresponding to 10 ml of blood emulating liquid for the finger (phalange) (see Figure 8.15) and 30 ml for the arm (radius) (Figure 8.14) and shin bones (tibia) (Figure 8.16). The length and diameter of the outer cylinder were 5 cm and 3 cm for the arm bone (radius), for

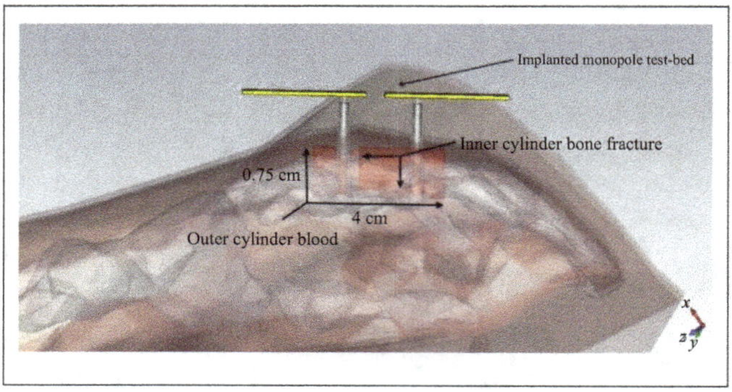

Figure 8.15 Simulated geometry of the bone fracture in the phalange bone (finger)

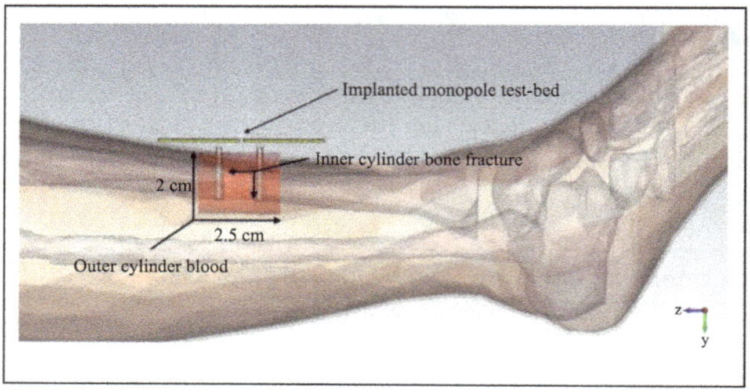

Figure 8.16 Simulated geometry of the bone fracture in the tibia bone

30 ml of blood hematoma of Figure 8.14. For each simulation, the outer cylinder's diameter and length reduced in size until it had equal diameter to the bone. In the final simulation, the fracture was considered healed, the inner cylinder had the dielectric properties of the bone and the outer blood emulating cylinder that corresponded to the inflammation was removed from the simulation.

8.6.4 Measurement setup using a three-material phantom

At first, to ensure correct implantation of the monopoles between the skin and the groundplanes the PLA insulators of the monopoles were first implanted into the middle of the bone layer. The coatings were separated by a gap of 2 cm from each other (centre to centre); then, the bone emulating tissue was placed inside the muscle phantom and its exposed top surface was carefully covered until all of the air cavities were filled with the muscle phantom (see Figure 8.17). The monopoles were then inserted into the insulating sleeves. The diameters of the bone tissues for all the phantoms presented in Sections 8.6.5–8.6.7, were given according to the average diameter of their equivalent bones provided by the voxel model of CST

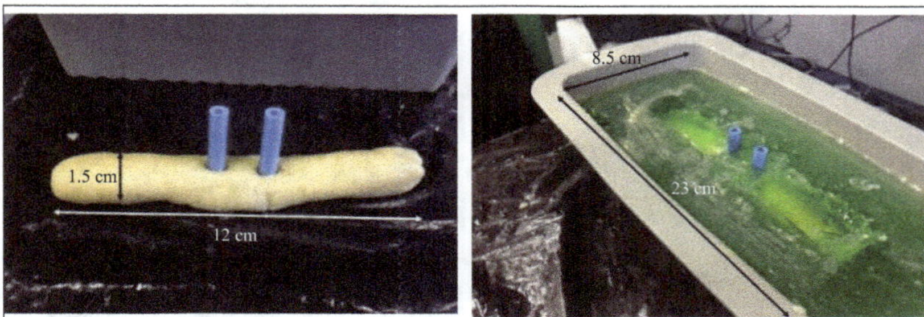

Figure 8.17 Bone layer and PLA insulators (left); forearm—radius phantom composed of bone and muscle emulating tissues (right) taken from [26]

Microwave Studio [9]. The corresponding diameters where 1.5 cm for the radius bone (arm), 0.75 cm for the phalange (finger) and 2 cm for the tibia (shin).

8.6.5 Forearm—radius bone measurement setup

The length of the monopoles for the measurement setup of the radius bone was 4 cm. The dimensions of the groundplanes were 6 × 5.6 cm. The groundplanes were attached to the exposed parts of the monopoles at the surface of the phantom (see Figure 8.17). The 0.4 cm gap between the groundplanes was used as an entry point for the injection of the blood emulating liquid via a sterilized syringe of 1 mm diameter. Although the envisioned 0.5 cm air gap between the phantom and the groundplanes was followed precisely, the groundplanes were also insulated using plastic sticky tape, thus eliminating the possibility of contact with the highly conductive muscle layer.

The diameter of the bone layer was approximately 1.5 cm according to the average diameter of the radius bone found in [30,31] and its length was 12 cm. The muscle layer occupied a plastic box with dimensions of 23 × 8.5 × 5 cm.

8.6.6 Finger—phalange bone measurement setup

For the phalange phantom, the bone layer had 0.75 cm diameter and 6 cm length and the muscle layer occupied a plastic box with dimensions of 9 × 6.5 × 4 cm (see Figure 8.18). The monopoles had 2 cm length and the dimensions of the groundplanes were 6 × 4 cm. Due to the smaller size of the phalange bone and its equivalent phantom, the maximum amount of blood-emulating liquid that was injected between the monopoles was 10 ml in five injections of 2 ml.

8.6.7 Leg—tibia bone measurement setup

For the measurements of the tibia phantom, the diameter of the bone layer was 2 cm and the length was 16 cm. The muscle layer occupied a plastic box with dimensions of 23 × 8.5 × 5 cm. The length of the monopoles was 2 cm and the dimensions of the groundplanes were 6 × 4 cm (see Figure 8.19).

Figure 8.18 Bone layer and PLA insulators (left); monopoles implanted into the finger—phalange phantom composed of bone and muscle emulating tissues (right) taken from [26]

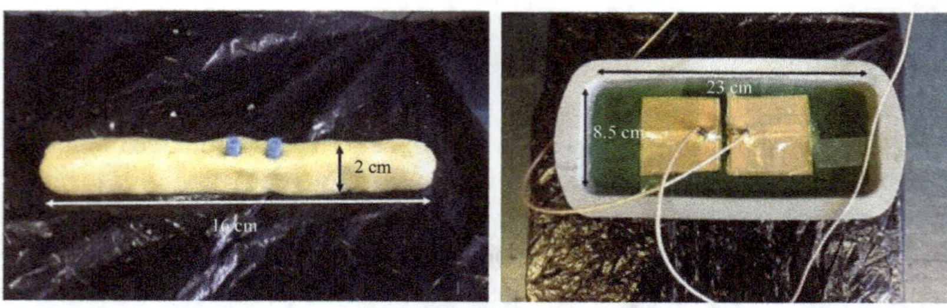

Figure 8.19 Bone layer and PLA insulators (left); monopoles implanted into the leg—tibia phantom (right) taken from [26]

For each phantom setup presented above, the S_{21} of the monopoles was measured using an Anritsu MS46524A vector network analyzer that was calibrated over the 1–4 GHz frequency range. The first measurement was conducted without injecting any blood emulating liquid into the phantom to emulate the fully healed condition of the bone. Each of the next measurements was taken for every 2 ml of blood emulating liquid that was injected into each phantom into the bone layer in the mid distance between the monopoles until the maximum number of milliliters was reached.

8.6.8 Ex vivo lamb joint measurement setup

For the ex vivo measurements, a lamb femur joint that was purchased from a supermarket was used. Two 0.4 cm diameter holes were drilled in the top surface of the bone for the monopole insulators to be implanted. The 2.5 cm length monopoles that were used in the phalange and tibia phantoms were selected for implantation due to the diameter of the bone joint being approximately 2 cm. The monopoles were coated using the 3D printed PLA coatings of Section 8.6.1. The distance between the holes was measured as 2 cm as shown in Figure 8.20. The first S_{21} measurement was conducted without fracturing the bone. For the S_{21} measurements that followed, a 4 mm diameter hole was drilled in the mid distance between the monopoles (position 1 in Figure 8.20, left) and 2 ml of blood emulating liquid was added for each step, up to the amount of 10 ml. Then, a second 4 mm diameter hole was drilled just to the right of the first hole (position 2 in Figure 8.20, left). The liquid injection process continued in 2 ml steps with an S_{21} measurement at each step until a total of 20 ml had been injected. Finally, a third 4 mm hole was drilled just to the left of the first hole (position 3 in Figure 8.20, left) and the injection and S_{21} measurement process continued in steps of 2 ml until a total of 30 ml of liquid was added to the fracture.

8.6.9 Measurement results of the three material phantoms and lamb joint test-beds

Dielectric measurements were taken in the centre of the fracture for the same amounts of injected blood emulating liquid using the coaxial probe of Chapter 4. The measurement was conducted in all three phantoms as an additional validation technique. The volume ratio of bone versus blood emulating liquid inside the fractured area was calculated based on the measured relative permittivity of the investigated area compared to the relative permittivity of the bone and the blood tissues. Radius fracture (arm) consisted of 24.1% blood at 10 ml, 63.7% blood at 20 ml, and 97.1% blood at 30 ml (see Figure 8.21). Phalange fracture (finger) consisted of 38.2% blood at 6 ml, 64.7% blood at 8 ml, and 90.0% blood at 10 ml (see Figure 8.22). Tibia fracture (shin) consisted of 36.9% blood at 10 ml, 51.7% blood at 20 ml, and 91.5% blood at 30 ml (see Figure 8.23). According to measurements conducted in [32], the measured relative permittivity of the trabecular bone callus that signals bone unification varies between 35 and 45 at 1 GHz. That range of values show the closest agreement with the 8 ml of injected blood

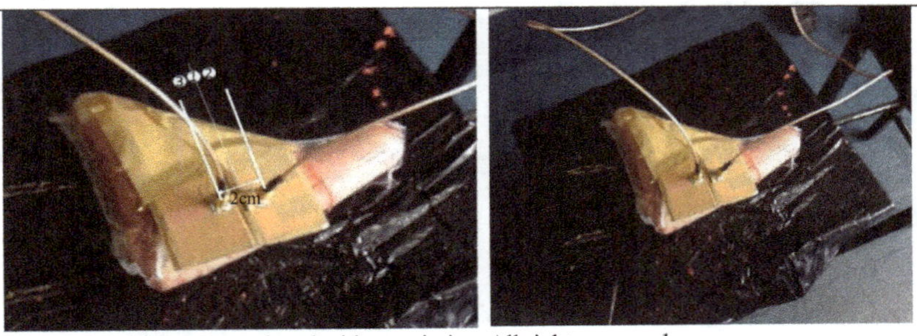

Figure 8.20 PLA insulators implanted into the lamb joint (left) and monopoles implanted into the lamb joint (right) taken from [26]

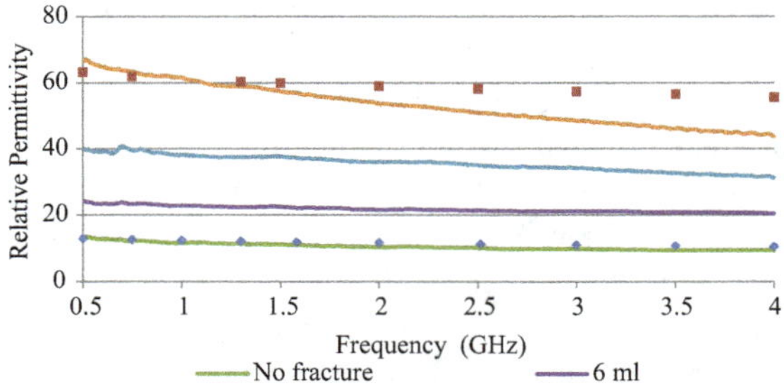

Figure 8.21 Coaxial probe dielectric measurements in the middle of the bone fracture in the radius phantoms taken from [26]

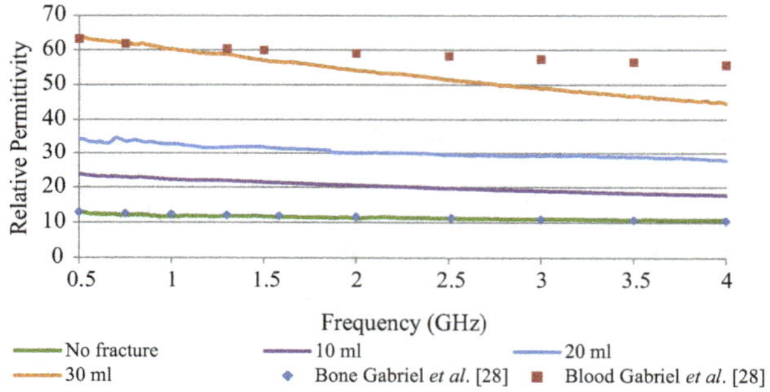

Figure 8.22 Coaxial probe dielectric measurements in the middle of the bone fracture in the phalange phantom taken from [26]

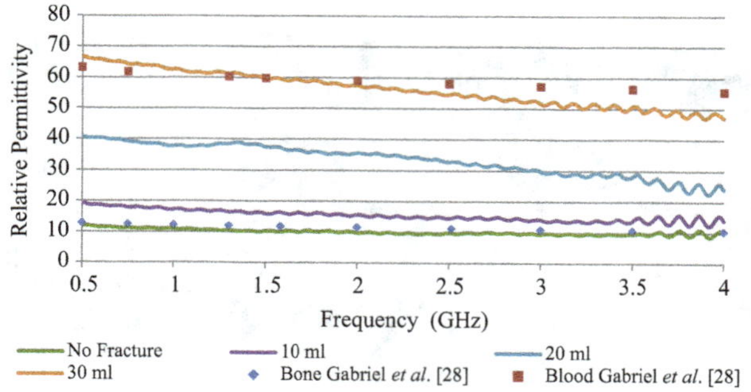

Figure 8.23 Coaxial probe dielectric measurements in the middle of the bone fracture in the tibia phantom taken from [26]

emulating liquid measurement for the phalange phantom and with the 20 ml for the tibia and radius phantoms.

The presented measurement results in Figures 8.24, 8.25 and 8.26 show that by gradually increasing the volume of blood emulating liquid between the monopoles, the magnitude of the S_{21} attenuated in the investigated frequency spectrum. Lamb joint measurements in Figure 8.27 confirm the S_{21} magnitude attenuation in ex vivo tissues. Several resonances created in the measurements compared to the smoother plots of simulations are due to the complex geometry of the dispersion of the blood emulating liquid inside the phantoms as shown in Figure 8.13. This effect is further analyzed at the end of this section with the visualization of the simulated electric fields of the monopoles inside a geometrical phantom simulated in CST Microwave Studio.

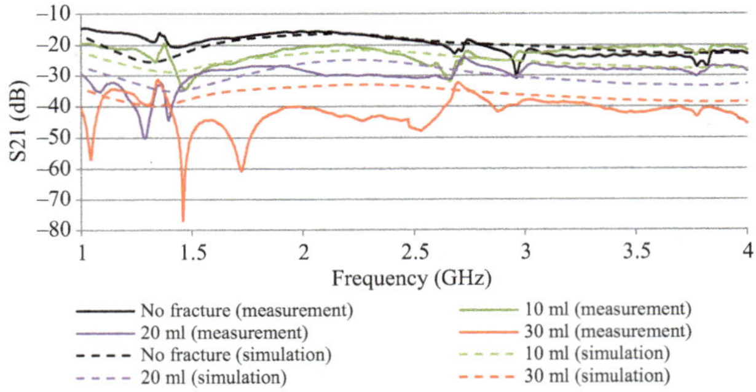

Figure 8.24 The S_{21} *measurement and simulation results for the radius phantom with different quantities of blood emulating liquid representing the hematoma taken from [26] © IOP Publishing. Reproduced with permission. All rights reserved*

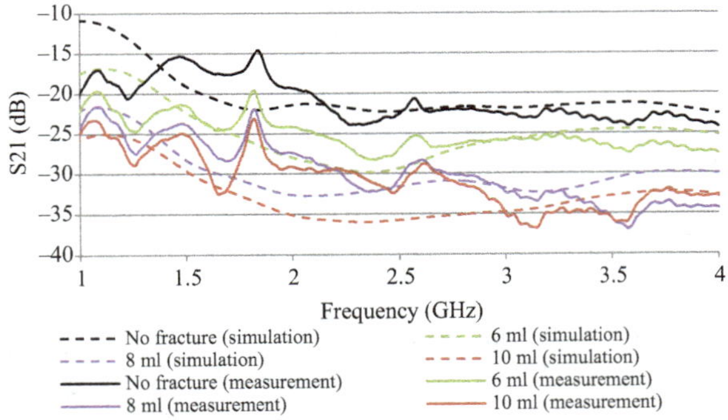

Figure 8.25 The S_{21} *measurement and simulation results for the phalange phantom with different quantities of blood emulating liquid representing the hematoma taken from [26] © IOP Publishing. Reproduced with permission. All rights reserved*

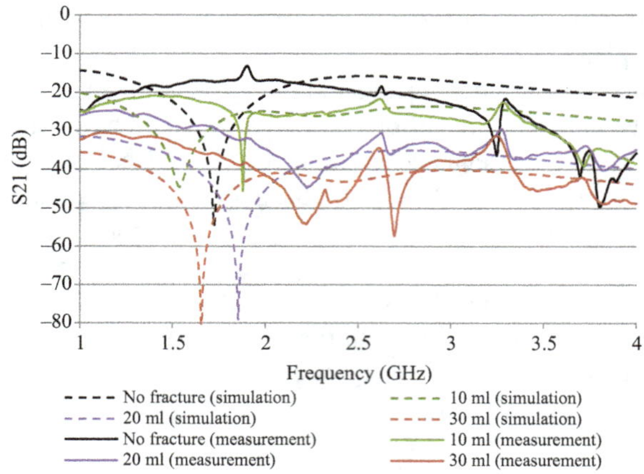

Figure 8.26 The S_{21} *measurement and simulation results for the tibia phantom with different quantities of blood emulating liquid representing the hematoma taken from [26] © IOP Publishing. Reproduced with permission. All rights reserved*

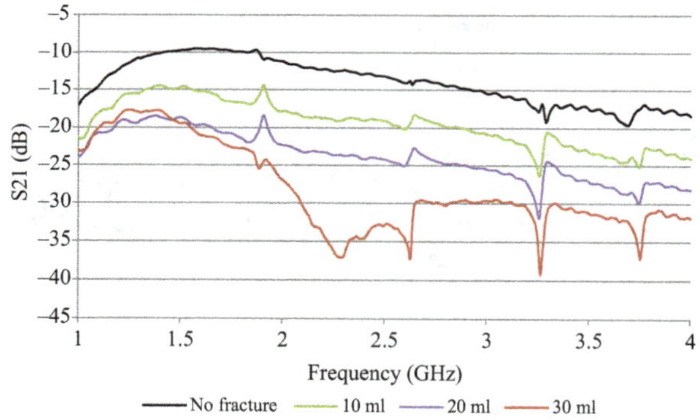

Figure 8.27 The S_{21} *measurement results for the lamb joint with different quantities of blood emulating liquid representing the hematoma taken from [26] © IOP Publishing. Reproduced with permission. All rights reserved*

The attenuation of the S_{21} due to the high relative permittivity and conductivity of blood can be additionally observed in Figure 3.28 where the electric fields of the monopoles are depicted inside a simulated multi-material geometrical phantom that features the exact dimensions of the measured tibia bone phantom of Section 8.6.7 with and without the blood cylinders representing the bone fracture.

The comparison between Figures 3.28 (*left*) and 3.28 (*right*) shows that the magnitude of the electric field is greatly reduced inside the muscle and in free space

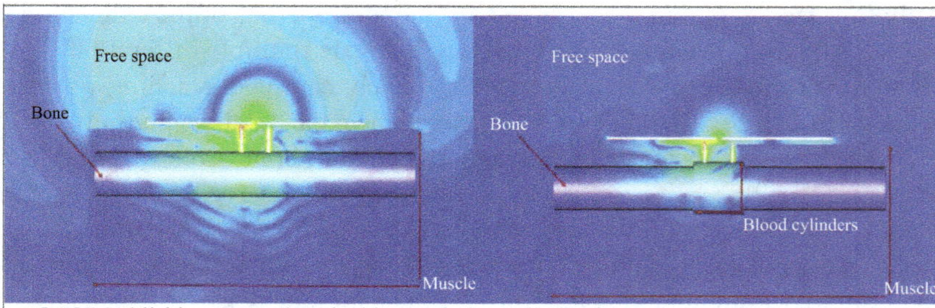

Figure 3.28 The electric field of the monopoles inside a simulated tibia phantom: without a blood cylinder (left); with a blood cylinder at 2.5 GHz (right) taken from [26]

when the blood cylinders are introduced by replacing the bone in the structure. The electric field of the monopoles in the Fresnel region ranges between $0.62\sqrt{\frac{D^3}{\lambda}} > R > \frac{2D^2}{\lambda}$ where D is the largest dimension of the antennas and λ the wavelength. For 4 cm monopoles, like the ones simulated in Figure 3.28, at the middle of the investigated frequency spectrum of 1–4 GHz, the Fresnel region of the electric field ranges between 1.4 and 2.6 cm. Therefore, the separation distance between the monopoles (2 cm throughout this chapter) is located in the Fresnel region of the electric fields. According to [33], the attenuation of the electric field in this region is described by: $e^{-j\vec{k}\vec{r}}/\vec{r}$ where $\vec{k}$ is the wavenumber and $\vec{r}$ is the distance between the field point (x, y, z) and the source.

The wavenumber k for lossy media depends on the relative permittivity, ε_r, and conductivity, σ, of each simulated tissue and is given by: $k = |\vec{k}| = \omega\sqrt{\mu_0\mu_r\varepsilon_0\left(\varepsilon_r - j\frac{\sigma}{\omega\varepsilon_0}\right)}$. The relative permittivity and conductivity of blood are higher than that of bone over the frequency range 1–4 GHz [17,18], i.e. ε_{rblood}= 58.1, $\sigma_{blood} = 2.6$ S/m and $\varepsilon_{rbone} = 11.3$, $\sigma_{bone} = 0.4$ S/m at 2.5 GHz. Therefore, k is larger when the blood is present in the separation distance between the monopoles, thus, the attenuation of the electric field is greater when the blood cylinder or blood emulating liquid is introduced inside the phantom for simulations and measurements, respectively, having as a result a worsening in the S_{21} magnitude as the volume of the blood increases (see Figure 8.27).

The S_{21} attenuation in the simulations was broadly validated in measurements as the amount of blood increased in volume inside the bone, one difference being the additional resonances in the S_{21} seen in the measurements. In the ex vivo measurements, it is not possible to create perfect geometric cylindrical shapes for the blood-emulating liquid. Creating a fracture with three circular holes only adds to the complexity of the bone's geometry. Furthermore, small bone fragments combined with the blood-emulating liquid and bone marrow of measurements are unlikely to perfectly mimic the homogenous blood properties of the simulation. As a result, the mixture inside the bone is likely to create multiple different dielectric layers to the propagating wave. These are the likely causes of standing waves and resonances in the S_{21} in the measurements. Since the relative permittivity and conductivity of blood is significantly higher than that of bone across the microwave frequency range, this behavior of S_{21} is likely to be observed for much of this frequency range. Therefore, the monopole lengths could easily be adapted to fit the diameter and depth of any human bone likely to have external fixators. However, it is important to bear in mind that microwave attenuation increases dramatically with increasing frequency. In order to obtain a measurable S_{21} above −40 dB, it would be advisable to conduct measurements in the 1–4 GHz frequency range. Finally, the reverse process where the magnitude of the S_{21} increases in the healing process after a bone fracture due to the reduction of blood, can provide an indication of the healing in the fractured area during the first four weeks after the trauma. Conversely, if the change in S_{21} slows or even stops, that would give a clear indication that the natural healing process has stalled.

8.7 Conclusions

The broadband recipes of the previous chapter have been used for the development of three multi-layered anatomical limb phantoms to monitor the healing of bones. A technique for monitoring the healing of severe bone fractures, using a pair of radiofrequency monopoles implanted in the bone has been investigated. The magnitude of change of the S_{21} measurements of two implanted monopoles in the radius, phalange, and tibia phantoms has shown to provide a good indication regarding the healing progress of different types of bone fractures that doctors use for external fixators for their treatment. These results have shown that, as the injected blood-emulating liquid's volume increased inside the fracture, the transmission between the two monopoles is severely attenuated. This attenuation occurred in simulations using a voxel model of radius, tibia, and phalange bones in CST Microwave Studio, and in measurements using three geometrical heterogeneous phantoms of the same bones. This was due to the attenuation of the electric field caused by the high relative permittivity and conductivity of the injected blood emulating liquid. It was shown that since the relative permittivity and conductivity of blood is significantly higher than that of bone across the 1–4 GHz frequency range, this behavior of S_{21} can be observed for much of this frequency range. The results were validated in an ex vivo measurement of a bone fracture in a lamb joint. The obtained results have shown that the feeding power of 0.3 W for the monopole test-bed complies with the SAR specifications for the general public exposure.

The proposed method could provide a fracture monitoring technique during the first four weeks after the trauma. In a clinical context, no two fractures would ever be identical, and it is the relative change that is important. The S_{21} that corresponds to a unified fracture 15 cm down from the top of the femur in an elderly male is unlikely to be the same as a fracture 25 cm down from the top of a femur in a young female. When the size of the fracture, the distance between the monopoles and their lengths is taken into account, there will be considerable variation in S_{21} between patients with consolidated fractures. Therefore, from this study at least, it would not be appropriate to equate an absolute value of S_{21} to consolidated fractures in general. Instead, the focus should be on the rate of change of the S_{21} magnitude over time. If a fracture is healing in the normal way, the blood haematoma will reduce over time and there will be an associated continual improvement in the S_{21}. If the healing time can be estimated by the clinicians, then one would expect the rate of change of S_{21} to slow as that period comes to an end. If the rate of change of S_{21} drops well in advance of the estimated healing time, this system would indicate healing has stalled and would warrant further investigation by the clinicians. The testing of this concept would not be possible without the low cost and easy to produce multi-layered anatomical limb phantoms developed here.

References

[1] IEEE Standards Coordinating Committee 28, IEEE C95. 1-1992: IEEE Standard for Safety Levels with Respect to Human Exposure to Radio Frequency Electromagnetic Fields, 3 kHz to 300 GHz, April 1999.

[2] "Specific absorption rate in Canada," Certification and Engineering Bureau, 2016.

[3] IEEE International Committee on Electromagnetic Safety (SCC39), *IEEE Standard for Safety Levels With Respect to Human Exposure to Radio Frequency Electromagnetic Fields, 3 kHz to 300 GHz*, April 2006.

[4] L. M. Marshall, T. F. Lang, L. C. Lambert, J. M. Zmuda, K. E. Ensrud, and E. S. Orwoll, "Dimensions and volumetric BMD of the proximal femur and their relation to age among older U.S. men," *J. Bone Miner. Res.*, vol. 21, no. 8, pp. 1197–206, 2006, doi: 10.1359/jbmr.060506.

[5] R. King, S. Prasad, and B. Sandler, "Transponder antennas in and near a three-layered body," *Microw. Theory Tech.*, vol. 28, no. 6, pp. 586–596, 1980.

[6] A. K. Skrivervik, "Implantable antennas: the challenge of efficiency," in *7th European Conference on Antennas and Propagation, EuCAP* 2013, 2013, pp. 3627–3631.

[7] J. M. Felício, C. A. Fernandes, and R. Jorge, "Wideband implantable antenna for body-area high data rate impulse radio communication," *IEEE Trans. Antennas Propag.*, vol. 64, no. c, pp. 1932–1940, 2016, doi: 10.1109/TAP.2016.2535500.

[8] H. Khodabakhshi and A. Cheldavi, "Irradiation of a six-layered spherical model of human head in the near field of a half-wave dipole antenna," *IEEE Trans. Microw. Theory Tech.*, vol. 58, no. 3, pp. 680–690, 2010, doi: 10.1109/TMTT.2010.2040410.

[9] "CST Homepage."

[10] C. Gabriel, "Compilation of the dielectric properties of body tissues at RF and microwave frequencies," *Environ. Heal.*, vol. 1, pp. 1–21, 1996, doi: Report N.AL/OE-TR- 1996-0037.

[11] M. Gudarzi, B. Marzban, H. Zamanian, and S. Amoozegar, "Investigation of the stress distribution in a bone due to screws pretensions of the bone plate," in *19th Iranian Conference on BioMedical Engineering (ICBME2012)*, 2012, pp. 1–7, doi: 10.13140/2.1.1814.8807.
[12] H. K. Uhthoff, P. Poitras, and D. S. Backman, "Internal plate fixation of fractures: short history and recent developments," *J. Orthop. Sci.*, vol. 11, pp. 118–126, 2006, doi: 10.1007/s00776-005-0984-7.
[13] G. Konrad, J. Bayer, P. Hepp, *et al.*, "Open reduction and internal fixation of proximal humeral fractures with use of the locking proximal humerus plate," *J. Bone Joint Surg.*, vol. 92, no. 1, pp. 85–95, 2010, doi: 10.2106/JBJS. I.01462.
[14] A. Goel and V. Laheri, "Plate and screw fixation for atlanto-axial subluxation," *Acta Neurochir. (Wien).*, vol. 129, pp. 47–53, 1994.
[15] S. Symeonidis, W. G. Whittow, and C. Panagamuwa, "Design and characterization of a three material anatomical bone phantom for implanted antenna applications," in *Loughborough Antennas & Propagation Conference (LAPC)*, 2017, pp. 4–8, doi: 10.1049/cp.2017.0253.
[16] S. Banik, S. Bandyopadhyay, and S. Ganguly, "Bioeffects of microwave—a brief review," *Bioresour. Technol.*, vol. 87, no. 2, pp. 155–159, 2003, doi: 10.1016/S0960-8524(02)00169-4.
[17] A. Hirata, H. Sugiyama, and O. Fujiwara, "Estimation of core temperature elevation in humans and animals for whole-body averaged SAR," *Prog. Electromagn. Res. PIER*, vol. 99, pp. 53–70, 2009, doi: 10.2528/ PIER09101603.
[18] N. Kuster and Q. Balzano, "Energy absorption mechanism by biological bodies in the near-field of dipole antennas above 300 MHz," *IEEE Trans. Veh. Technol.*, vol. 41, no. 1, pp. 17–23, 1992, doi: 10.1109/25.120141.
[19] K. Y. Yazdandoost and R. Miura, "SAR studies for UWB implanted antenna for Brain-Machine-Interface application," 2016 *10th European Conference on Antennas Propagation, EuCAP* 2016, pp. 3–6, 2016, doi: 10.1109/ EuCAP.2016.7481770.
[20] W. G. Whittow, R. M. Edwards, C. J. Panagamuwa, and J. C. Vardaxoglou, "Effect of tongue jewellery and orthodontist metallic braces on the SAR due to mobile phones in different anatomical human head models including children," *2008 Loughborough Antennas Propagation Conference LAPC*, March, pp. 293–296, 2008, doi: 10.1109/LAPC.2008.4516924.
[21] "ICNIRP Guidelines for limiting exposure to time-varying electric, magnetic and electromagnetic fields," *Int. Comm. Non-Ionizing Radiat. Prot. Heal. Phys.*, vol. 74, no. 4, pp. 494–*522*, 1998, doi: 10.1097/HP.0b013e3181f06c86.
[22] A. Cihangir, W. G. Whittow, C. J. Panagamuwa, *et al.*, "Feasibility study of 4G cellular antennas for eyewear communicating devices," *IEEE Antennas Wirel. Propag. Lett.*, vol. 12, pp. 1704–1707, 2013, doi: 10.1109/ LAWP.2013.2287204.
[23] K. Gupta, P. Gupta, G. Singh, S. Kumar, R. Singh, and R. Srivastava, "Change in electrical properties of bone as diagnostic tool for measurement

of fracture healing," *Hard Tissue*, vol. 2, no. 1, pp. 1–8, 2013, doi: 10.13172/2050-2303-2-1-316.

[24] M. Khan, S. P. S. M. A. Sirdeshmukh, and K. Javed, "Evaluation of bone fracture in animal model using bio-electrical impedance analysis," *Perspect. Sci.*, vol. 8, pp. 567–569, 2016, doi: 10.1016/j.pisc.2016.06.022.

[25] T. Yoshida, W.-C. Kim, Y. Oka, N. Yamada, and T. Kubo, "Assessment of distraction callus in rabbits by monitoring of the electrical impedance of bone," *Acta Orthop.*, vol. 81, no. 5, pp. 628–633, 2010, doi: 10.3109/17453674.2010.519910.

[26] S. Symeonidis, W. G. Whittow, M. Zecca, and C. Panagamuwa, "Bone fracture monitoring using implanted antennas in the radius, tibia and phalange heterogeneous bone phantoms," *Biomed. Phys. Eng. Express*, vol. 4, no. 4, pp. 1–36, 2018, doi: 10.1088/2057-1976/aab974.

[27] D. Kerns, "Plane-wave scattering-matrix theory of antennas and antenna-antenna interactions: formulation and applications," *J. Res. Natl. Bur. Stand. (1934).*, vol. 80B, no. 1, pp. 5–51, 1976, doi: 10.1109/MAP.1979.27388.

[28] Y. Ramot, M. Haim-Zada, A. J. Domb, and A. Nyska, "Biocompatibility and safety of PLA and its copolymers," *Adv. Drug Deliv. Rev.*, vol. 107, pp. 153–162, 2016, doi: 10.1016/j.addr.2016.03.012.

[29] R. A. D. Carano and E. H. Filvaroff, "Angiogenesis and bone repair," *Drug Discov. Today*, vol. 8, no. 21, pp. 980–989, 2003, doi: 10.1016/S1359-6446(03)02866-6.

[30] Y. Sheu, F. Amati, A. V. Schwartz, *et al.*, "Vertebral bone marrow fat, bone mineral density and diabetes: The Osteoporotic Fractures in Men (MrOS) study," *Bone*, vol. 97, pp. 299–305, 2017, doi: 10.1016/j.bone.2017.02.001.

[31] M. Doblaré, J. M. García, and M. J. Gómez, "Modelling bone tissue fracture and healing: a review," *Eng. Fract. Mech.*, vol. 71, no. 13–14, pp. 1809–1840, 2004, doi: 10.1016/j.engfracmech.2003.08.003.

[32] R. M. Irastorza, S. Valente, F. Vericat, and E. Blangino, "Dielectric properties in fresh trabecular bone tissue from 1 MHz to 1000 MHz: a fast and non destructive quality evaluation technique," *Mater. Sci. Forum*, vol. 638–642, pp. 730–735, 2010, doi: 10.4028/www.scientific.net/MSF.638-642.730.

[33] M.-K. Hu, "Fresnel region field distributions of circular aperture antennas," *J. Res. Natl. Bur. Stand. (1934).*, vol. 65D, no. 2, pp. 137–147, 1961, doi: 10.1109/TAP.1960.1144853.

Chapter 9
Wearable antennas and sensing

Vigyanshu Mishra[1] *and Asimina Kiourti*[1]

9.1 Introduction

The term "wearable antennas" refers to antennas that are placed upon the human body by embedding them into fabrics and other accessories (e.g., belts, shoes). Typically, the purpose of such antennas is to serve as the wireless interface of a wearable device that has the antenna embedded in it. For example, wearable antennas may enable data transfer from wearable sensors to a personal digital assistant (such as a smart phone or laptop), they may collect information from an underlying ingestible device, or they may send wireless power to remotely charge a patient's implant. As would be expected, extensive research work has been pursued in terms of designing wearable antennas, optimizing their performance to account for loss associated with biological tissues, miniaturizing their size for seamless operation, and enabling flexible and robust implementations. This above has already been extensively reviewed [1–3].

In this chapter, we instead focus on the use of wearable antennas for sensing applications. That is, we discuss sensors whose operation relies on how a wearable antenna system performs as a function of the sensing parameter of interest. An example application that has recently been reviewed in this regard and would not be repeated here, lies in the use of wearable antennas to monitor human body kinematics in real-world environments [4]. For instance, electrically small resonant loop antennas can be worn above and below the joint with their transmission coefficient being monitored: as the joint flexes or rotates, relative position of the antennas changes and their transmission coefficient changes accordingly [5]. Another such example entails wearable antenna-based breast cancer detection systems [6,7] that have also been reviewed in the past and will not be repeated here [8]. Along similar lines, several research groups have relied on the use of wearable antennas to empower applications as diverse as: respiration/breath monitoring, cardiac monitoring, cardiopulmonary monitoring, body fluid monitoring, and steotatic liver detection. This chapter reviews these works in an attempt to provide

[1]ElectroScience Laboratory, Department of Electrical and Computer Engineering, The Ohio State University, Columbus, OH, USA

a thorough coverage of the state-of-the-art and to inspire future research in these areas and beyond.

9.2 Respiration/breath monitoring

Continuous monitoring of respiration/breathing can help in the early detection, prevention and treatment of conditions such as sleep apnea, cardiac arrest, chronic obstructive pulmonary disease (COPD) or asthma, lung cancer, and ventilator associated pneumonia (VAP), among other respiratory conditions [9].

Though it is possible to monitor breathing rate (BR) indirectly using electrocardiography (ECG) and pulse signals [10–12], these methods are complex, obtrusive, and lack in accuracy [13]. Electrical impedance measurement techniques can quantify respiration by measuring changes in the chest electrical impedance and lung volume [13]. This is a non-invasive and safe approach, but results are prone to corruption [13]. Another method entails respiratory inductive plethysmography (RIP) which has good interference cancellation ability, but the associated equipment is cumbersome and expensive [13]. Spirometry is a clinical device that is quite accurate and often used as reference for validating various other sensors [13,14]. However, this approach requires attentive participation on behalf of the patient as well as forced breathing maneuvers, making it impractical for continuous monitoring or for patients suffering from asthma [14]. Other alternatives include nasal probes which are uncomfortable [9] and require calibration to provide lung volume estimates [14]; strain or inductance belts that require reasonable tension and hence are uncomfortable [14]; local strain sensors which are comfortable but require tight skin contact [14]; and physical examination by clinicians that is not quantitative [14].

Recently, several wearable antenna-based solutions have been proposed to monitor various breathing parameters. The vision is to empower continuous and unobtrusive mechanisms for respiration monitoring, while circumventing limitations of current technologies. Some of these recent developments are covered here, including RFID antenna-based technologies, smart T-shirts integrated with spiral fiber sensor arrays, and radar-based methods.

9.2.1 RFID antenna-based

Radio-frequency IDentification (RFID) technology is commonly used in areas such as inventory and warehouse management, transportation, libraries, and more [15]. The approach consists of a tag and a reader, wherein the tag can be either passive or active. Both reader and tag contain their own antennas and communicate with each other. Due to the passive nature of the tags, RFID finds tremendous applications in wearables and the healthcare sector in general. A major challenge in this direction is that the tag antenna needs to be designed to operate efficiently on the human body that is known to be complex from an electromagnetic standpoint. Here, we discuss two types of RFID antenna-based technologies that utilize two different approaches for monitoring respiration parameters.

- *Graphene oxide sensor* [9]: This work presents a flexible RFID sensor along with a nanomaterial sensor made from graphene oxide integrated into a facemask to monitor: (a) inhalation/exhalation cycles and (b) abnormal patterns of respiration, like apnea. Graphene is preferred for this application given its biocompatible nature, low cost, and easy biological/chemical functionalization. The sensor operates by sensing the change in humidity levels associated with inhalation/exhalation or abnormal patterns of breathing. Graphene oxide can detect such changes by means of changes in resistance, which can be recorded and relayed using RFID. The RFID tag antenna utilized is a meandered line antenna manufactured on 50 μm thick flexible Kapton substrate. It utilizes the Radio6sense Radio-board (see Figure 9.1(a)) to integrate the graphene oxide sensor and form the complete RFID integrated sensor, which is further connected to an AMS SL900A microchip for analog to digital conversion. This sensor is attached to a textile facemask (see Figure 9.1(b)) and measurements are carried out using a portable CAEN RFID qIDmini Keyfob Bluetooth UHF (ultra-high frequency) RFID reader. The sensor is exposed to environmental humidity to form a baseline and then the change in resistance is recorded during breathing cycles while the subject is wearing the sensor-integrated facemask (see Figure 9.1(c)). A maximum dynamic range of 82 Ω was observed which is well above the noise level range of 4 Ω. During exhalation, the air is highly humidified (relative humidity > 60%) and hence the resistance increases and vice versa for inhalation, Figure 9.1(c). Sleep apnea can also be detected using this method. We remark that this work is at nascent stages of development. We also note that the approach has further been implemented via epidermal technology that can be placed close to the nose, thereby enhancing wearability and user comfort [16].
- *Compression sensor* [15]: Tajin *et al.* proposed a wearable, knitted conductive yarn based belly patch antenna (see Figure 9.2(a)) as a compression sensor to monitor respiration. The antenna operates in the Ultra High Frequency (UHF) 902-928 MHz band and is fabricated on a flexible substrate that allows change of the substrate height (a key factor for compression sensing). The antenna is designed for optimal performance on the human body and can be placed on the belly part of the human torso. During respiration, the substrate height changes, altering the radiation efficiency. This change is depicted as change in the received signal strength indicator (RSSI) due to different levels of backscattered signal to the RFID reader, Figure 9.2(b). This is then utilized to monitor the compression caused on the antenna due to breathing, in turn monitoring the breathing pattern. Experiments were conducted on the SimBaby mannequin placed in a crib (e.g. Figure 9.2(c)) and on human subject participants. In both cases, breathing patterns could be deciphered using the RSSI (see Figure 9.2(d) and (e)). The maximum on-body read range of the sensor was found to be 5.8 m, which is important both from mobility and safety point of view. We remark that this technique has only recently been proposed and is in nascent stages of development. In the future, authors of the work intend to further increase the read range of the sensor and incorporate machine learning for estimating the breathing rate.

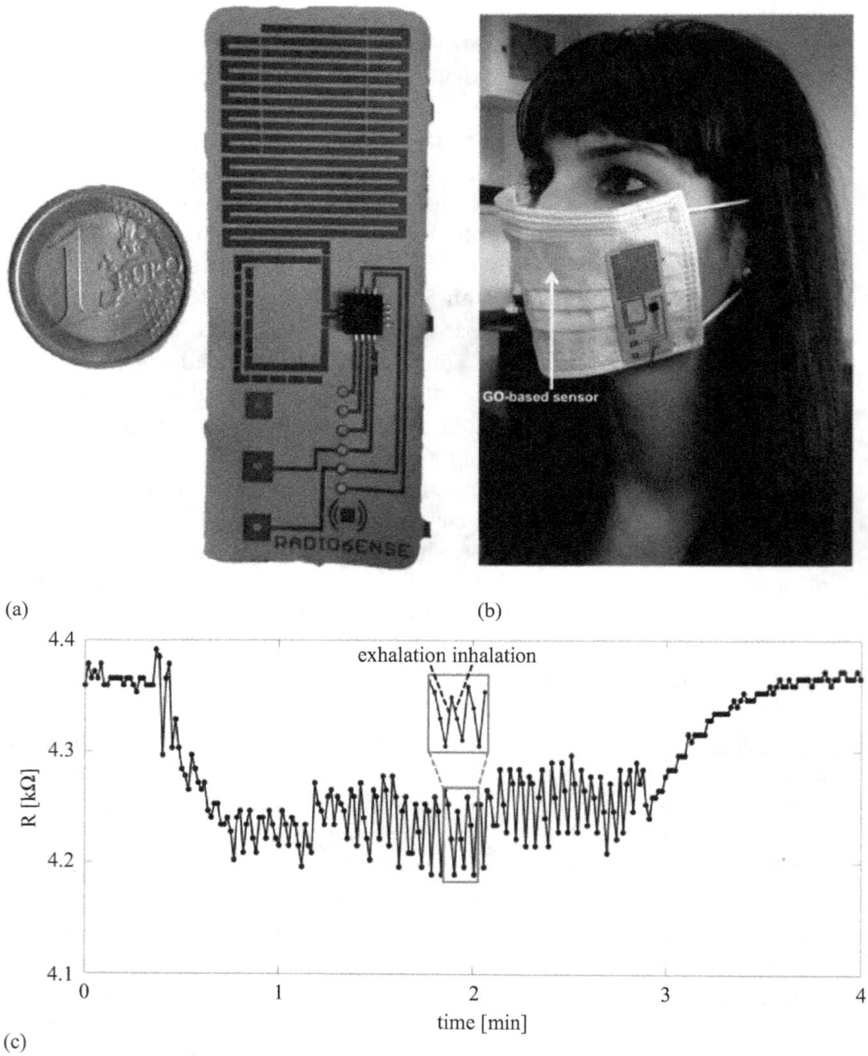

Figure 9.1 (a) Flexible antenna, (b) antenna board with graphene oxide sensor integrated on facemask, and (c) results capturing inhalation and exhalation via change in resistance of graphene oxide sensor. © [2017] IEEE. Reprinted, with permission, from (9)

9.2.2 Smart T-shirt: spiral fiber sensor array [13]

Roudjane *et al.* recently proposed a smart textile wearable stretching T-shirt integrated with an array of six contactless and non-invasive sensors. These sensors are made using spiral shaped multi-material fiber antennas operating at 2.4 GHz. The T-shirt can monitor various breathing parameters such as expiration and inspiration

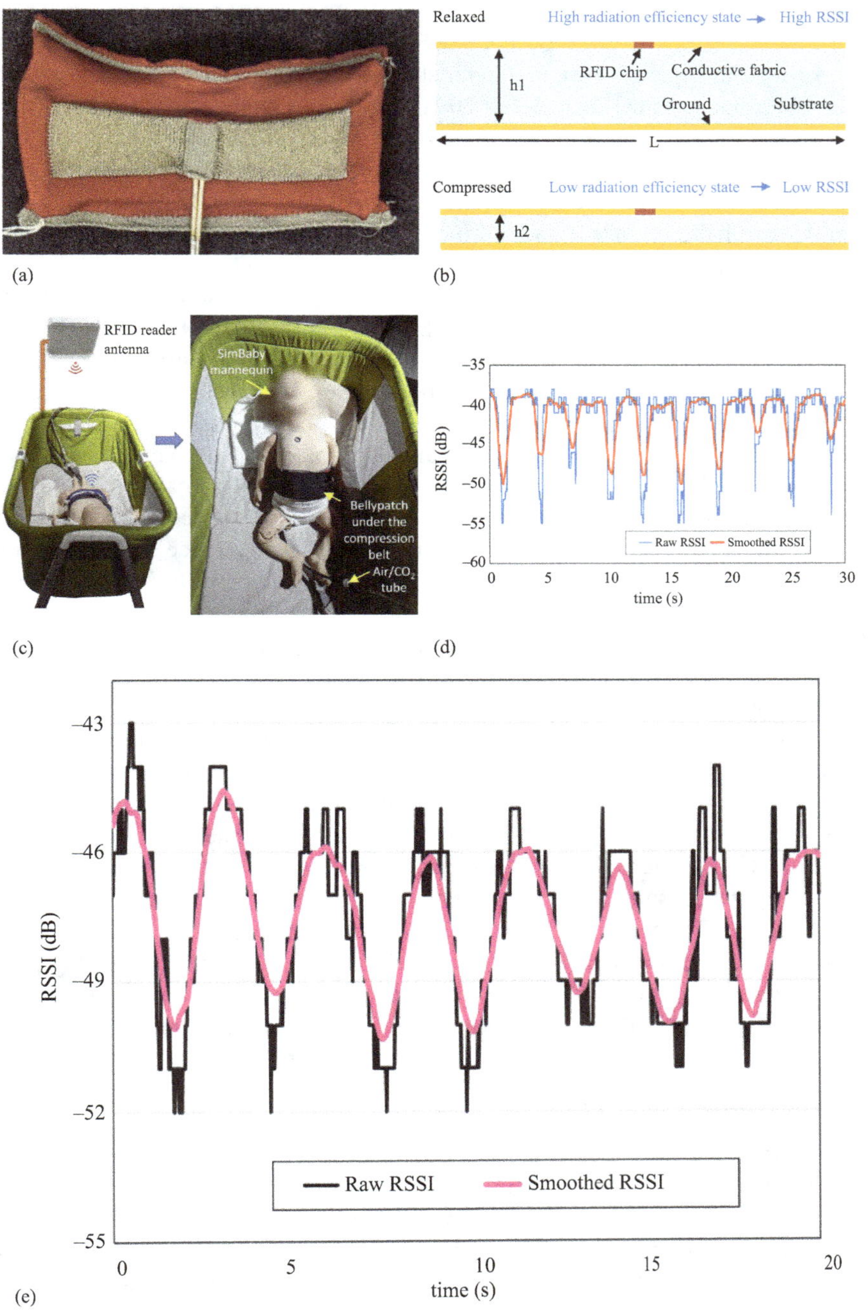

Figure 9.2 (a) Belly patch antenna, (b) schematic describing operating principle (compression→low radiation efficiency→low RSSI and vice versa), (c) experimental set-up on a SimBaby mannequin, (d) results of RSSI for the set-ups in (c), and (e) results demonstrating respiration monitoring on human abdomen. © [2021] IEEE. Reprinted, with permission, from (15)

times, breathing period, and breathing rate. These can then be wirelessly transmitted via Bluetooth technology to a portable base station.

As seen in Figure 9.3(a), six spiral shaped antennas (made from hollow-core silica fibers having inner diameter of 200 μm and outer diameter of 362 μm, coated with 18 μm thick polyimide layer, with silver thin layer on the inner surface of a hollow core-fiber) operating at 2.4 GHz are located across the thoraco-abdominal walls. Such placement allows capturing of signals from different areas, hence improving reliability of the monitoring process. During breathing, the antenna radius changes, leading to an average shift of 12 MHz/mm in the central frequency of the antenna. In addition, the relative permittivity of the inflated lung is 20.51 while it is 48.45 for the deflated lung, which further shifts the frequency. In their previous work [17,18], authors demonstrated that frequency shifts between 10 and 250 MHz are possible depending on the breathing frequency. This change is recorded in terms of the RSSI transmitted via a Bluetooth module to a portable base station (laptop in this case but can be any other portable device such as smart phone/tablet etc.), Figure 9.3(a) and (b). The above RSSI signal contains information of different breathing parameters. A spirometer, a device utilized to measure breathing parameters in clinical settings, is used as reference for comparison.

Experiments are performed on seven human subjects (six healthy males and one healthy female volunteers) with the experimental set-up (without subjects) shown in Figure 9.3(b). The result obtained for one of the volunteers is shown in Figure 9.3(c), demonstrating the successful capture of different breathing parameters including breathing period (BP), breathing rate/breathing cycle, and inhalation and expiration times. Furthermore, Bland–Altman statistical analysis was performed, demonstrating good agreement between results obtained via the reference and the proposed sensor. Specifically, the mean difference was on the orders of millisecond and 0.01 bpm (breaths per minute). Sleep apnea was also emulated by volunteers as enabled by pauses during breathing. It was also successfully detected by the sensors, with similar results obtained via the spirometer.

Overall, this technology has been tested on human subjects with successful results and can be potentially useful under clinical settings after performing more rigorous tests. However, one evident limitation is the usage of a tight T-shirt which can be uncomfortable for some people and may need to be addressed in the future.

9.2.3 Radar-based [19]

Radar stands for radio detection and ranging and utilizes both transmitter and receivers (along with corresponding antennas) to detect different aspects of the target, such as distance and velocity. This is typically achieved by deciphering the information captured in the reflected signal (also known as echo) from the target. During the past several years, radars have been utilized to detect diverse vital signals, including respiration [20–22]. Specifically, ultrawideband (UWB) Radars offer better performance compared to continuous-wave (CW) radars in terms of signal to noise ratio, and in multipath channels among others, making UWB radars a preferable choice [19]. For example, Pittella *et al.* demonstrated a wearable UWB

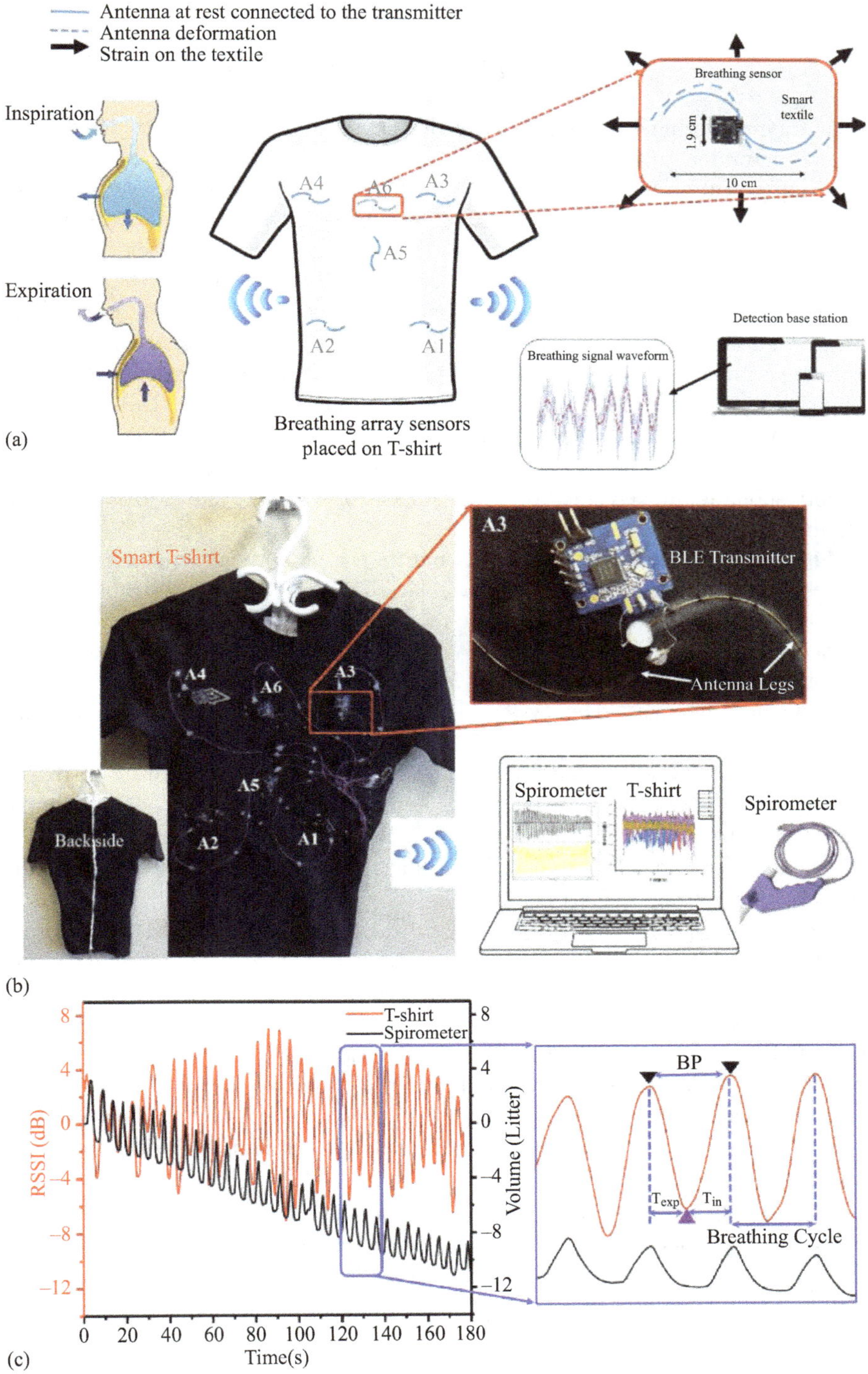

Figure 9.3 (a) Schematic of smart T-shirt with six integrated antennas (sensors) demonstrating the working principle used to capture breathing parameters during inhalation and exhalation, (b) experimental set-up with T-shirt showing wireless connectivity to record data and spirometer as reference, and (c) corresponding results for four breathing parameters. © [2020] IEEE. Reprinted, with permission, from [13]

radar for monitoring breathing activity that is a critical step towards monitoring respiration on the go [19].

In a typical arrangement of UWB radars, respiration can be sensed by monitoring the changing distance of the target (tissues in this case) which is captured in the received voltage due to different time-of-arrival of the reflected echo. For fixed UWB radars, breathing activity can be captured in the signal reflected from the superficial skin layer. However, for wearables, the signal reflected from the skin cannot be used to monitor respiration. This is because the reflected signal comes back superimposed with the incident signal which does not change with thorax movement given that the radar itself is attached to the thorax [23]. However, after some time, a small echo (late-in-time echo) is detected which carries the information of different breathing activity. The latter comes from the lung boundary and its corresponding movements, forming the basis of wearable radars for monitoring breathing parameters.

This work utilized numerical and two different experimental methods to verify the results. Numerical simulations are conducted via CST Microwave Studio with a multilayered planar tissue model of the thorax to get the baseline results. The first experimental set-up included a NVA 620 CMOS impulse UWB radar from Novelda that has 10 dB bandwidth of 3.1–5.6 GHz, connected to two UWB half-heart shaped antennas. The system is controlled by dedicated software. Two antennas are placed at a distance of 3 cm from each other, located on the thorax, and at a distance of 1 cm from the human subject's T-shirt (see Figure 9.4(a)). The second set-up utilized the indirect time-domain reflectometry (TDR) technique with 10 dB frequency bandwidth of 0.5–7 GHz, and employed the same heart-shaped UWB antenna along with a network analyzer and LabVIEW to perform the experiments (see Figure 9.4(b)).

Numerical simulations detected three different phases of the lung during respiration, (a) deflated lung (resting state), (b) extended lung, and (c) inflated lung (tidal breath) (Figure 9.4(c)). The Impulse Novelda radar confirmed the late-in-time echo signal which arrived 8 ns later, compared to 7 ns from the numerical analysis. The TDR experiment, on the other hand, was shown to detect the end expiration and end inspiration phases of respiration (Figure 9.4(d)). Statistical analysis was performed via a campaign of measurements conducted on two different subjects using the TDR method. Student's *t*-test was shown to be statistically significant with a 99% confidence interval.

Overall, this method demonstrated that wearable radars placed in a breast pocket can be utilized to detect different respiration phases by detecting reflections from the lung boundaries. Real-time data processing capabilities to implement the design in real-world scenarios would provide impetus to this technology.

9.3 Cardiac monitoring

This section reviews the work on sensor developed for cardiac monitoring in [24]. Blood pressure (BP) and heartbeat dynamics capture some of the most crucial

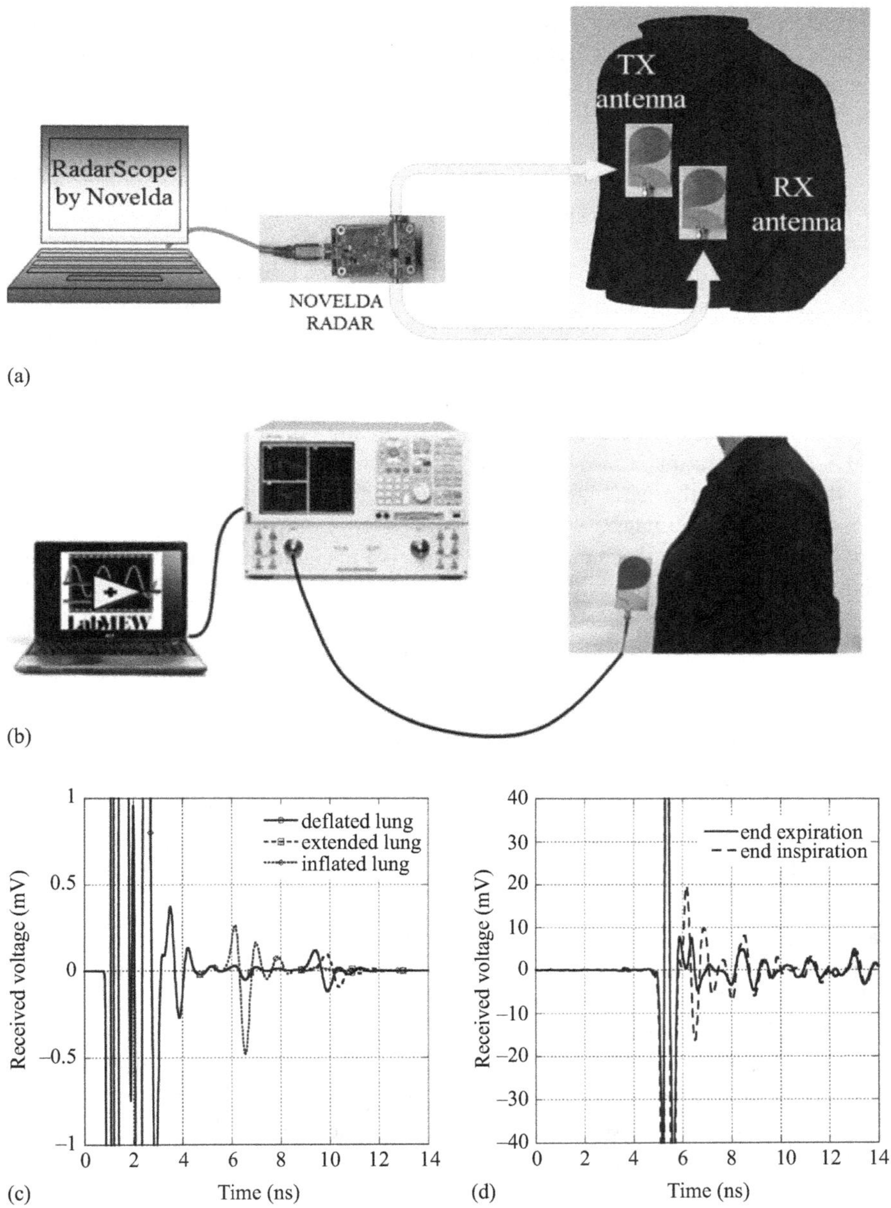

Figure 9.4 Experimental measurement set-ups (a) using CMOS impulse radar system from Novelda, and (b) using indirect TDR UWB system. Results obtained for respiration parameters using (c) numerical simulation and (d) the TDR experimental set-up shown in (b). © [2016] IEEE. Reprinted, with permission, from [19]

information related to cardiac activity and are the focus of this section. Systolic and diastolic blood pressure is commonly estimated using arm-cuff sphygmomanometers via branchial vascular occlusion. However, these can disturb sleep and interrupt circadian rhythm. In addition, heartbeat dynamics during central aortic and atrioventricular cycles, as well as information on pulmonary circulation, cannot be captured via this method. Nevertheless, these factors provide crucial information for accurate evaluation of cardiac activity. Invasive catheters are available to measure these [24,25], but are obtrusive and uncomfortable. Blood pressure can also be obtained indirectly via pulse transit time (PTT), an approach that is more comfortable to users [26,27]. For example, PTT can be measured using photoplethysmography (PPG) [28,29] to estimate the systolic and diastolic pressure. However, PTT sensors are typically required to be placed at two disjoint body positions, synchronized with sub-millisecond precision, making it cumbersome for long-term ambulatory monitoring. Also, here, a single parameter, i.e. pulse transit time, is utilized to obtain two parameters (systolic and diastolic pressure), which is not ideal. As an alternative, ultrasound, computed tomography (CT) scans, and magnetic resonance imaging (MRI) can be utilized to retrieve geometrical features of the heart. Although these can be used to derive pressure quantities indirectly, the accuracy and certainty of values obtained are not quite reliable.

Hui *et al.* report a near-field coherent sensing (NCS) method which utilizes multiple transmit and receive antenna pairs coupled via near-field interactions and placed over the cloth in the heart region. The heartbeat dynamics with even small vibrations modulate the radio-frequency (RF) signals coupled between the antenna pairs. This information is then utilized to accurately predict cardiac activity parameters, such as detailed central blood pressure transients and heart dynamics. Specifically, to capture information, this work utilizes a multi-point (4-point) mechanism. The NCS antenna pairs are located at the four conventional stethoscope points around the heart region and are used to listen to the sounds of four valves of the heart (aortic, mitral, pulmonary, and tricuspid) (see Figure 9.5(a)). Each NCS pair consists of two antennas and each pair can operate at different frequencies [(see Figure 9.5(b)). The system is connected to software defined radios to control the measurement via software implemented in LabVIEW.

Operation can be summarized as follows. The NCS system pumps RF energy in the UHF band (300 MHz–3 GHz) into the heart region. Radiation levels are very low and are within safety limits set by the occupational safety and health administration (OSHA). The dielectric boundary movement of the heart modulates the magnitude (any movement occurring relative to the antenna) and phase (any movement occurring along with the antenna) of the RF signal. Frequency division multiple access (FDMA) is used to multiplex signals at the transmitters. A carrier frequency of 1.82 GHz is used, with transmitters 1–4 having baseband offsets of 0.71 MHz, 1.22 MHz, 1.71 MHz, and 2.33 MHz. The NCS system is synchronized with ECG and stethoscope to obtain the required parameters. Overall, the NCS system is shown to capture quite rich information with raw data requiring further processing to derive meaningful results. Results extracted for aortic blood pressure and pulmonary cycle dynamics are shown in Figure 9.5(c) and (d), respectively.

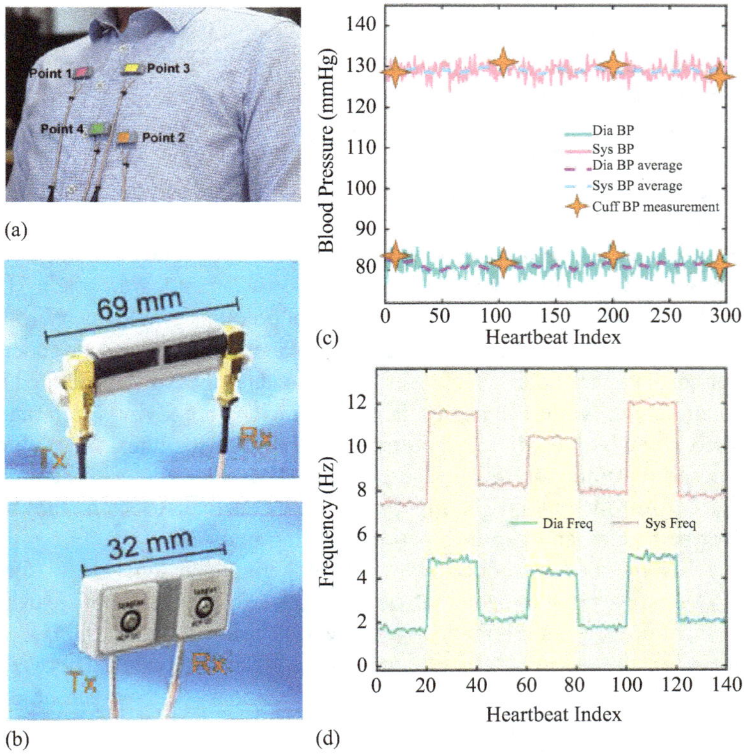

Figure 9.5 (a) Two types of NCS antenna pairs operating at 1.8 GHz band (top) and at 900 MHz and 5 GHz bands (bottom). (b) NCS antenna pairs placed at designated 4 points to capture heart dynamics. Results extracted for (c) aortic blood pressure along with reference arm-cuff measurement, and (d) pulmonary cycle dynamics (diastolic and systolic pressure frequencies) [24]. [2020] IEEE. CC by license

Arm-cuff BP used as a reference for aortic BP is in agreement with the one obtained via proposed NCS system.

This wearable technology is suitable for clinical and at-home diagnostics and can be worn over clothing. However, in the present study, the experiment is performed with the subject sitting still and, hence, the use is limited to static scenarios. The study is still at an early stage of development and more efforts are needed on refining the hardware and understanding the signal characteristics, processing and calibration.

9.4 Cardiopulmonary monitoring

This section reviews the work on wearable sensor developed for cardiopulmonary monitoring reported in [14]. Cardiopulmonary monitoring involves monitoring of both the heart and lung (pulmonary) functions simultaneously using the same sensor. This is clearly advantageous as compared to using two different sensors to monitor the above. Conventional technologies pertinent to pulmonary functions of this section are covered in Section 9.2. However, conventional technologies for cardiac activity specifically related to heart rate (focus of this section) were not covered in Section 9.3. For such cardiac activity, ECG is one of the most common technologies, but it requires multiple electrodes attached to the skin. Pulse oximetry is another less invasive method, but still requires stable and direct contact with the skin [30,31]. There are other non-invasive and non-skin-touch sensing technologies, but again they come with their own limitations. Seismocardiography [32–34] and ballistocardiography [32,35,36] are such technologies which measure surface movements caused by heartbeat in terms of displacement, velocity or acceleration to monitor cardiac parameters. However, artifacts can be easily introduced in such measurements, e.g., while the person is on a wheelchair, moving, or in an ambulance. To combine both cardiac and pulmonary functions together, smart shirts [37] were proposed that integrate ECG, chest belts, and accelerometers, but these have large device form-factor and require direct skin contact. As such, they are not suitable for bedridden and geriatric patients having bed sores [38].

Sharma *et al.* also introduce near-field coherent sensing (NCS) similar to Section 9.2 to develop a sensor that has potential to measure both cardiac and pulmonary parameters simultaneously, while overcoming shortcomings of prior technologies. In this approach, continuous wave (CW) RF signals between NCS antenna pairs couple to the internal dielectric boundary movement of different organs via near field coupling. This leads to measurement of lung, heart, and diaphragm movement directly, instead of relying on surface movements, and helps capture cardiopulmonary related parameters such as respiratory rate (RR), respiratory volume (RV), and heart rate (HR).

The sensor consists of transmit (Tx) and receive (Rx) antenna pairs connected to software-defined radio (SDR) transceiver (see Figure 9.6(a)) to perform the measurements. Transmit and receive antennas are coupled in near field that capture the internal dynamics. Near-field interaction reduces interference and improves signal-to-noise ratio (SNR). Two sensors are utilized, one below the xiphoid process close to the diaphragm, and the other close to the heart (see Figure 9.6(b) and (c)). Placement is optimized for direct Tx–Rx coupling and to enhance the sensitivity of signals reflected from the organs. Both sensors operate in the UHF frequency band because of reasonable penetration in the tissue dielectrics. Specifically, the thorax and abdomen sensor, utilize 1.82 GHz (RF power $= -12.84$ dBm) and 1.9 GHz (RF power $= -10.42$ dBm), respectively [(see Figure 9.6(c)) that allows for frequency division multiplexing of the sensors. The employed RF power is less than -10 dBm for both and is within safety limits set by the occupational safety and health administration (OSHA). Experiments are performed on 20 healthy participants (14 females). For reference measurements, BIOPAC sensors are utilized which consist of 3-lead ECG SS2LB (heart parameters), two torso belts SS5LB (respiration parameters), and PTM SS11LB (type of spirometer used for airflow rate L/s). The schematic and set-up are shown in Figure 9.6(b) and (d), respectively. Results

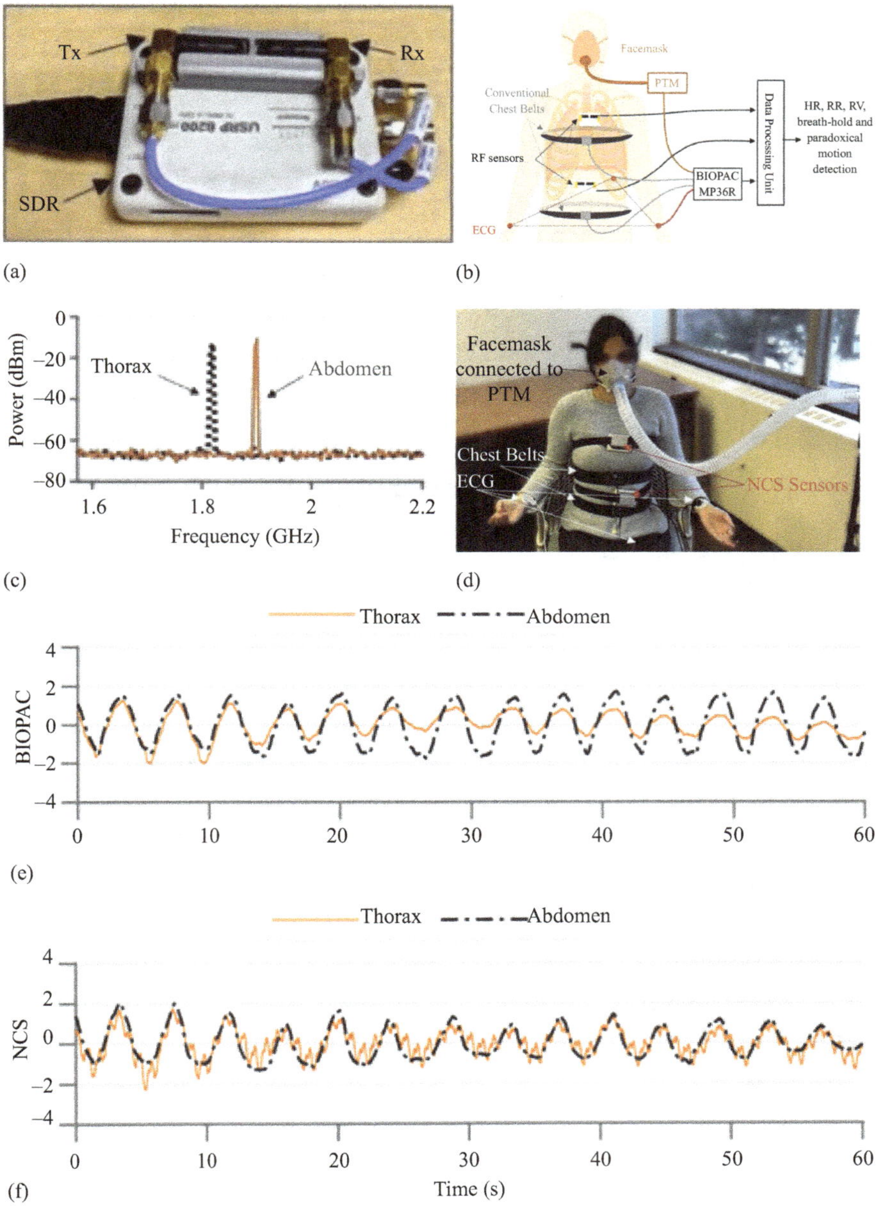

Figure 9.6 (a) RF sensor with transmit (Tx) and receive (Rx) antennas connected with software defined radio (SDR), (b) schematic representing the entire set-up with sensors and reference measurement system, (c) two carrier frequencies for two sensors (located on thorax and abdomen), and (d) experimental set-up with human subject. Results: normalized signal showing respiration waveform during tidal (normal breathing) using (e) reference chest belt (BIOPAC) and (f) NCS (unfiltered signal), (g) instantaneous volume from reference (PTM and chest belt) and NCS, and (h), (i), and (j) show different breathing and heart parameters obtained from the analysis of (e) and (f) [14]. [2020] Nature. CC by license

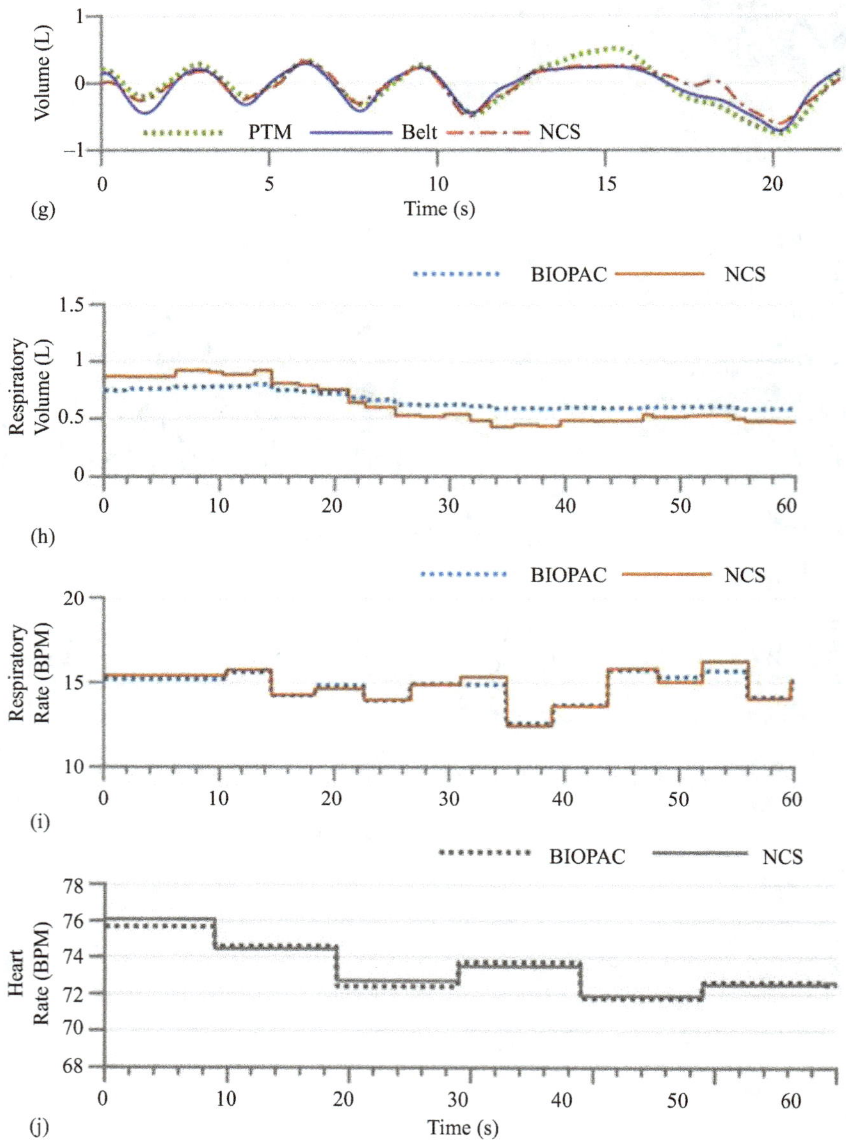

Figure 9.6 (*Continued*)

obtained are depicted in Figure 9.6(e–j) with quite good agreement compared to the reference. Furthermore, statistical analysis was performed for data collected in different postures and under different breathing exercises on 20 participants, accounting for a total of 100 experiments over 590 minutes. Satisfactory correlation is obtained between the NCS sensor and the reference, specifically, heart rate ($r = 0.95$), respiratory rate ($r = 0.93$), and respiratory volume ($r = 0.84$). Detection of breath-hold periods was achieved with an accuracy of 96%.

Overall, this is a promising technology as it has been tested on multiple human subjects with satisfactory results. Notably, environmental changes and interference have little effect due to near-field coupling of the antennas. In addition, due to the small form-factor of the sensor, it can be integrated with fabric and worn over multiple clothing layers, without the need for skin contact (similar to Section 9.2). Furthermore, the study showed that the NCS system can be utilized to record data across the body for different types of vitals by using multiple sensors simultaneously via frequency division multiplexing. However, there are limitations in the technology, such as signal artifacts produced during subject movement, making it unsuitable for monitoring during dynamic tasks. Although still suitable for static/quasi-static scenarios (similar to the NCS system discussed in Section 9.2), the technology has potential and can be further advanced in the future.

9.5 Body fluid monitoring

There are different ways in which the human body secretes fluids. This section focuses on two types of fluid monitoring, viz. (a) urine and (b) excessive face mask wetness caused due to moisture in breath and sweat produced by the body.

9.5.1 Urine monitoring [39]

Incontinence refers to partial or complete inability of bladder control, leading to urine leakage, and is most commonly found in infants or the elderly. While for infants, it is a natural process, for elderly it can be caused due to nerve deterioration or conditions such as Alzheimer's and dementia. Diapers are often used in such cases, but – if left unattended – they can cause rashes, discomfort, and bacterial infection, particularly when patients cannot change diapers on their own. In turn, the solution lies in diapers being integrated with urine monitoring capable sensors to alert caregivers on the need for a diaper change.

In terms of the state-of-the-art, there are mainly two types of diapers integrated with sensors that can monitor urine leakage and convey the sensed information wirelessly to alert the caregiver. These include (a) external [40–42], and (b) internal types [43–46]. External sensors are placed outside the diaper after it is worn. These typically utilize external batteries and utilize wireless technologies such as Bluetooth and Wi-Fi for transmission of the information. This technology allows for quite large read range (theoretically up to 30 m) for communicating the information. However, the complete module is relatively bulky and has to be placed on diapers using adhesive tapes. Furthermore, the sensors need to be removed and cleaned after contamination. Although suitable for single patient care, these are not ideal for hospital settings with multiple patients. Internal types utilize sensors that are passive and hence do not require batteries, making them easy to use and at larger scale. However, they are limited by read range (e.g. <30 cm). In few instances, internal type sensors are reported with better read range (30 cm–5 m), but these are bulky and the response time is more than 5 min.

Sen *et al.* utilize passive RFID tags made in conjunction with super absorbent polymer (SAP) (a subclass of hydrogels which is body-safe and also acts as a flexible sensing element) to build an internal type of sensor. Note that SAP is already used as a primary absorbent material in diapers, which is an added advantage. A bow-tie antenna design is utilized for the RFID tag due to its robust and stable performance over a wide frequency range in the UHF band. The novelty of the sensor lies in a hybrid RFID tag which is made using a combination of metal and SAP (see Figure 9.7(a)). Conductivity of SAP increases (up to orders of 10^{-2} S/m) when wet, in-turn increasing the size of the antenna. This changes/increases backscattered power to the RFID reader, thereby changing/increasing the received signal strength indicator (RSSI). The latter can then be utilized to distinguish wet state from dry state. This antenna is sandwiched between the diaper's absorption (third) and bottommost layer (see Figure 9.7(b)) to avoid any interference with the water retention functionality of the diaper itself, while simultaneously detecting the wet state of the diaper. Experiments are performed by capturing RSSI in dry state and wet state (emulated by adding 60 ml of 0.9% saline solution). Results obtained are shown in Figure 9.7(c) that depict the change in RSSI as the sensor moves from dry to wet state. The dark line represents the average over 10 samples, while the lighter boundary depicts the standard deviation across 10 samples. The system takes 30 s to reach a stable reading and there is an increase in signal level by 13 dB between dry and wet state. The read range is found to be 1 m which is satisfactory. In addition, an unequal variances *t*-test is conducted which further statistically confirms the higher strength of the signal in wet state compared to dry state. A complete picture of the entire system and its operating principle are summarized in Figure 9.7(d).

This sensor is passive, and the tag antenna uses only 20% metal (rest utilizes SAP material). As such, the sensor is low cost, relatively flexible and comfortable to use, and with a satisfactory read range of ~1 m. This is a promising solution that can also facilitate the bulk manufacturing of sensor-integrated diapers. Further improvement in read range can provide an extra boost to its potential. Authors of the work intend to further improve the accuracy of the sensor to also detect the amount of urine passed each time rather than just distinguishing between dry and wet states.

9.5.2 *Facemask moisture detection [47]*

Usage of facemask or filtering facepiece respirators (FFRs) have increased tremendously across the world with the recent onset of the COVID-19 pandemic. The World Health Organization advises replacement of FFRs once they become too damp, as attributed to their reduced effectiveness [47–49]. The most common reasons for such wetness in masks include moisture in breath and sweat. However, currently there is no practical and quantitative way to monitor masks' wetness.

Bianco *et al.* propose an RFID-based method that can identify excessively wet masks from commonly used ones. To implement this, a thin-wire tag antenna with

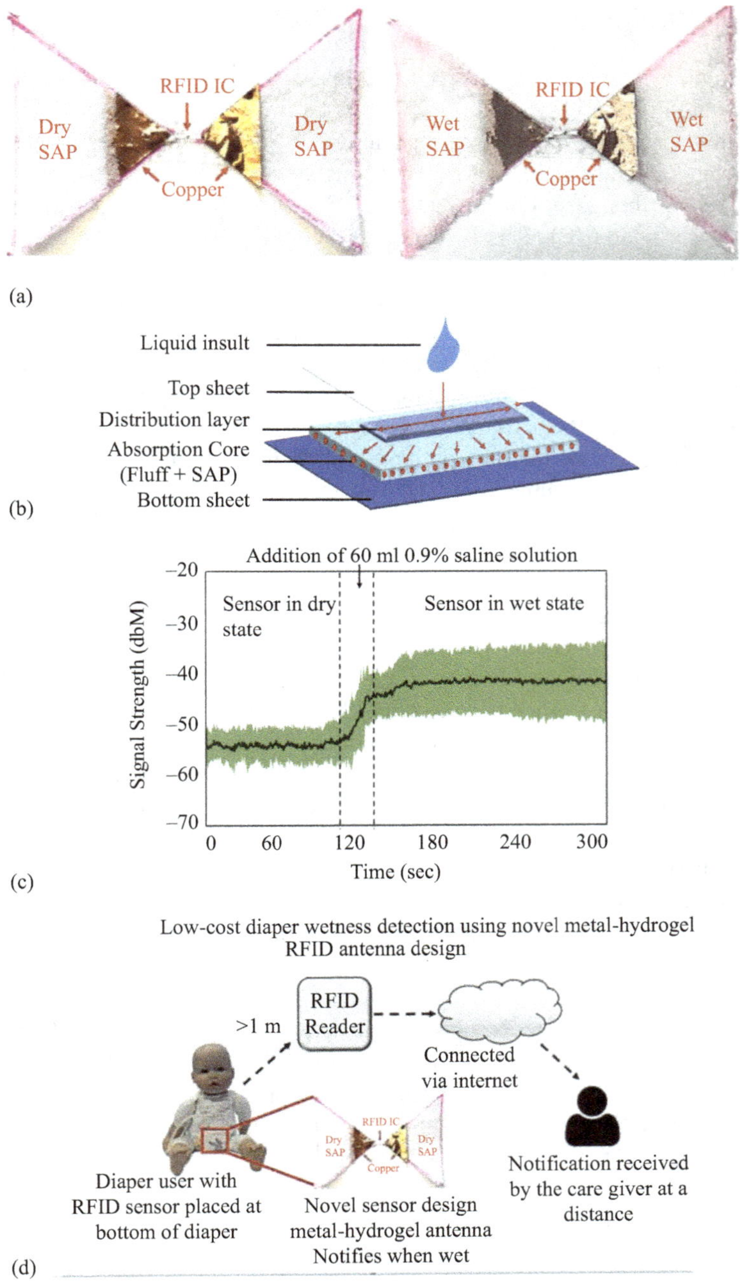

Figure 9.7 *(a) Hybrid bow-tie antenna (dry-left and wet-right) used as RFID tag made using 20% metal and rest using super absorbent layer (SAP) whose conductivity increases when wet thereby changing size of antenna. (b) Cross-section of diaper layers wherein antenna in (a) is placed between bottom and third layer. (c) Results depicting the sensing ability to distinguish dry and wet state using signal received from tag. (d) Complete system and its operation. © [2020] IEEE. Reprinted, with permission, from [39]*

an auto-tuning integrated circuit (IC) is utilized. The sensor operates by utilizing the auto-tuning capability of the IC. These family of ICs have the capability to dynamically maintain impedance matching for maximum power transfer (thus maintaining the read range) even when the tag antenna's impedance changes due to any reason (in this case, due to moisture). The IC utilized here returns a digital metric, sensor code (SC). This is related to the tuning process of the IC and hence can be utilized to sense the change in moisture levels. Ambient moisture conditions are used as a baseline and subtracted from the value of SC obtained, making change in sensor code (ΔSC) the actual sensing parameter.

The antenna utilized is a symmetric loop-matched meandered dipole (see Figure 9.8(a)). It is built on a substrate that can collect moisture produced by breath and is a twill woven fabric (50% cotton, 50% nylon) of size 70 mm × 25 mm × 1 mm. The antenna design is optimized using the CST Microwave Studio numerical solver. Simulation utilized a numerical homogeneous head phantom ($\varepsilon_r = 42.7$, and $\sigma = 0.99$ S/m) and N95 FFR made of polypropylene ($\varepsilon_r = 2.2$, and $\tan\delta = 0.002$), Figure 9.8(b). The sensor operates in the UHF frequency band.

For experiments, three prototypes were fabricated. Figure 9.8(c) depicts a mask integrated with a tag, and Figure 9.8(d) depicts the tag for one of the prototypes. Three sets of experiments were conducted. In the first experiment, the prototypes were placed inside a Binder MKF 56 climatic chamber and the ambient humidity was kept at 10%, which was then increased to 100% and kept for 10 min. The measured value of ΔSC for three tags along with simulation results is shown in Figure 9.8(e) (left), confirming the feasibility of the proposed idea. Furthermore, the time lag for the sensor is measured by spraying 200 mg of physiologic solution to induce the step response. The response time was found to be ~0.8 s, Figure 9.8(e, right). In the second experiment, three human subjects wore FFRs for 8 h continuously while performing everyday activities, such as working on computers or walking to a shop, etc. The results obtained are shown in Figure 9.8(f). As seen, variation in the sensing parameter is modest and does not demonstrate any significant behavior. Finally, in the third experiment, subjects performed intense physical exercise while wearing the mask (see Figure 9.8(g)). Results are depicted in Figure 9.8(h) which clearly demonstrates a significant change in the sensing parameter and change in realized gain of the antenna (that is also affected by the moisture).

The proposed sensor has potential to monitor excessive wetness of facemask and is suitable for hectic work environments. However, the sensor only provides qualitative information, works mainly in intense environments, and its sensitivity can be further improved. Authors of the work intend to improve dynamic range of SC (and hence sensitivity) and integrate temperature measurement to better predict the ageing of masks in the future.

9.6 Steatotic liver detection

This section reviews sensor for steotatic liver detection covered in [50]. Hepatic steatosis is a condition of lipid accumulation within major liver cells known as

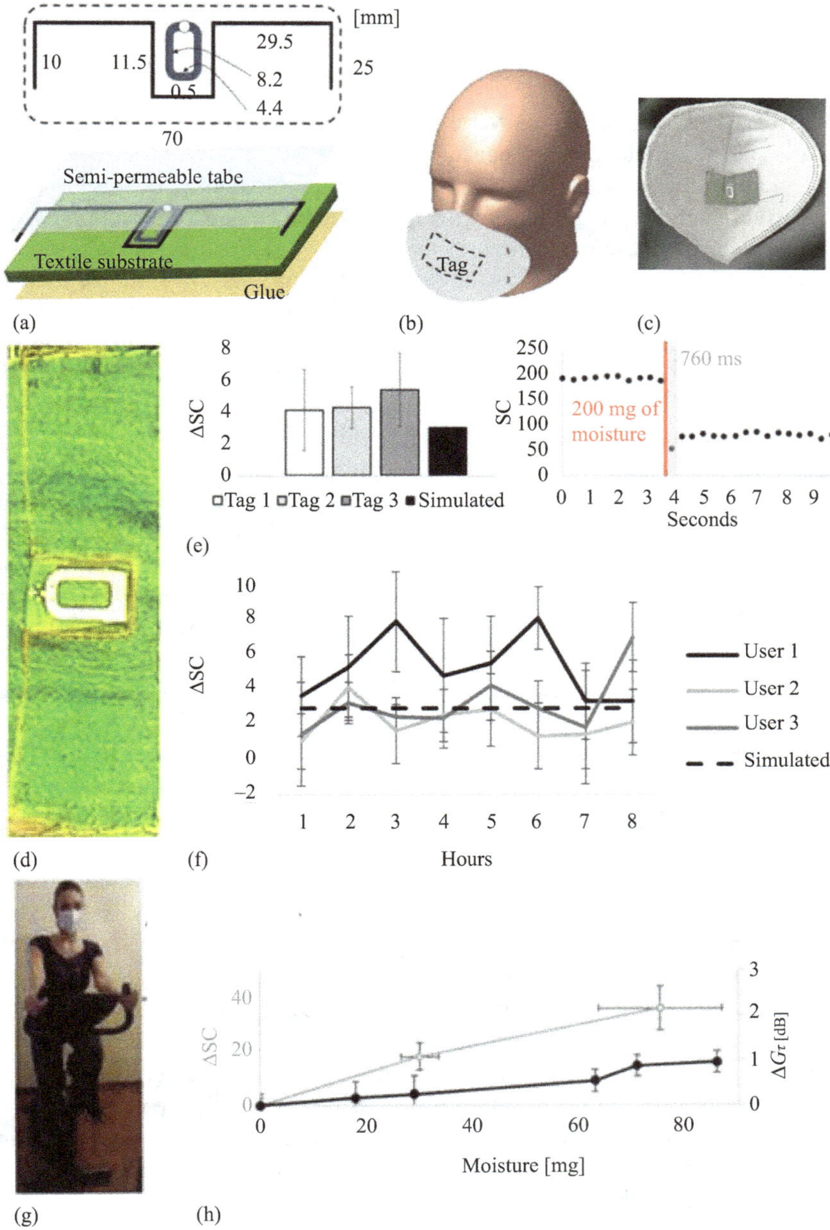

Figure 9.8 *(a) Tag antenna design, (b) numerical simulation set-up of tag antenna on N95 mask with human head model, (c) fabricated prototype of tag antenna on the mask, and (d) fabricated tag antenna prototype. Results (e) for masks placed in climatic chamber while humidity level is changed from 10% to 100% (left), and ~0.8 s of time lag in sensing response measured by spraying 200 mg of moisture to induce step response (right). (f) Results on human subjects while performing routine activities for continuous 8 h with no significant change observed in sensing parameter. (g) Subject performing intense physical exercise with mask, and (h) result obtained depicting feasibility of detecting excessively wet mask. © [2021] IEEE. Reprinted, with permission, from [47]*

hepatocytes. This can occur in several clinical situations such as obesity, type 2 diabetes, and hazardous alcohol consumption. In case of obesity, it is known as non-alcoholic fatty liver disease (NAFLD) and is quite common in western countries. Current methods include several imaging-based methods for detection of such condition, such as ultrasound, computed tomography (CT), and magnetic resonance imaging (MRI). However, these are limited in terms of sensitivity, utilize ionizing radiation (CT) or strong magnetic field (MRI) which restricts long-term usage.

Rezaeieh *et al.* propose a wearable electromagnetic belt imaging system that can be placed around the lower chest region for detection of steatotic liver. The accumulation of fat in the liver creates a dielectric contrast (e.g., $\varepsilon_r = 46$ (healthy liver), and $\varepsilon_r = 36$ (steatotic liver) at 1 GHz) which can be utilized to build such a system. The components of such system include antennas on the hardware part, and signal processing/image creation algorithm on the software part. Operating frequency is kept around 0.5–2 GHz as a tradeoff between reasonable penetration of radiation inside tissue and better resolution. The antenna designed is a 3D hybrid loop-dipole that is matched to the human body (see Figure 9.9(b) and (c)). It does not need a reflector and is miniaturized using Ω-shaped meandered arcs with an operating frequency of 0.58–1.6 GHz. The size-adjustable belt is made using 12 such antennas placed on a pair of elastic bands. The belt is made such that antennas are located symmetrically around the left (L) and right (R) sides of the torso (see Figure 9.9(a)). The system operates based on the fact that left and right sides of the torso have similar dielectric properties around the liver. Now, for healthy liver, high degree of similarity is expected, while for the steatotic liver, there will be a considerable difference due to low permittivity of accumulated fat. This similarity value is calculated using the multivariate energy statistics method.

Both simulation (using CST) and experiments are utilized to validate the results. The schematic and the experimental set-up are shown in Figure 9.9(d) and (e), respectively. Tissue mimicking mixtures are used to emulate different tissues. Data are collected by using one antenna as a transmitter and others as receivers and repeating this for all 12 antennas to obtain a complete scan. Measurement for each antenna is performed four times that took less than 10 s. Similarity value (obtained via above-mentioned statistical method) of 77.2% was obtained in simulation and 41.6% in experiments for healthy liver ($\varepsilon_r = 45$ at 1 GHz) as compared to 35.5% in simulation and 35.3% in experiments for highly steatotic liver ($\varepsilon_r = 32$ at 1 GHz). Though there is significant difference between simulation and experiment for healthy liver, results establish the feasibility of using this approach for diagnosis of steatotic liver or NAFLD. Furthermore, repeatability on different sized phantoms is also verified.

Overall, this is a new method in the area of steatotic liver detection and has potential. The method draws its inspiration from other traditional electromagnetic imaging-based methods that have been researched over the past several years, such as breast cancer or head imaging. The results obtained in simulation and experiments have considerable difference and future efforts will be needed to build a more concrete ground for this promising technology.

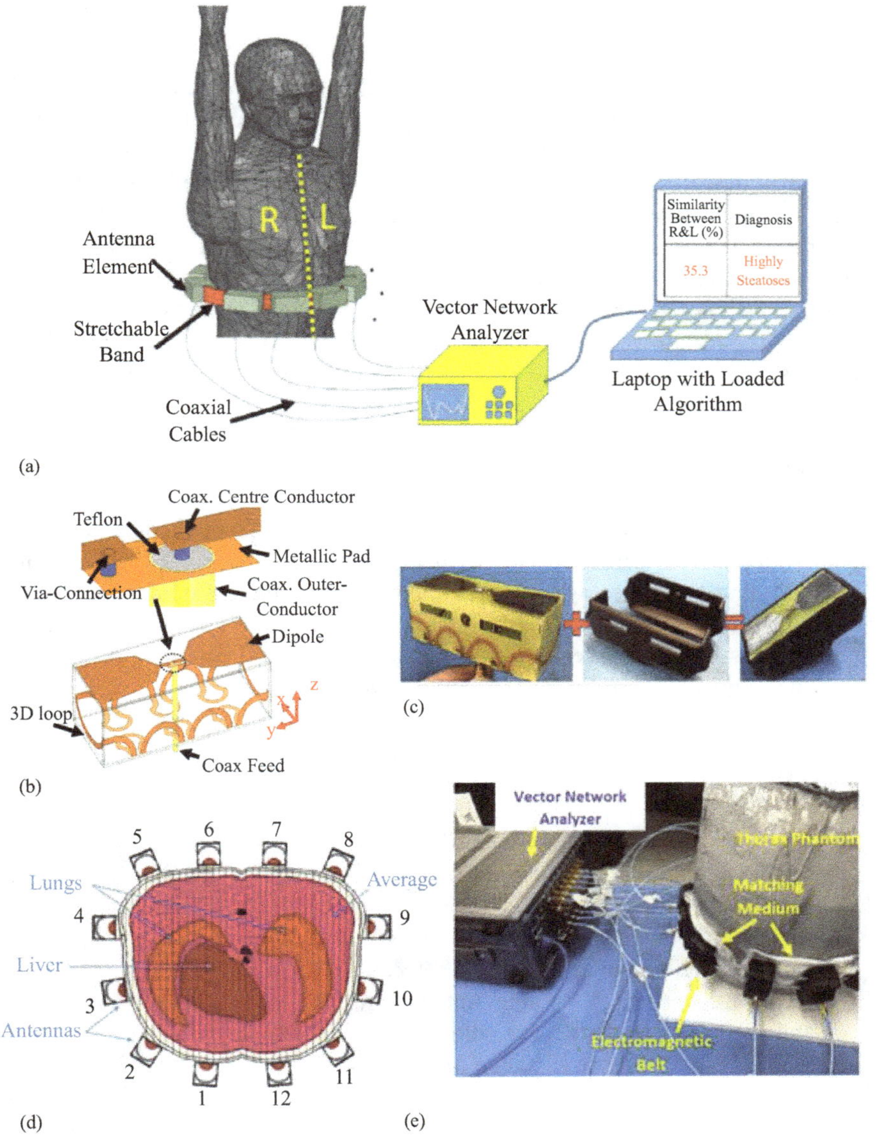

Figure 9.9 *(a) Schematic of the proposed system. (b) Simulated and (c) fabricated antenna prototypes. (d) Schematic of 12 antenna array system to perform simulation and measurements. (e) Experimental set-up to conduct measurement [50]. [2020] IEEE. CC by license*

9.7 Conclusion

This chapter has covered a brief review of recent wearable antenna-based technologies that are being developed for sensing different vitals, including respiration/breath, cardiac and cardiopulmonary activities, body fluid, and steatotic liver. We remark that, though some of the most recent methods are presented, the list is by no means comprehensive. Several other wearable antenna-based technologies are being developed for other vital monitoring, e.g. head/brain [51,52], breast cancer [6,7], and motion capture [5,53].

The most common hardware utilized across different methods is RFID-based, demonstrating its strong potential in healthcare. Received signal strength indicator (RSSI) is one of the most common parameters used to sense different vitals. Recently emerging near-field coherent sensing (NCS) applied for detection of different vitals also shows great promise. The approach can retrieve quite rich information across different parts of the human body simultaneously, from the modulated transmission data between antennas.

It is worth noting that radar-based and imaging-based methods that rely on antennas to monitor human vitals are being developed for several years. Most recently, this area is marching toward building a wearable ecosystem that can allow monitoring on the go. Most of the technologies discussed in this chapter are at the nascent stages of development and demonstrate substantial potential to meet various future clinical requirements once fully developed.

References

[1] Corchia L, Monti G, Tarricone L. Wearable antennas: nontextile versus fully textile solutions. *IEEE Antennas Propag Mag.* 2019;61(2):71–83.

[2] Paracha KN, Abdul Rahim SK, Soh PJ, Khalily M. Wearable antennas: a review of materials, structures, and innovative features for autonomous communication and sensing. *IEEE Access.* 2019;7:56694–712.

[3] Nepa P, Rogier H. Wearable Antennas for Off-Body Radio Links at VHF and UHF Bands: challenges, the state of the art, and future trends below 1 GHz. *IEEE Antennas Propag Mag.* 2015;57(5):30–52.

[4] Mishra V, Kiourti A. Wearable sensors for motion capture. In: *Antenna and Sensor Technologies in Modern Medical Applications [Internet].* New York, NY: John Wiley & Sons, Ltd; 2021 [cited 2021 Jul 6], pp. 43–90. Available from: https://onlinelibrary.wiley.com/doi/abs/10.1002/9781119683285.ch3.

[5] Mishra V, Kiourti A. Wearable electrically small loop antennas for monitoring joint flexion and rotation. *IEEE Trans Antennas Propag.* 2020; 68(1):134–41.

[6] Bahramiabarghouei H, Porter E, Santorelli A, Gosselin B, Popović M, Rusch LA. Flexible 16 antenna array for microwave breast cancer detection. *IEEE Trans Biomed Eng.* 2015;62(10):2516–25.

[7] Porter E, Bahrami H, Santorelli A, Gosselin B, Rusch LA, Popović M. A wearable microwave antenna array for time-domain breast tumor screening. *IEEE Trans Med Imaging*. 2016;35(6):1501–9.

[8] Misilmani HME, Naous T, Khatib SKA, Kabalan KY. A survey on antenna designs for breast cancer detection using microwave imaging. *IEEE Access*. 2020;8:102570–94.

[9] Caccami MC, Mulla MYS, Di Natale C, Marrocco G. Wireless monitoring of breath by means of a graphene oxide-based radiofrequency identification wearable sensor. In: 2017 11th European Conference on Antennas and Propagation (EUCAP), 2017, pp. 3394–6.

[10] Cretikos M, Chen J, Hillman K, Bellomo R, Finfer S, Flabouris A. The objective medical emergency team activation criteria: a case–control study. *Resuscitation*. 2007 Apr 1;73(1):62–72.

[11] de Waal CG, Kraaijenga JV, Hutten GJ, de Jongh FH, van Kaam AH. Breath detection by transcutaneous electromyography of the diaphragm and the Graseby capsule in preterm infants. *Pediatr Pulmonol*. 2017;52(12): 1578–82.

[12] Liu H, Allen J, Zheng D, Chen F. Recent development of respiratory rate measurement technologies. *Physiol Meas*. 2019;40(7):07TR01.

[13] Roudjane M, Bellemare-Rousseau S, Drouin E, *et al.* Smart T-shirt based on wireless communication spiral fiber sensor array for real-time breath monitoring: validation of the technology. *IEEE Sens J*. 2020;20(18): 10841–50.

[14] Sharma P, Hui X, Zhou J, Conroy TB, Kan EC. Wearable radio-frequency sensing of respiratory rate, respiratory volume, and heart rate. *Npj Digit Med*. 2020;3(1):1–10.

[15] Tajin MAS, Amanatides CE, Dion G, Dandekar KR. Passive UHF RFID-based knitted wearable compression sensor. *IEEE Internet Things J*. 2021; 8(17):13763–73.

[16] Caccami MC, Miozzi C, Mulla MYS, Di Natale C, Marrocco G. An epidermal graphene oxide-based RFID sensor for the wireless analysis of human breath. In: 2017 IEEE International Conference on RFID Technology Application (RFID-TA), 2017, pp. 191–5.

[17] Roudjane M, Khalil M, Miled A, Messaddeq Y. New generation wearable antenna based on multimaterial fiber for wireless communication and real-time breath detection. *Photonics*. 2018;5(4):33.

[18] Guay P, Gorgutsa S, LaRochelle S, Messaddeq Y. Wearable contactless respiration sensor based on multi-material fibers integrated into textile. *Sensors*. 2017;17(5):1050.

[19] Pittella E, Pisa S, Cavagnaro M. Breath activity monitoring with wearable UWB radars: measurement and analysis of the pulses reflected by the human body. *IEEE Trans Biomed Eng*. 2016;63(7):1447–54.

[20] Pisa S, Bernardi P, Cavagnaro M, Pittella E, Piuzzi E. A circuit model of an ultra wideband impulse radar system for breath-activity monitoring. *Int J Numer Model Electron Netw Devices Fields*. 2012;25(1):46–63.

[21] Nijsure Y, Tay WP, Gunawan E, *et al.* An impulse radio ultrawideband system for contactless noninvasive respiratory monitoring. *IEEE Trans Biomed Eng.* 2013;60(6):1509–17.

[22] Aardal Ø, Paichard Y, Brovoll S, Berger T, Lande TS, Hamran S-E. Physical working principles of medical radar. *IEEE Trans Biomed Eng.* 2013;60(4):1142–9.

[23] Cavagnaro M, Pittella E, Pisa S. UWB pulse propagation into human tissues. *Phys Med Biol.* 2013;58(24):8689–707.

[24] Hui X, Conroy TB, Kan EC. Multi-point near-field RF sensing of blood pressures and heartbeat dynamics. *IEEE Access.* 2020;8:89935–45.

[25] Avolio A. Central aortic blood pressure and cardiovascular risk. *Hypertension.* 2008;51(6):1470–1.

[26] Mukkamala R, Hahn J-O, Inan OT, *et al.* Toward ubiquitous blood pressure monitoring via pulse transit time: theory and practice. *IEEE Trans Biomed Eng.* 2015;62(8):1879–901.

[27] Hui X, Kan EC. Monitoring vital signs over multiplexed radio by near-field coherent sensing. *Nat Electron.* 2018;1(1):74–8.

[28] Allen J. Photoplethysmography and its application in clinical physiological measurement. *Physiol Meas.* 2007;28(3):R1–R39.

[29] Holz C, Wang EJ. Glabella: continuously sensing blood pressure behavior using an unobtrusive wearable device. Proc ACM Interact Mob Wearable Ubiquitous Technol. 2017;1(3):58:1–58:23.

[30] Biswas D, Simões-Capela N, Van Hoof C, Van Helleputte N. Heart rate estimation from wrist-worn photoplethysmography: a review. *IEEE Sens J.* 2019;19(16):6560–70.

[31] Zheng Y-L, Ding X-R, Poon CCY, *et al.* Unobtrusive sensing and wearable devices for health informatics. *IEEE Trans Biomed Eng.* 2014;61(5):1538–54.

[32] Inan OT, Migeotte P-F, Park K-S, *et al.* Ballistocardiography and seismocardiography: a review of recent advances. *IEEE J Biomed Health Inform.* 2015;19(4):1414–27.

[33] Taebi A, Solar BE, Bomar AJ, Sandler RH, Mansy HA. Recent advances in seismocardiography. *Vibration.* 2019;2(1):64–86.

[34] Xia Z, Shandhi MdMH, Inan OT, Zhang Y. Non-contact sensing of seismocardiogram signals using microwave doppler radar. *IEEE Sens J.* 2018;18(14):5956–64.

[35] Pinheiro E, Postolache O, Girão P. Theory and developments in an unobtrusive cardiovascular system representation: ballistocardiography. *Open Biomed Eng J.* 2010;4:201–16.

[36] Kim C-S, Ober SL, McMurtry MS, *et al.* Ballistocardiogram: mechanism and potential for unobtrusive cardiovascular health monitoring. *Sci Rep.* 2016;6(1):31297.

[37] Inc (Hexoskin) CT. Hexoskin Smart Shirts – Cardiac, Respiratory, Sleep & Activity Metrics [Internet]. Carre Technologies inc (Hexoskin). [cited 2021 Oct 11]. Available from: https://www.hexoskin.com/

[38] Bansal C, Scott R, Stewart D, Cockerell CJ. Decubitus ulcers: a review of the literature. *Int J Dermatol.* 2005;44(10):805–10.

[39] Sen P, Kantareddy SNR, Bhattacharyya R, Sarma SE, Siegel JE. Low-cost diaper wetness detection using hydrogel-based RFID tags. *IEEE Sens J.* 2020;20(6):3293–302.

[40] Amazon.com: Opro9 Smart Wearable Humidity Sensor Instant Alert Prevent Baby Diaper: Baby [Internet]. [cited 2021 Oct 11]. Available from: https://www.amazon.com/OPRO9-Wearable-Humidity-Notification-Reminder/dp/B07934MXY9/ref=sr_1_3?keywords=diapersens&qid=1562380477&s=gateway&sr=8-3&th=1

[41] Amazon.com: Geecare Thin Smart Baby Wet Diaper Alarm Monitoring Baby Urine Regularity and Lost Prevent with Mobile Phone APP: Health & Household [Internet]. [cited 2021 Oct 11]. Available from: https://www.amazon.com/Geecare-Diaper-Monitoring-Regularity-Prevent/dp/B06XTQW33B/ref=pd_sbs_121_2/140-3278811-0984137?_encoding=UTF8&pd_rd_i=B06XTQW33B&pd_rd_r=9c8e6891-9f96-11e9-ae2b-152fe49d8d51&pd_rd_w=yIWSG&pd_rd_wg=cpFtJ&pf_rd_p=588939de-d3f8-42f1-a3d8-d556eae5797d&pf_rd_r=A4QCT3Z2ANJBC7QHQECF&psc=1&refRID=A4QCT3Z2ANJBC7QHQECF

[42] Khan T. A Smart Wearable Gadget for Noninvasive Detection and Notification of Diaper Moister. In: 2018 IEEE International Conference on Electro/Information Technology (EIT), 2018, pp. 240–4.

[43] Tanaka A, Yamanaka T, Yoshioka H, Kobayashi K, Douseki T. Self-powered wireless urinary incontinence sensor for disposable diapers. In: 2011 IEEESENSORS, 2011, pp. 1491–94.

[44] Tanaka A, Suematsu R, Sakamoto H, Douseki T. Self-powered wireless urinary-incontinence sensor determines time for diaper change from spacing between sensing signals. In: 2016 IEEESENSORS, 2016, pp. 1–3.

[45] Siden J, Koptioug A, Gulliksson M. The "smart" diaper moisture detection system. In: *2004 IEEE MTT-S International Microwave Symposium Digest (IEEE Cat No04CH37535)*, vol. 2, 2004, pp. 659–662.

[46] Ziai MA, Batchelor JC. Smart radio-frequency identification tag for diaper moisture detection. *Healthc Technol Lett.* 2015 Feb 2;2(1):18–21.

[47] Bianco GM, Marrocco G. Sensorized facemask with moisture-sensitive RFID antenna. *IEEE Sens Lett.* 2021;5(3):1–4.

[48] Gao S, Kim J, Yermakov M, *et al.* Performance of N95 FFRs against combustion and NaCl aerosols in dry and moderately humid air: Manikin-based study. *Ann Occup Hyg.* 2016;60(6):748–60.

[49] Pacitto A, Amato F, Salmatonidis A, *et al.* Effectiveness of commercial face masks to reduce personal PM exposure. *Sci Total Environ.* 2019;650: 1582–90.

[50] Rezaeieh SA, Brankovic A, Janani AS, *et al.* Wearable electromagnetic belt for steatotic liver detection using multivariate energy statistics. *IEEE Access.* 2020;8:201847–60.

[51] Alqadami ASM, Nguyen-Trong N, Mohammed B, Stancombe AE, Heitzmann MT, Abbosh A. Compact unidirectional conformal antenna based on flexible high-permittivity custom-made substrate for wearable

wideband electromagnetic head imaging system. *IEEE Trans Antennas Propag.* 2020;68(1):183–94.

[52] Alqadami ASM, Zamani A, Trakic A, Abbosh A. Flexible electromagnetic cap for three-dimensional electromagnetic head imaging. *IEEE Trans Biomed Eng.* 2021;68(9):2880–91.

[53] Mishra V, Kiourti A. Wrap-around wearable coils for seamless monitoring of joint flexion. *IEEE Trans Biomed Eng.* 2019;66(10):2753–60.

Chapter 10

Epidermal and conformal electronics for bio sensing applications

John Batchelor[1], *Viktorija Makarovaite*[1] *and Robert Horne*[1]

This chapter is concerned with low power epidermally or conformally mounted flexible sensing tags that use passive backscattered communications links with the UHF Gen2 RFID protocol (865–928 MHz) [1]. Very low-profile polymer or fabric substrate skin mountable UHF RFID tag designs were introduced by [2], with proposed fabrication by inkjet printing [3], and incorporation of sensing [4], while alternatively, integrated, flexible silicon NFC sensing systems at 13.56 MHz have been pioneered by [5].

The high permittivity and loss of human tissue [6] together with the requirement for epidermal tags to stretch and allow the skin to breathe, mean that RFID tags must factor in mechanical as well as electromagnetic considerations [7]. Appropriate substrate materials must be used, with conductors able to flex and stretch, while reliable component mounting needs to be achieved in a system that protects the skin from soreness and exposure to irritants or toxic materials. There are several challenges, including reliability, energy consumption and scalability to overcome before implementing UHF RFID technology within the healthcare sector. Beyond UHF, RFID-5G combination using antenna arrays has been considered in [8] to compare read range and data rates at 3.6, 28 and 60 GHz and proposing that above 6 GHz, arrays offer longer read range and beamsteering advantages, while 3.6 GHz designs compare well with UHF performance for single element antennas. Some such arrays have been described in [9,10], which showed that on-body application of tags at 5G bands could achieve a 6 m read range in passive mode with performance comparable between a UHF antenna, a single dipole at 5.8 GHz and a 23-element array at 60 GHz. Finally, utilising rigid circuit board based battery-assisted UHF RFID (866 MHz) tags as wearable 9-axis accelerometry sensors has also been demonstrated for pelvis and torso mounting [11].

Section 10.1 will consider the design and initial testing of a UHF tag to sense the onset of biofilm growth resulting from a fungal infection on a throat valve prosthesis, and Section 10.2 will describe how low power streaming data can be

[1]School of Engineering, University of Kent, Kent, UK

obtained over backscattered links using the Gen2 protocol. Section 10.3 concludes the chapter.

10.1 Throat prosthesis biofilm sensing tag design

Here we describe an antenna directly mounted onto an implant as part of a proposed fungal infection management system, Figure 10.1. The antenna is part of a sensing tag located within the oesophagus at neck level. The passive tag is activated by an external reader antenna mounted a short distance from the throat. The UHF RFID implant antenna was originally designed for direct epidermal mounting on the arm based on [12]. Figure 10.2 shows the original 30 mm diameter skin mounted antenna design (dimensions provided in Table 10.1), which was a curved and convoluted dipole centred on a proximity coupled shunt capacitance matching network. The tag used a Higgs-3 RFID transponder chip with an input impedance $Z_c = 31 - j216$.

Adjusting the arm-mounted original antenna substrate thickness between 1.1 and 1.5 mm in Figure 10.2 (B) reduces aperture field coupling with the skin and improves efficiency for individuals with high muscle content (permittivity above 30). This helps reduce the need for tailored sensor antenna geometry adjustments to suit electrical parameter variations between different users. The antenna was tested on the skin of three volunteers with different body types and with the measured tissue electrical characteristics indicated in Table 10.2. The objective was to

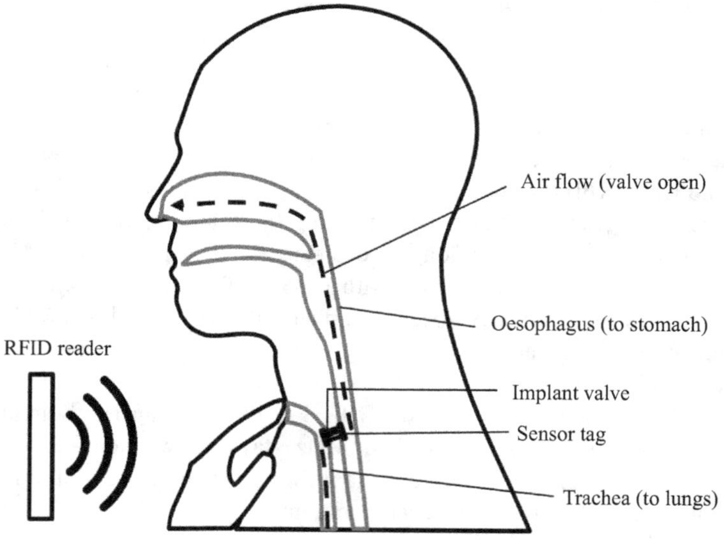

Figure 10.1 Voice prosthesis placement after a tracheostomy allowing speech to occur with blockage of the stoma (shown by utilising a finger to block stoma)

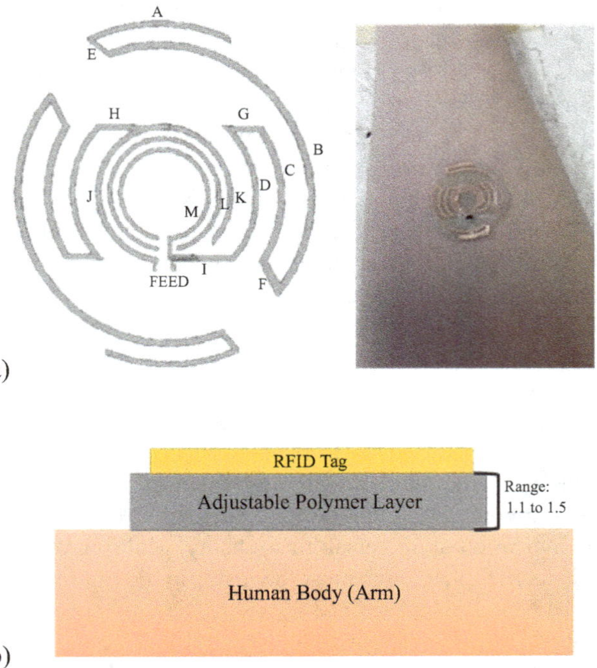

(a)

(b)

Figure 10.2 Epidermal skin UHF RFID tag design. (A) Copper etched mylar tag (0.2 mm thickness) on body with an adjustable thickness polymer backing. (B) Representation of 1.1–1.5 mm adjustable polymer substrate between UHF RFID tag and body

Table 10.1 Antenna dimensions for Figure 10.2 in mm

A	**B**	**C**	**D**	**E**	**F**	**G**	**H**	**I**	**J**	**K**	**L**	**M**
15	33	12	12.5	3.3	3.7	3.5	5	6	13	13.5	23	20

Table 10.2 SPEAG probe dielectric measurement of volunteer forearms related to BMI and substrate thickness required to achieve 50 cm read range taken at 865 MHz

Volunteer	**BMI**	**Relative permittivity (294.25 K)**	**Conductivity (S/m)**	**Polyurethane substrate thickness (mm)**
1 (thin female)	20.6	22	0.27	1.1
2 (overweight male)	27.2	30	0.42	1.1
3 (muscular male)	24.6	41	0.67	1.5

All readings represent averaged values taken in triplicate.
Source: data from [14].

determine the thickness of the antenna polyurethane layer required to achieve at least 50 cm passive backscatter read range on each individual volunteer.

A Voyantic TagformancePro calibrated system was used to indicate that an antenna substrate of just 1.1 mm is required for individuals with lower permittivity skin tissue, though an increase to 1.5 mm is needed where the permittivity rises to higher values (Figure 10.3). It is known that the high permittivity value of muscle causes greater detuning and attenuation in skin mounted antenna performance for muscular individuals than for both low and higher fat body types where there is less musculature [13] and this means that tag selection to match different individuals cannot be based on BMI alone, but should also be informed by body type.

The antennas must be tuned to account for the strong loading effect of the skin and the matched range is severely reduced when they are not attached (in air), Figure 10.3. The 'air' measurement was taken to determine consistency between the replicate tags and to account for any etching variability (leading to differing impedance mismatch and ohmic losses); no noticeable variability was seen between replicate tags when they are not attached to users. Therefore, the differences observed in read range in Figure 10.3 are accounted for by variation in user tissue composition, and any epidermal tag design characterised on a single volunteer is not an adequate indication of universal functionality. Multiple volunteers or a range of simulated dielectric phantoms must be used to demonstrate wider antenna performance.

The simulated radiation efficiencies of the arm-mounted antenna (Figure 10.2) on the 1.1 mm thick substrate were found to be −7 dB and −14 dB for volunteer 1 (thin female, $\varepsilon_r = 22$, $\sigma = 0.27\ \mathrm{Sm^{-1}}$) and 3 (muscular male, $\varepsilon_r = 41$, $\sigma = 0.67\ \mathrm{Sm^{-1}}$), respectively. These values are in the range expected for low-profile skin mounted antennas where efficiencies can fall as low as −20 dB. An efficiency lower than −20 dB is regarded as acceptable as the human body is a

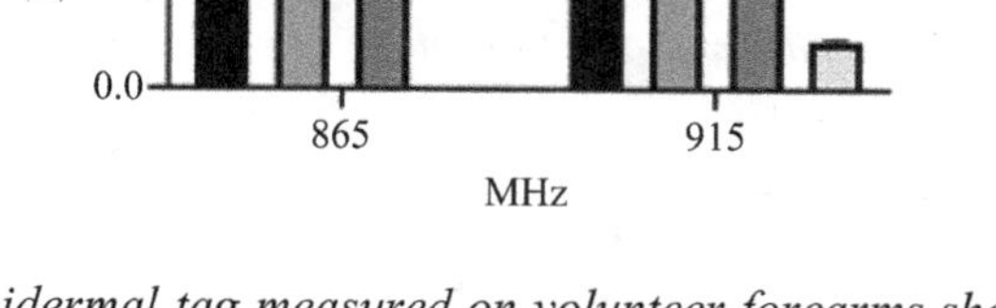

Figure 10.3 Epidermal tag measured on volunteer forearms showing that 50 cm minimum read range can be achieved for the three volunteers, thin to muscular body types (Table 10.2), with a slight adjustment in tag backing thickness as to decouple the tag from skin. Measured data represents triplicate repeats (source data from [14])

highly lossy material. However, if the tag efficiency is below −30 dB, the design will need improvement as it is unlikely to be medically useful at the chosen frequency. This is largely due to impedance mismatch and maximum read range falling below 20 cm for passive on-body UHF tags. A simple fix for impedance mismatch, to a certain degree, could be to scale the tag dimensions. This however will often have minimum effect on ohmic losses and adjustment to the tag geometry, material, adding lumped matching network components (i.e. shunting [15]) or increasing the tag separation from the lossy human tissue will be necessary [16].

Having established the antenna design to function on skin, it was reduced in diameter to 22 mm and retuned for direct mount on a PDMS elastomer throat prosthesis as indicated in Figure 10.4 with the dimensions in Table 10.3 [17]. It was initially necessary to include a 2.2 pF capacitor across the feed gap in order to tune the reduced size antenna for simulation purposes; however, during measurements,

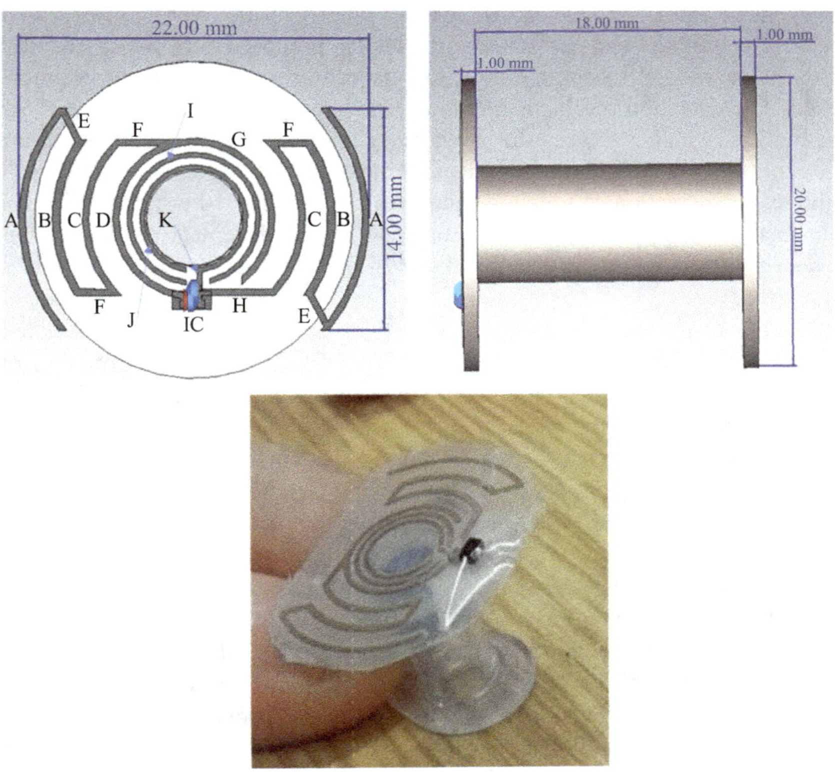

Figure 10.4 *UHF RFID voice prosthesis tag design for real-time detection of microbial detection while within a body (phantom). The tag prototype consisted of a 22 mm by 14 mm meandered circular dipole etched in copper on a standard polymer voice prosthesis*

Table 10.3 Antenna dimensions for Figure 10.4 in mm

A	B	C	D	E	F	G	H	I	J	K
15	8	8	12	2.8	2.5	16	4.5	23	18	2.3

it was found that capacitive loading of the gelatin phantom reduced the resonant frequency eliminating the need for the tuning component.

The purpose of the prosthesis tag is to indicate the presence of a biofilm associated with an establishing infection of the micro-organism *Candida albicans*. This fungal infection is commonly associated with the premature failure of voice prosthesis implants and can require surgical intervention for replacement [18]. Therefore, early detection of a newly establishing fungal colony improves the chances of medically treating the infection before a mature biofilm forms that would compromise the mechanical function of the implant, usually by impeding the movement of a phalange which switches air movement between the two states of speech or breathing. The passive sensing system proposed here would enable the prosthesis user to self-check the infection status and seek medical intervention as needed. The detection mechanism exploits the hydrophobic nature of the *C. albicans* biofilm, which displaces the saliva layer that would typically cover an uninfected prosthesis. Exploiting the displacement of the high permittivity saliva facilitates the detection of immature biofilms of around 5–10 μm thickness which occur in the first 4 h of a new infection. The sensor placement on the prosthesis between a substrate and superstrate, the biofilm, and a saliva layer are shown in Figure 10.5.

The antenna was covered with a 10 μm thick polymer superstrate to isolate the conductive material from the user tissue. Therefore, before the formation of a biofilm, the saliva layer lies 10 μm from the antenna and couples strongly with its aperture fields. The loading effect on the polymer mounted antenna is significant as saliva has respective relative permittivity and conductivity of around 40 and 0.3 Sm^{-1} at UHF frequencies. However, the aperture field-saliva coupling diminishes as an emerging biofilm pushes the saliva further from the antenna layer. The sensor functions through tracking the frequency detune that occurs as the saliva is displaced. This is manifest as a change in the tag transponder energy required from the reader, which correlates to biofilm thickness.

The sensing process is demonstrated using a physical salt and gelatin neck phantom with an embedded throat implant. For testing purposes, the hydrophobic biofilm is represented by applying 10–100 μm layers of polyurethane tape over the surface of the tag antenna. An upper limit of 100 μm was chosen as it approaches the maximum thickness of a *C. albicans* biofilm before shedding starts to occur, while 0–50 μm thickness presents normal growth over 48 h [6]. A 10 μm saliva layer was added by a pipette to form the top layer. Figure 10.6 shows the neck phantom at a fixed 30 cm from the external reader antenna. The implant has been inserted into the centre of the phantom at a point where a vertical air column

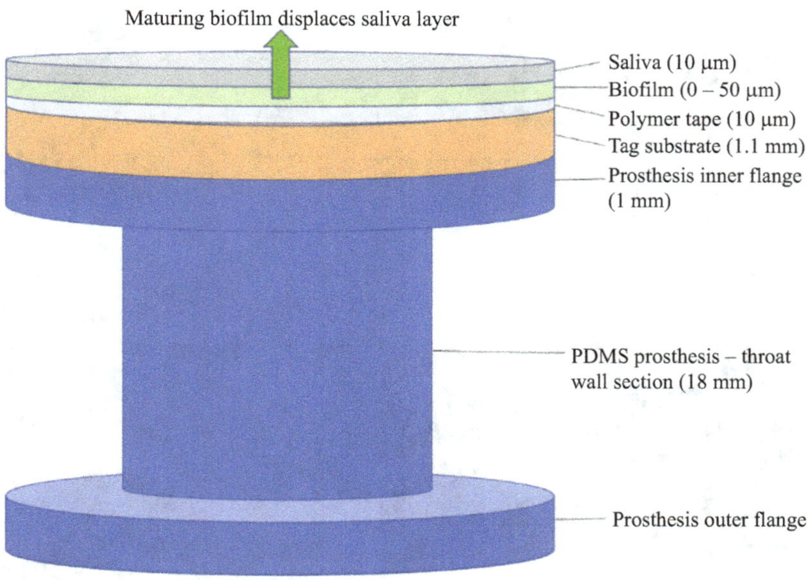

Figure 10.5 UHF RFID voice prosthesis sensing structure and biofilm. The design consisted of a 10 μm polymer superstrate over the tag substrate which was placed on the upper portion of the voice prosthesis body (facing away from the stoma opening). An adjustable 'biofilm layer' demonstrates the tag can detect fungal biofilm growth by decoupling the uppermost saliva layer from the antenna over time

representing the trachea intersects with a horizontal air tube representing the surgical stoma in the throat. As stated, the biofilm was represented by layers of polyurethane tape which reasonably models the characteristics of the polysaccharide-based biofilm matrix.

The human neck has relatively stable electrical characteristics as neck musculature is not strongly correlated with body type [19]. This is confirmed through dielectric probe measurement of four volunteers with BMI ranging between 20.5 and 44.9, Table 10.4. These relatively stable permittivity and conductivity values across human subjects have the benefit that unlike for other mounting points on the body, a single neck phantom can be utilised making a single tag design suitable for a wide variety of users.

10.1.1 Bio film sensor measurements

The Voyantic system demonstrated the sensing tag achieved read ranges of 65 and 80 cm at the 865 and 920 MHz bands, respectively, with good stability over triplicate tests. Figure 10.7 shows, that as the biofilm thickness increases, the reduction in read range at the two considered UHF bands corresponds to an increase in required

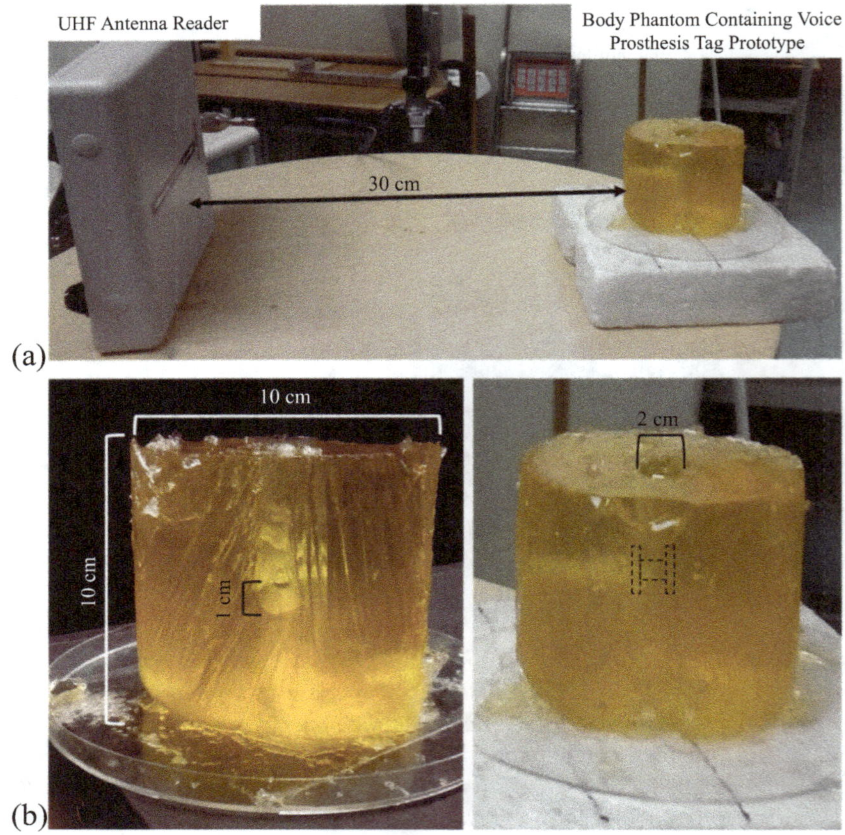

Figure 10.6 *UHF RFID voice prosthesis prototype: (A) within a gelatin and salt neck phantom at 30 cm from the UHF antenna reader. (B) The 10 cm by 10 cm gelatin phantom was designed as a near human muscle replacement (58 permittivity and 1 S/m conductivity). Phantom contains a 2 cm vertical opening (modelling trachea) and 1 cm horizontal opening (modelling tracheostomy stoma). Prototype (dashed lines) placement is represented at the intersection between the two openings to module surgical placement of a standard voice prosthesis. Adapted with permission from [17]*

Table 10.4 *Electrical parameters of volunteer necks taken at 865 MHz*

BMI	Relative permittivity (293.75 K)	Conductivity (S/m)
20.8	40.30 ± 0.19	0.63 ± 0.007
44.9	40.29 ± 0.68	0.60 ± 0.200
26.1	42.70 ±0.77	0.72 ±0.017
20.5	36.75±1.14	0.58±0.030
Average	**40.1±2.31**	**0.64±0.058**

All readings represent averaged values taken in triplicate.
Source: data from [17].

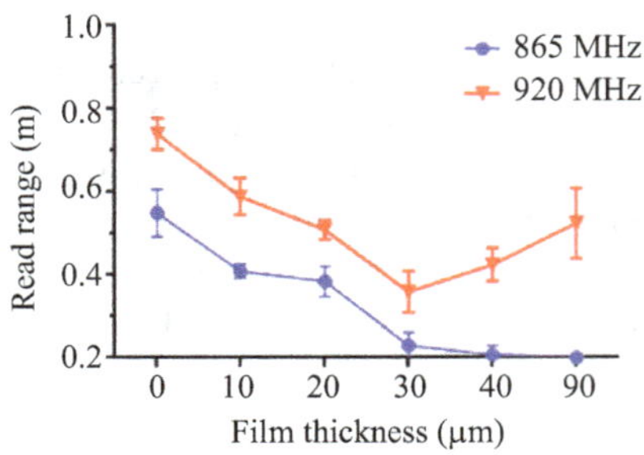

Figure 10.7 Emulated sensor response at EU and US RFID bands. Biofilm growth has been simulated by adding additional thicknesses of polyurethane to decouple the saliva layer (high dielectric constant) from the tag. It can be seen that as emulated biofilm growth increases (change in film thickness), the impedance mismatch (change in capacitance) shifts the frequency allowing for simple detection for any single frequency within the band of interest (source data from [17])

transponder activation power, and this is the sensor transfer response. The original tag tuning allows for unambiguous detection at both bands up to 30 μm biofilm thickness, and up to 40 μm for 865 MHz. Measuring films below this thickness is critical as they are associated with immature biofilms which might be treated with antifungal medication before they become mature and harden, leading to prosthetic valve failure. As this was a prototype, improved system robustness could be accomplished by utilising an RFID transponder capable of directly sensing changes in the antenna input reactance rather than relying on the required tag activation power which is affected by channel variations. However, this is one of the first UHF RFID designs to show detection of microbial growth while within human tissue.

In summary, a passive sensing tag has been shown to indicate the presence and thickness of a newly establishing fungal biofilm on a throat prosthesis and to be able to communicate wirelessly to an external RFID reader that does not need to be directly pressed against the skin. There is an advantage in the throat location of the implant as biological tissue parameters are much less of a function of body type than for other areas and this reduces the need to provide a comprehensive range of tuned tags. While this first prototype indicates biofilm thickness via link power parameters, future refinements could be made using capacitance sensing RFID transponders that digitally code the extent of detuning that they experience. The technology is proposed as a cost-effective means for a prosthesis user to self-manage the health of their device and reduce the need for surgical replacement.

10.2 Streaming bio data over Gen 2 RFID link

While the biofilm infection monitoring described in Section 10.1 is achieved through low sampling rates that suit the constraints of the passive UHF RFID Gen2 protocol, there are epidermal sensors measuring parameters such as ECG (electrocardiography), EMG (electromyography), EEG (electroencephalography), PPG (photoplethysmography) or accelerometery where there is a requirement to stream data. Skin patches worn for long periods require low profile flexible batteries. These cells have low capacities, and therefore power management becomes critical in low profile epidermal devices. A technique to achieve reliable streamed data transmission using Battery Assistive Passive (BAP) RFID communication is discussed in this section.

To harness the ultra-low power functionality of UHF RFID, while providing continuous data transmission with low latency, it is necessary to understand the available memory technologies and access procedures within the UHF RFID EPC Gen 2 protocol.

Typical RFID communications are based around static data being sent as an identifier. This means that items of merchandise, that have a static Electronic Product Code (EPC) [1] can be identified, logged, and located. Therefore, bursty low data rates are tolerated (in the region of 8–16 bytes of data), however the underlining fundamentals of the technology do allow for streaming dynamic data transfer at rates suitable for biological measurement, with examples including devices such as acceleration (60–200 Hz, 3 axis, 8–10 bit per axis) [20], Heart Rate data such as PPG [21], or ECG [22] in the range of 250 Hz at 8–10 bit resolution [23] (Figure 10.8). Typically, when looking at popular fitness devices within the

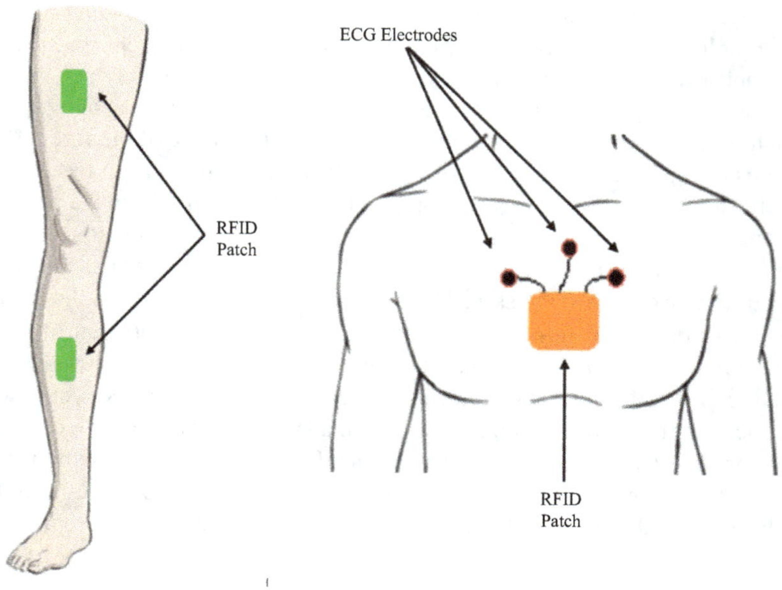

Figure 10.8 Mounting positions of two examples of streamed data RFID sensing modalities on the body, with dual attachment on the leg enabling three-axis acceleration and singular chest attachment for EEG heart waveform

consumer market, transmission of this data off the body is achieved using low power versions of Bluetooth [24], however, UHF RFID has the capability to send this data reliably and at a much lower power consumption [25].

Normal read requests access the tag transponder EEPROM (electrically erasable programmable read-only memory), which is a static memory type with a limited lifespan (100,000 read/write cycles) and moderate latency in performing functions (20–50 ms). For static data, or applications where data is not going to change frequently, this storage medium is more than sufficient, though when considering data that could be changing hundreds of times per second, its suitability diminishes. However, some RFID UHF chips offer a small region of memory located within the RAM registers. This is volatile memory and thus data is not retained when the device powers off, though compared to EEPROM, RAM has much lower read/write latency and the endurance is rated at 10,000,000,000 read/write cycles. However, RAM has much lower memory per unit area than EEPROM, and devices which employ this memory type typically only have 16 Bytes available within the RAM.

When dealing with dynamic data, hardware additional to the RFID transponder is required including a microcontroller (MCU) to capture and process data from sensors and relay the information seamlessly to the RFID chip. One of the major roles of the MCU is to ensure that there are no data clashes when the remote reader is trying to communicate with the RFID Chip whilst the microcontroller is attempting to write new data. This would corrupt the data which then proves troublesome to validate, and results in either invalid information or a significant reduction in the overall operating speed of the device.

To overcome data clashing, a flag pin of the transponder chip can be used to regulate access to its volatile memory. The flag indicates when the tag transponder is communicating with the RFID reader and signals to the microcontroller that it should wait before attempting to access the transponder memory. This allows for a state machine to be defined that avoids clashes in the read/write cycles. When timed correctly, it facilitates low latency real-time data streaming from the sensing tags with no data clashes.

Figure 10.9 shows that there are essentially four prominent actors in the sampling of live data. To harness the full speed of a UHF RFID link, each stage has a prescribed set of instructions and purpose throughout the capture of raw data to its presentation at the remote computer.

To evaluate the system performance, an EM4325 RFID transponder from EM Microelectronics was connected to an MSP430FR5969 MCU from Texas Instruments. A high-performance computer was attached to the reader (Intel i9 7960x with 64 GB RAM) to reduce any latency that could be introduced into the system by slow USB polling. Figure 10.10 outlines the basic circuit implementation used to integrate all of the component parts of an example system to stream data. Figure 10.11 shows the physical hardware assembled with a monopole antenna and a large stitched ground plane. The hardware was moved in increments of 5 cm away from the reader to establish read speeds at varying ranges. A ThingMagic M6E reader was used with a circular polarised antenna pointed at the monopole antenna.

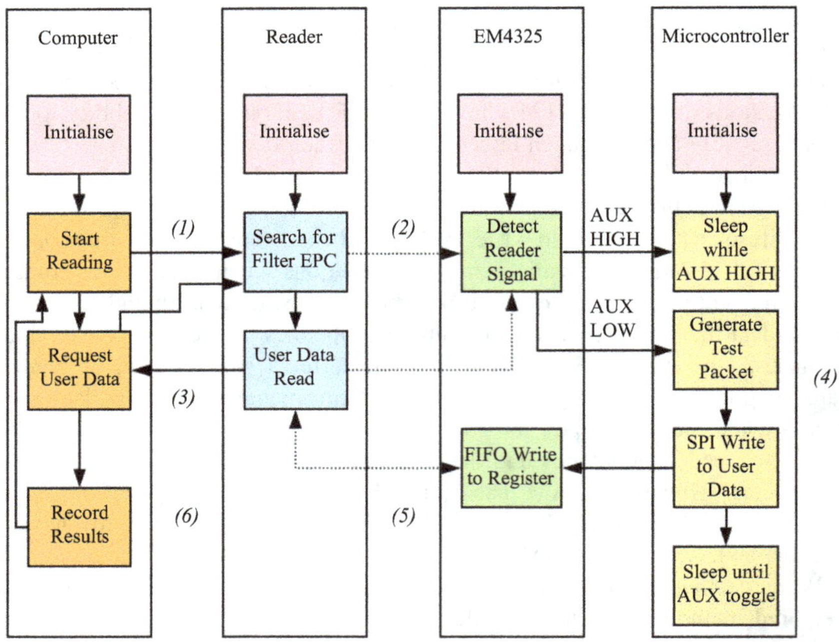

Figure 10.9 System diagram of each stage for a single transaction of valid data, mimicking the transmission of body centric data from an RFID device. Sourced from [26]

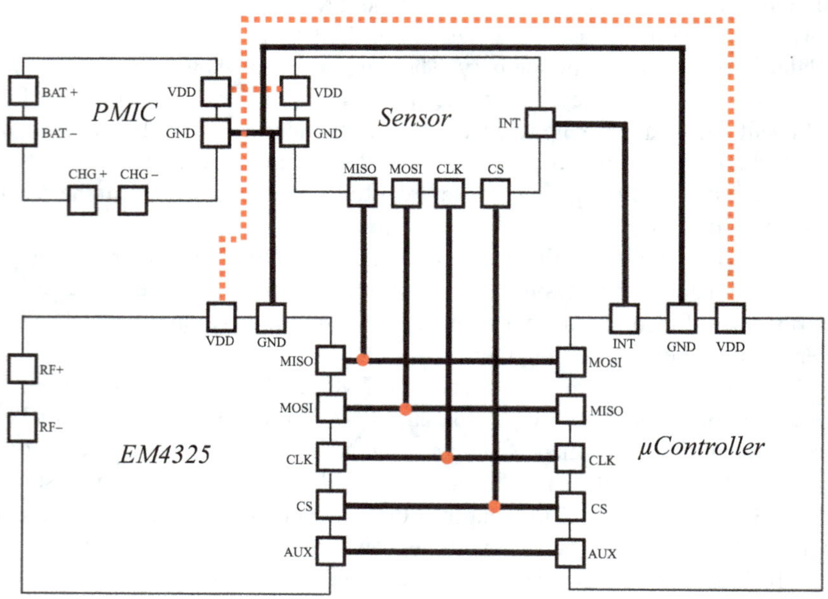

Figure 10.10 System block diagram showing the interconnects between the power management integrated circuit (PMIC), sensor, EM4325 RFID chip and the microcontroller

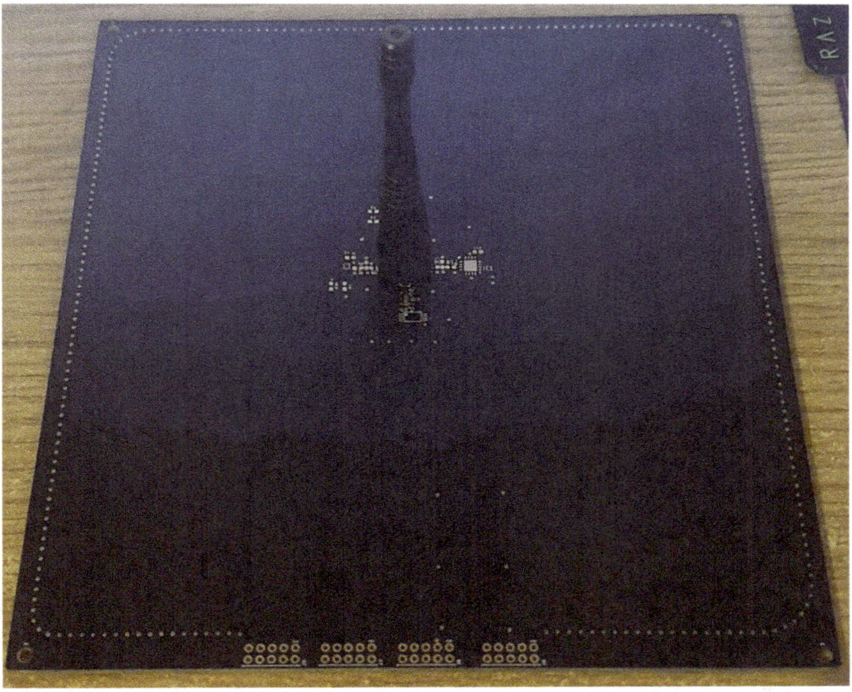

Figure 10.11 Dedicated RFID testing board with SMA connector for monopole antenna attachment and large ground plane. Sourced from [26]

The six main stages of the data management operation are discussed below:

1. Stage One: To begin, the computer would establish a serial connection with the reader via USB. Once the connection had been confirmed, settings for the reader would be sent detailing the use of battery assisted passive mode, fast search, *Q*-factor (0), BSK frequency (640 kHz), Hop table (2 hops) and the polling frequency (Maximum).
2. Stage Two: The reader would attempt to energise tags within the field, interrogating any tags with the specific EPC filter defined from the computer. This is to ensure the tag under test is the only one being engaged, as the performance can be degraded if other tags are within the field during the interrogation. The EM4325 then responds with its EPC and begins to toggle its internal AUX state.
3. Stage Three: Once the reader has established that the desired tag is within the field and able to return an EPC, a further request for tag data is sent. This request for data is pointed at the specific register section of the EM4325 for 16 bytes. The memory within this area is indexed in logical word addresses, resulting in a query of hex words 0×104 to $0 \times 10B$.
4. Stage Four: As the EM4325 transponder has detected the reader and toggled its AUX pin, the microcontroller can be brought out of a low power idle mode and

returned to normal operating speed. A test packet is then generated, which iterates upon every generation and is transmitted through SPI to the high speed 16 byte register of the EM4325. This transmission can only occur when the EM4325 AUX pin denotes that the reader is not actively communicating. This technique is used to stop both the microcontroller and the reader from accessing the same memory space simultaneously, causing a data deadlock. Once this transmission has occurred, the microcontroller can then resume its low power state to conserve battery power.

5. Stage Five: After the microcontroller has loaded the register, the reader will access the register, causing the AUX pin to flag that the register is unavailable.
6. Stage Six: Once the reader has captured the register data, it is then sent to the controlling computer which carries out a direct comparison with the previous frame of data to check whether the iteration of the frame contents matches. If it did not match, then it would be flagged and results in an error being presented to the tester, as well as being recorded. The data is logged into a .csv file for further offline processing.

Data sent via the communications line needs to be correctly packaged. The frame described in [26] was designed for data safety and only intended to send acceleration data at 60 Hz. It included frame number and CRC for added data integrity. However, the frame could be reconfigured and simplified to try and improve performance by only reading 7 bytes of data.

The frames were found to have an average latency of roughly 3 ms but with delay spikes of 6 ms and with occasional values above 10 ms. These spikes are observed for times when the reader is conforming to the EPC GEN2 standard. Read range, as shown in Figure 10.12 has good performance up until 2.5 m, whereby the transfer fails. It can be seen to recover at 2.55 m which is suspected to be the device moving out of a localised null, but the signal again fails and never recovers as distance increases. There is a minor penalty for using the full 16 bytes of the memory which can be seen in Figure 10.13 (A), but this is only around 10% of the update rate. It can also be seen in Figure 10.13 (B) that the effects of null regions

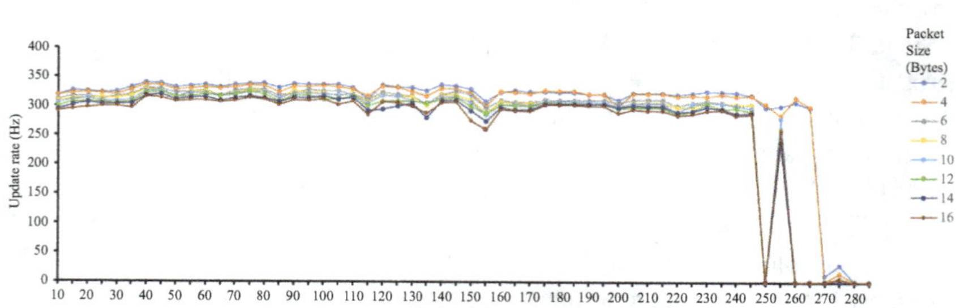

Figure 10.12 Update rate versus read range showing consistent speeds until 250 cm of EPC + memory data in the eight packet sizes tested. Adapted from [26]

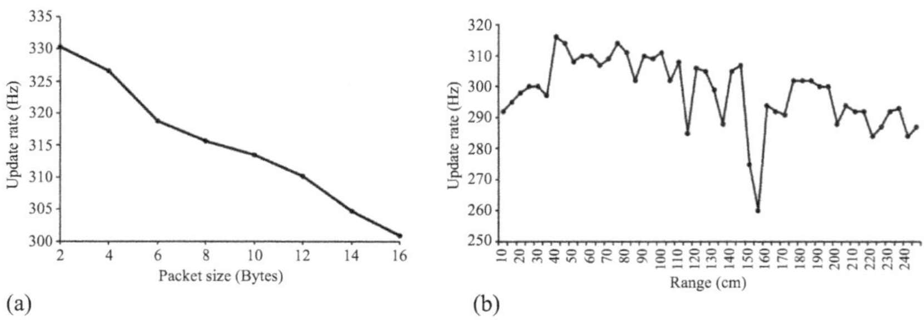

Figure 10.13 *(a) Update rate versus packet size showing degradation in performance with increased packet size. (b) Update rate versus range with 16 byte packets showing effects of null regions on update rate. Adapted from [26]*

can cause significant swings in data transfer rates (±15%) and that the distance from the reader can cause further degradation.

Whilst the bulk of the results shown here are for in-air measurements, the indication of performance for body-worn applications which may degrade the Read Range cannot be ignored. A link speed of 16 bytes at 300+ times a second fulfils the EMG, ECG, PPG and accelerometry requirements with additional overhead to allow for error correction and recovery.

Furthermore, given the latency of around 3 ms per transmission, UHF data transfer offers significant improvement against Bluetooth low energy transfers with typical minimum values of around 5–6 ms with high likelihoods of large spikes in the 20 ms range [27].

10.3 Summary

This chapter has discussed two aspects of low power epidermally or conformally mounted UHF RFID sensing tag applications. First, colonising infections on voice prostheses can be detected using an entirely passive system by exploiting the thickening biofilm which decouples a saliva layer from the sensor tag antenna aperture fields. Obtaining passive backscattered UHF communications is challenging, though this voice prosthesis application benefits from both the relative stability in throat tissue electrical parameters in association with the low bulk density of the neck.

Second, for systems such as skin mounted accelerometry or ECG measuring tags, a microcontroller-based system can achieve streaming data at rates of 5.2 kB per second using RAM rather than EEPROM and avoiding clashes between the microcontroller and the RFID transponder chip. This technique has the potential for enabling low cost, compact and body conformal electronics in areas such as rapid diagnostics, where low overhead and instant response sensors are vital. Future investigations into this technique will aim to focus on optimised read/write cycles to improve read range and data throughput to the extent that more advanced multicombination sensor suites might be achievable.

References

[1] Protocol U. EPC UHF Gen2 Air Interface Protocol | GS1 [Internet]. Gs1.org. 2022 [accessed 6 January 2022]. Available from: https://www.gs1.org/standards/rfid/uhf-air-interface-protocol

[2] Ziai M, Batchelor J. Temporary on-skin passive UHF RFID transfer tag. *IEEE Transactions on Antennas and Propagation*. 2011;59(10):3565–3571.

[3] Sanchez-Romaguera V, Wünscher S, Turki B, *et al.* Inkjet printed paper based frequency selective surfaces and skin mounted RFID tags: the interrelation between silver nanoparticle ink, paper substrate and low temperature sintering technique. *Journal of Materials Chemistry C*. 2015; 3(9):2132–2140.

[4] Camera F, Miozzi C, Amato F, Occhiuzzi C, Marrocco G. Experimental assessment of wireless monitoring of axilla temperature by means of epidermal battery-less RFID sensors. *IEEE Sensors Letters*. 2020; 4(11):1–4.

[5] Kim D, Lu N, Ma R, *et al.* Epidermal electronics. *Science*. 2011; 333 (6044):838–843.

[6] Kellomaki T. On-body performance of a wearable single-layer RFID Tag. *IEEE Antennas and Wireless Propagation Letters*. 2012; 11:73–76.

[7] Ray T, Choi J, Bandodkar A, *et al.* Bio-integrated wearable systems: a comprehensive review. *Chemical Reviews*. 2019; 119(8):5461–5533.

[8] Amato F, Occhiuzzi C, Marrocco G. Epidermal backscattering antennas in the 5G framework: performance and perspectives. *IEEE Journal of Radio Frequency Identification*. 2020;4(3):176–185.

[9] Amato F, Amendola S, Marrocco G. Upper-bound performances of RFID epidermal sensor networks at 5G frequencies. In 2019 IEEE 16th International Conference on Wearable and Implantable Body Sensor Networks (BSN), 2019.

[10] Hughes J, Occhiuzzi C, Batchelor J, Marrocco G. Miniaturized grid array antenna for body-centric RFID communications in 5G S-band. In 2020 50th European Microwave Conference (EuMC), 2021.

[11] Colella R, Turmolo MR, Sabina S, *et al.* Design of UHF RFID sensor-tags for the biomechanical analysis of human body movements. *IEEE Sensors Journal*. 2021;13(21):14090–14098.

[12] Makarovaite V, Hillier A, Holder S, Gourlay C, Batchelor J. Passive wireless UHF RFID antenna label for sensing dielectric properties of aqueous and organic liquids. *IEEE Sensors Journal*. 2019;19(11):4299–4307.

[13] Oyeka D, Batchelor J, Ziai A. Effect of skin dielectric properties on the read range of epidermal ultra-high frequency radio-frequency identification tags. *Healthcare Technology Letters*. 2017;4(2):78–81.

[14] Makarovaite V, Gourlay C, Hillier A, Batchelor J, Holder S. Adjustable passive RFID skin mounted sticker. In: 2019 IEEE 16th International Conference on Wearable and Implantable Body Sensor Networks (BSN), 2019.

[15] Zuffanelli S, Aguila P, Zamora G, Paredes F, Martin F, Bonache J. An impedance matching method for optical disc-based UHF-RFID tags. In:

2014 IEEE International Conference on RFID (IEEE RFID), Apr 2014; pp. 15–22.

[16] Sidén Johan, Nilsson H-E. RFID antennas – possibilities and limitations. In: Radio Frequency Identification Fundamentals and Applications. London: INTECH Open Access Publisher; 2010; pp. 69–92.

[17] Makarovaite V, Hillier A, Holder S, Gourlay C, Batchelor J. Passive UHF RFID voice prosthesis mounted sensor for microbial growth detection. *IEEE Journal of Radio Frequency Identification.* 2020;4(4):384–390.

[18] Talpaert M, Balfour A, Stevens S, Baker M, Muhlschlegel F, Gourlay C. Candida biofilm formation on voice prostheses. *Journal of Medical Microbiology.* 2015;64(3):199–208.

[19] Kamibayashi L, Richmond F. Morphometry of human neck muscles. *Spine.* 1998;23(12):1314–1323.

[20] Taraldsen K, Chastin SF, Riphagen II, Vereijken B, Helbostad JL. Physical activity monitoring by use of accelerometer-based body-worn sensors in older adults: a systematic literature review of current knowledge and applications. *Maturitas.* 2012;71(1):13–19.

[21] Fortino G, Giampà V. PPG-based methods for non-invasive and continuous blood pressure measurement: an overview and development issues in body sensor networks. In: 2010 IEEE International Workshop on Medical Measurements and Applications, New York, NY: IEEE; Apr 30 2010; pp. 10–13.

[22] Wong AC, McDonagh D, Omeni O, *et al.* Sensium: an ultra-low-power wireless body sensor network platform: design & application challenges. In: 2009 Annual International Conference of the IEEE Engineering in Medicine and Biology Society, New York, NY: IEEE; Sep 3 2009; pp. 6576–6579.

[23] Daskalov I, Christov I. Improvement of resolution in measurement of electrocardiogram RR intervals by interpolation. *Medical Engineering & Physics.* 1997;19(4):375–379.

[24] Kim H, Kim YS, Mahmood M, *et al.* Fully integrated, stretchable, wireless skin-conformal bioelectronics for continuous stress monitoring in daily life. *Advanced Science.* 2020; 7(15):2000810.

[25] Horne R, Batchelor J, Taylor P, Balaban E, Casson A. Ultra-low power on skin ECG using RFID communication. In: 2020 IEEE International Conference on Flexible and Printable Sensors and Systems (FLEPS), New York, NY: IEEE; Aug 16 2020; pp. 1–4.

[26] Horne R, Batchelor JC. A framework for a low power on body real-time sensor system using UHF RFID. *IEEE Journal of Radio Frequency Identification.* 2020;4(4):391–397.

[27] Rondon R, Gidlund M, Landernas K. Evaluating Bluetooth low energy suitability for time-critical industrial IoT applications. *International Journal of Wireless Information Networks.* 2017;24:278–290.

Chapter 11

Radar for healthcare technology

Panagiota Kontou[1], *Souheil Ben Smida*[1] *and Dimitris E. Anagnostou*[1]

The application of monitoring systems in the field of healthcare can be of considerable value to patients and caregivers. The implementation of a system that collects data such as heartbeat rates, respiration, or various other indications of a patient round the clock could enable a practitioner to access this data to monitor the patient. This could be beneficial in preventing, predicting, and tracking progression of illnesses.

There are products available in the market that help keep track of and monitoring a person's heart rate and other personal data, however they are mostly wearable devices. It is becoming more important to monitor and possibly treat patients without physical contact mainly due to maintain comfort. Additionally, there are cases where this would be the only way, such as an increasing risk of contamination from viruses in the hospitals and post-surgery monitoring. Other advantages from this method are to allow for high mobility, less stress for the patient, convenience in monitoring breathing patterns when patient is asleep, and finally, making domestic healthcare more feasible.

Non-contact methods can be categorized in optical, thermal, and radar methods. Optical methods use cameras that detect chest movement and are a good choice especially for activity or gesture recognition. However, with cameras people stress levels increase, which would affect their heart and breathing rate.

Moreover, there are many privacy issues concerning cameras, as well as security issues (i.e., hacking). Thermal methods include thermo sensors to monitor the temperature of the breathing airflow to detect the respiratory rates [1]. However, they are not able to detect heartbeat unless they are in contact with the body, usually in the mouth [2].

Radars have been used in many applications, mostly to determine distance from an aircraft or to measure the speed of a vehicle. This is achieved by measuring the frequency or phase shift in the reflected signal [3]. The same principle can be used to detect the vital signs. Microwave radars have been used to detect these tiny

[1]VisionRF (www.visionrf.com), and Institute of Sensors, Signals and Systems (ISSS), Heriot-Watt University, Edinburgh, UK

physiological movements and this application has been successfully demonstrated by researchers in the past.

The first Doppler radar was developed back in the 1970s [4], FMCW radar in the 1990s [5] and ultrawide-band (UWB) radar in the early 2000s [6]. Since then, there has been progress not only in understanding the theory of the continuous wave radar but also to develop and optimize the current designs, to fit numerous applications in the daily life, to get FDA/MHRA approvals and to potentially substitute traditional more expensive cardiopulmonary assessment devices.

Without any sensors attached to the body, the medical radar is very appealing to both industry and academia and over the last decades, many efforts and state-of-the-art work has been made from both parties in this field.

11.1 Physiological monitoring

Vital signs are clinical measurements that provide information on a patient's essential body functions such as respiration rate, blood pressure, temperature and heart rate. The classic vital signs also include height and weight. Measuring these vital signs not only helps evaluate an individual's physical health but also achieve prognosis of a disease and monitoring until full recovery.

11.1.1 Heart monitoring

The heart is a muscular organ that pumps blood to the organs, tissues and the lungs. In humans, it is divided into four chambers: upper left and right atria and lower left and right ventricles [7]. The heart is situated in the middle of the thorax. During each heartbeat, the heart contracts, pauses then relaxes. This is called the cardiac cycle and is demonstrated in Figure 11.1. The period where the heart relaxes is known as the Diastole phase and the period where the heart contracts is known as the Systole phase. The contractions of the heart are determined by a group of pacemaking cells that generate electrical impulses. Due to that, the atria contract at about the same time, followed by the ventricles [8]. The peak-to-peak motion due to the heartbeat is about 0.5 mm.

One of the most common ways to currently measure heart activity is through an electrocardiograph (ECG), which uses electrodes on the chest and limbs to measure the electrical current generated, as shown in Figure 11.2. The ECG has several disadvantages, as it requires electrodes to be attached to the skin via wires. These restrict

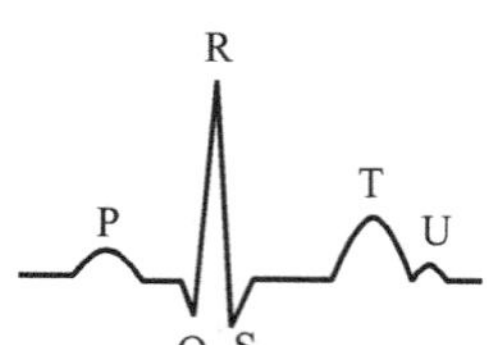

Figure 11.1 Cardiac cycle as correlated to ECG

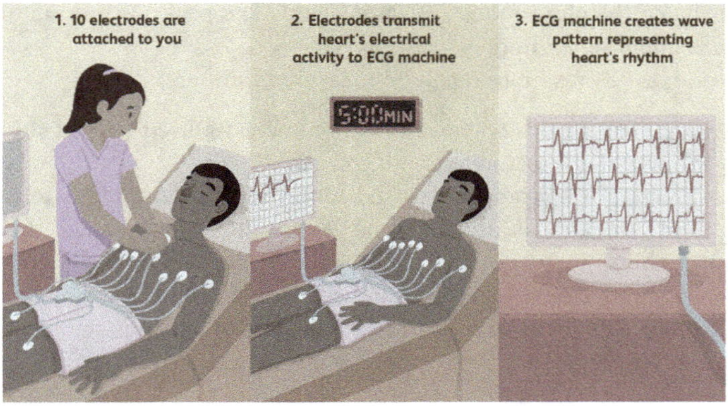

Figure 11.2 Electrocardiogram (ECG) [9]

a patient's mobility and range of movement causing discomfort, especially in the case of burn victims. Incorrect results can be displayed due to unclean skin and body hair blocking the electrical contact between the skin and the electrodes.

Another method of measuring heart activity is via polar straps, which are electric chest straps. They work by calculating the pulse rate when an R-wave (see Figure 11.1) is encountered and using the timing of the generated pulses. The main disadvantage with polar straps is that the strap under the skin must be moistened enough underneath to deliver accurate readings.

11.1.2 Respiration monitoring

Respiration is the action of breathing during which oxygen from the outside environment is inserted in the lungs and carbon dioxide is removed from the lungs. Respiration causes the chest wall to move due to contracting muscles as seen in Figure 11.3. Specifically, when air is inhaled, the chest moves outwards and when

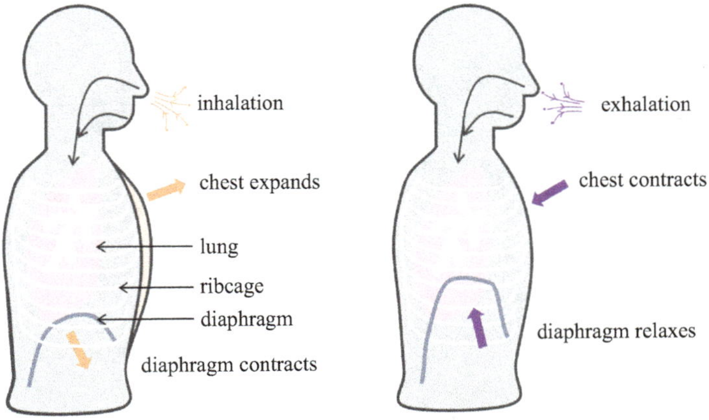

Figure 11.3 Respiration description

the air is exhaled the chest moves inwards again. The peak-to-peak motion due to respiration is about 4–12 mm.

According to [8] respiration can be measured through:

- The oxygen saturation, where a pulse oximeter is used to find the levels of hemoglobin which is saturated with oxygen.
- Airflow, where a spirometer is used along with a face mask to record the volume of air that is inhaled and exhaled, which makes this method uncomfortable.
- Respiratory movement can be achieved by measuring the body volume or impedance plethysmography with ECG electrodes which is the most common method in hospitals.

11.2 Radar principles

Radars use electromagnetic (EM) waves which are transmitted and are reflected if they meet any surface. If there is an object in the direction of propagation, a small portion of the reflected energy (echo) returns. A radar system is shown in Figure 11.4.

EM energy travels through air at a constant speed, the speed of light. That allows the determination of the distance between the reflecting objects (airplanes, ships or cars) and the radar site by measuring the running time of the transmitted pulses. A radar system can determine the distance, direction and height of the reflecting object.

11.2.1 Doppler radar

Doppler effect or Doppler shift is the change in frequency of a wave in relation to an observer who is moving relative to the wave source.

The Doppler effect (Figure 11.5) can be observed for any wave type such as water, sound, and light. It can be described as follows: a moving source of waves will

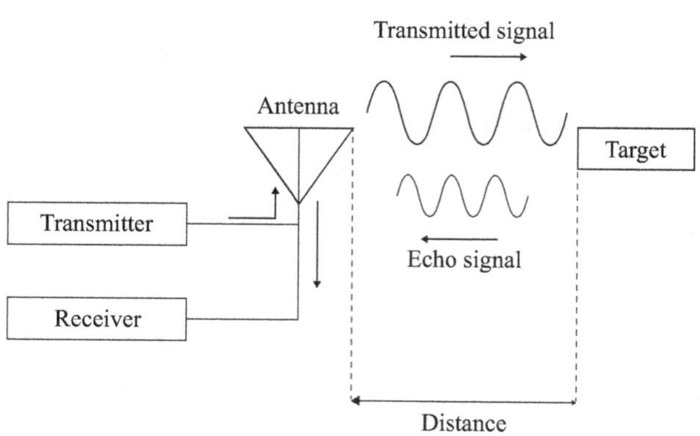

Figure 11.4 Basic radar principle. The transmitted signal will be reflected by the target, resulting in an echo signal

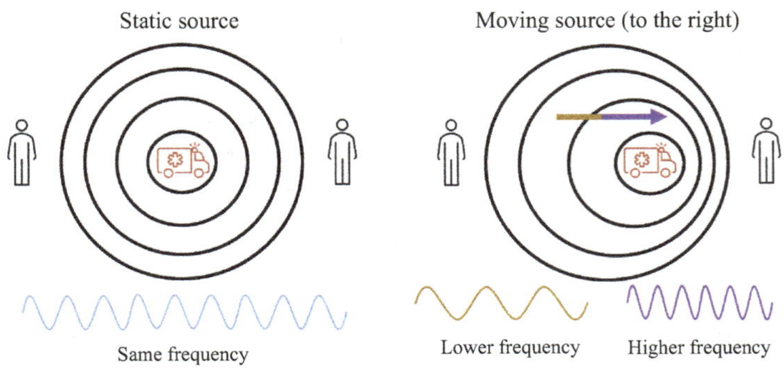

Figure 11.5 The Doppler effect

cause a frequency increase to the observer toward whom the source is approaching, and a frequency decrease to the observer from whom the source is receding.

In the radar application, the waves that are transmitted from the radar to a target that is approaching the radar, will result in reflected waves with greater frequency. Inversely, a target receding from the radar, will result in waves with lower frequency.

If R is the distance between the target and the radar, then the waves are travelling a distance of $2R$. The wavelength is λ, therefore the total number of wavelengths is $2R/\lambda$. The total angular excursion φ made by the EM wave to and from the target is $4\pi R/\lambda$. The motion of the object causes both R and phase to vary. A change in phase with respect to time is equal to frequency. The angular Doppler frequency ω_d is given by [3]:

$$\omega_d = 2\pi f_d = \frac{d\varphi}{dt} = \frac{4\pi}{\lambda}\frac{dR}{dt} = \frac{4\pi v_r}{\lambda} \tag{11.1}$$

where v_r is the relative velocity of the target with respect to the radar. Consequently, the Doppler frequency shift is:

$$f_d = \frac{2v_r}{\lambda} = \frac{2v_r f_0}{c} \tag{11.2}$$

where f_0 is the transmitted frequency. The velocity of the target can be found by rearranging (11.2) and using the frequency obtained by the receiver.

11.2.2 Radar types

A radar can be differentiated regarding their transmitted waveform, as can be seen in Figure 11.6 and explained below:

1. *Continuous wave (CW) radar:* operates by transmitting a continuous signal.
   - *Unmodulated CW radar:* detects stationary objects using one transmitting and one receiving antenna. Measures the speed of the target, not the distance.

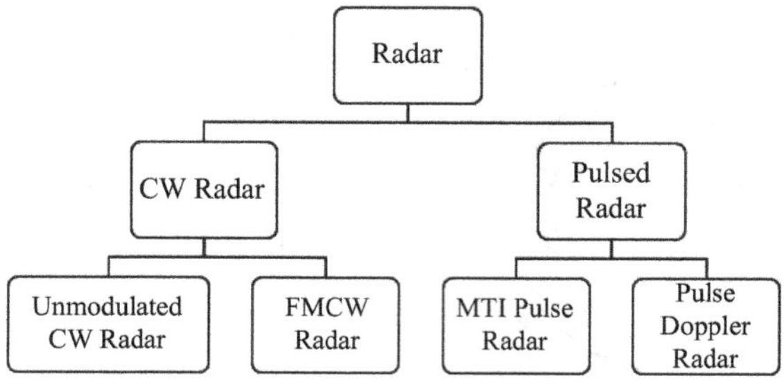

Figure 11.6 Types of radar depending on transmitted pulse

- *Frequency modulated continuous wave (FMCW) radar:* has the ability of frequency modulating the transmitting signal which allows for detection of the distance. It also requires one transmitting and one receiving antenna.

2. *Pulsed radar:* operates by transmitting a pulsed signal.
   - *Moving target indication (MTI) pulse radar:* uses one transmitting antenna and a duplexer for receiver. Based on Doppler effect it can distinguish stationary from non-stationary targets.
   - *Pulse Doppler radar:* similarly uses one transmitting antenna and a duplexer for receiver. The antenna transmits a pulse in a rate, which must be such that the echo signal is received before the next pulse is transmitted.

   Pulse radar is not appealing in human health monitoring applications due to its broadband nature, system integration and low-power operation. Additionally, it is not that effective in short-range applications because of the short backscattering trip. This would require a very high-speed, power-hungry analog-to-digital converter (ADC).

   CW and FMCW radars are attractive to human health monitoring applications. This is mainly due to higher level of system integration and lower power operation. Although FMCW has the ability of determining the distance and the relative movement, in most applications, based on literature, CW is preferred due to higher measurement accuracy, less complex hardware architecture and simpler signal processing approaches.

11.2.3 CW radar for healthcare applications

A CW radar is capable to detect vital signs of a person from a distance. The radar can sense the physiological movement through clothing and bedding [10]. A person breathing is considered an object with zero net velocity and periodic movement according to Doppler theory. A low-power transmitted RF signal will be reflected by the person's body and will be phase modulated. It contains information from the respiration and heartbeat. In other words, the radar will detect the movement of the chest due to cardiopulmonary activity.

The basic components of a radar system are usually an RF signal generator, power dividers, mixers, amplifiers, antennas and analog to digital convertors.

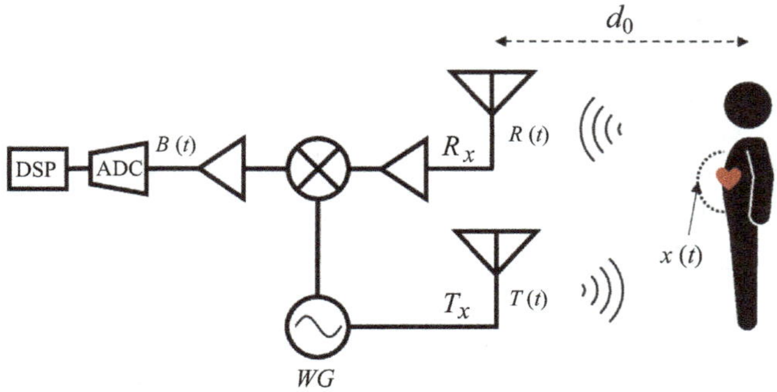

Figure 11.7 Block diagram of a Doppler radar system for vital signs detection

As shown in Figure 11.7, an unmodulated signal $T(t)$ is generated by the waveform generator (WG) and transmitted via a *Tx* antenna to the subject at a distance d_0. The unmodulated signal can be written as:

$$T(t) = A_T \cos(2\pi f t) \tag{11.3}$$

where f is the carrier frequency and A_T is the power amplitude.

The area of interest for radar-based vital sign detection in the chest wall. The subject's chest wall movement $x(t)$, from respiration and heartbeat, modulates the RF signal and reflects a part of it back towards the radar. The total distance travelled by the radar wave (signal) is:

$$2d(t) = 2d_0 + 2x(t) \tag{11.4}$$

where d_0 is the nominal distance between the subject and the radar. The signal received at the *Rx* antenna of the radar is:

$$R(t) = A_R \cos\left[2\pi f\left(t - \frac{2d(t)}{c}\right)\right] \tag{11.5}$$

where c is the propagation velocity of the signal, and the carrier wavelength is given from:

$$\lambda = \frac{c}{f} \tag{11.6}$$

Substituting (11.4) and (11.6) in (11.5), the received signal is:

$$R(t) = A_R \cos\left[2\pi f t - \frac{4\pi d_0}{\lambda} - \frac{4\pi x(t)}{\lambda}\right] \tag{11.7}$$

Comparing (11.3) and (11.7), it is evident that the received signal resembles the transmitted one with an added time delay defined by the distance of the subject and its phase is modulated by the periodic chest movement of the subject. The information about this periodic movement can be demodulated if the received signal is mixed with an LO from the same source as the transmitted.

The received and LO signals are mixed and then low-pass filtered. The result is a baseband signal:

$$B(t) = \cos\left[\theta + \frac{4\pi x(t)}{\lambda}\right] \tag{11.8}$$

where

$$\theta = \frac{4\pi d_0}{\lambda} + \theta_0 \tag{11.9}$$

is the constant phase shift dependent on the nominal distance d_0. Factors that affect θ_0 include the phase shift at the reflection surface (near 180°) and any distance between the mixer and the antenna.

11.3 Challenges

11.3.1 Range correlation effect

Phase noise is defined as random fluctuations within the oscillator in the frequency caused by time-domain instabilities (Figure 11.8). Any signal source suffers from phase noise and reducing this phase noise is a challenge in vital sign monitoring with radar-based systems. Some RF sources have a very low-phase noise and others have a larger phase noise. The RF phase noise will become baseband noise after frequency down conversion which will determine the baseband signal-to-noise ratio (SNR) [9,11].

Heart and respiration rates are between 0.1 and 10 Hz, where phase noise is close its peak. The residual phase noise would appear as additive noise in the baseband signal. Heart and respiration motion create very small phase variations which would not be feasible to detect, if two different signal generators with uncorrelated phase noise were used to transmit and receive.

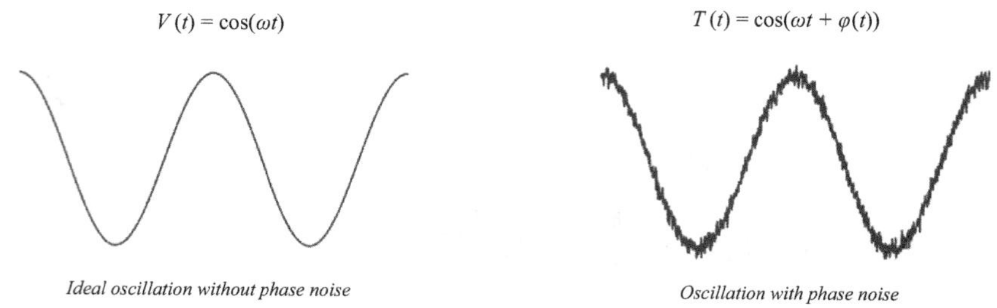

Figure 11.8 Phase noise in a signal

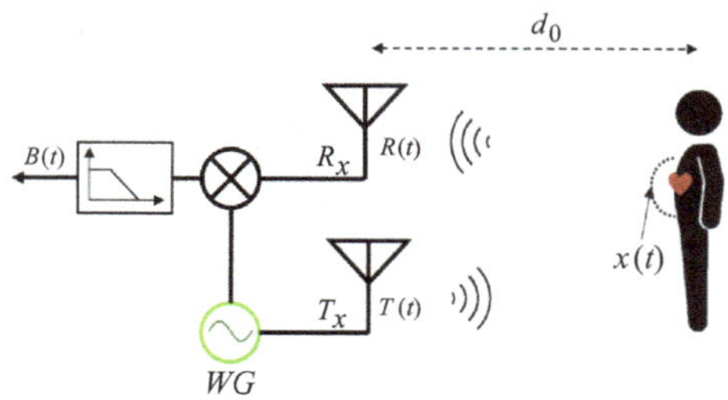

Figure 11.9 Range correlation effect: the same source is used to generate the transmit signal and as LO to down-convert the received signal

A solution proposed by [12] is to use the same local oscillator (LO) for transmitting and receiving, as depicted in Figure 11.9. Then, the phase noise of the two signals would be correlated. The level of correlation would be dependent on the time delay between the two signals which in radar application, is proportional to the target range. Due to that, this effect is called range correlation effect and reduces the baseband noise significantly.

11.3.2 Null points

11.3.2.1 Problem definition

Theorem 1 *(Small angle approximation):* *Small angle approximation is used to approximate values of main trigonometric functions, provided the angle is small and measured in radians. Therefore, when $\theta \approx 0$:*

$$\sin\theta \approx \theta, \quad \cos\theta \approx 1 - \frac{\theta^2}{2}, \quad \tan\theta \approx \theta \tag{11.10}$$

Looking at the baseband signal in (11.8), if θ is odd multiple of $\pi/2$, then the baseband output is:

$$B(t) = \sin\left[\frac{4\pi x(t)}{\lambda}\right] \tag{11.11}$$

The small angle approximation is valid for $x(t) \ll \lambda$, and the baseband output is approximately:

$$B(t) \approx \frac{4\pi x(t)}{\lambda} \tag{11.12}$$

Hence, phase-demodulation sensitivity is achieved as the output is proportional to chest displacement $x(t)$ summed with the residual phase noise.

If θ is integer multiples of π, the baseband output is:

$$B(t) = \cos\left[\frac{4\pi x(t)}{\lambda}\right] \tag{11.13}$$

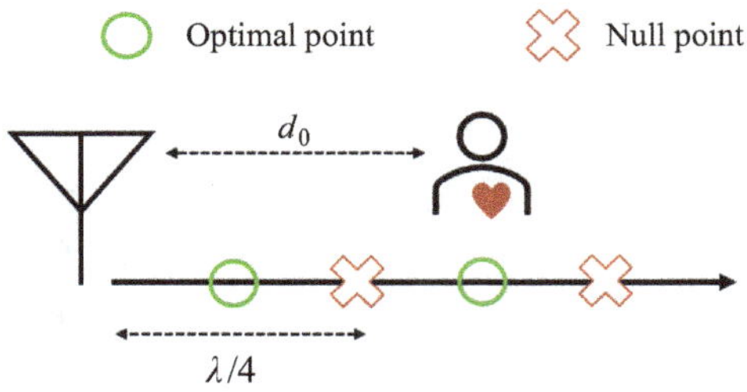

Figure 11.10 Null and optimal points demonstration in a receiver system

The small angle approximation is valid for $x(t) \ll \lambda$, and the baseband output is approximately:

$$B(t) \approx 1 - \left[\frac{4\pi x(t)}{\lambda}\right]^2 \tag{11.14}$$

In this case, phase-demodulation sensitivity is decreased, and the output is not proportional to chest displacement $x(t)$.

This happens if LO $T(t)$ and the received $R(t)$ signals are in-phase or 180° out-of-phase. The angle θ depends on the distance from the subject d_0. Therefore, the null point appears with a subject distance $\lambda/4$ from the radar as shown in Figure 11.10. For a 2.4 GHz radar null points occur every 3 cm. This leads to poor signal and incorrect measurement values and would be almost impossible to overcome this by changing the position of the system.

11.3.2.2 Solution

First approach is a quadrature receiver [13], where they measured heart and respiration simultaneously with digital rate detection algorithm. In [14], Droitcour *et al.* used quadrature receiver to design the first CMOS/BiCMOS Doppler radar chips. A quadrature receiver would benefit the performance of the radar with two receiver chains of $\pi/2$ apart. The quadrature receiver architecture is demonstrated in Figure 11.11. The transmit signal $T(t)$ which comes from the LO, is split to feed the two mixers. The received signal $R(t)$ is mixed with the transmit to generate the

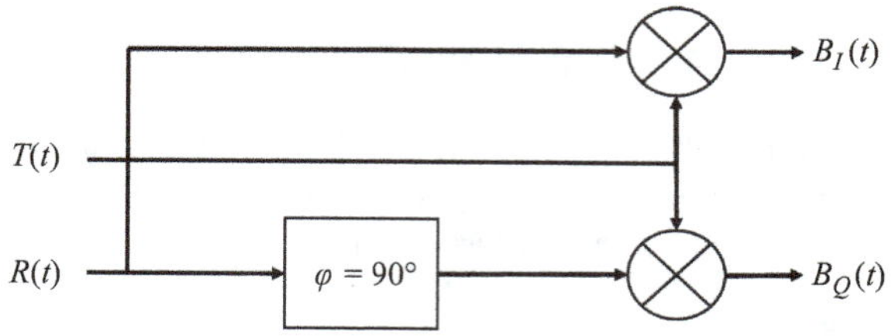

Figure 11.11 Quadrature receiver

in-phase I component. The $R(t)$ signal is going through a 90° phase shifter and is mixed with the transmit to create the quadrature Q component.

The in-phase B_I and quadrature B_Q outputs are:

$$B_I(t) = A_I \cos\left(\theta_0 + \frac{4\pi x(t)}{\lambda}\right) \tag{11.15}$$

$$B_Q(t) = A_Q \cos\left(\theta_0 + \frac{4\pi x(t)}{\lambda} - \frac{\pi}{2}\right) = A_I \sin\left(\theta_0 + \frac{4\pi x(t)}{\lambda}\right) \tag{11.16}$$

The chest displacement $x(t)$ is hidden in both signals. With this receiver, when one output is at a null point, the other will be at the optimal phase demodulation point.

The worst case is when θ is an integer multiple of π. Then, both baseband outputs will be:

$$B_I(t) = B_Q(t) \approx \frac{1}{\sqrt{2}} - \frac{1}{\sqrt{2}}\left[\left(\frac{4\pi x(t)}{\lambda}\right) + \frac{1}{2}\left(\frac{4\pi x(t)}{\lambda}\right)^2\right] \tag{11.17}$$

On condition that $x(t) \ll \lambda$, the linear term of (11.17) is greater than the squared term, thus the heart signal can be detected.

Based on [12], a single chain receiver in a null point gave a heart rate detection of an accuracy less than 40%. The implementation of a quadrature receiver improved the heart rate detection accuracy from 40% to approximately 80%. Quadrature architecture has been widely used; some examples would be [15–18].

Other work to avoid the null point: in [19] the writers proposed a heterodyne Ka band radar, where a double-sideband transmission is implemented. They managed to eliminate the DC offset as well. The same technique at 4–7 GHz was also used by [20]. Null point can also be avoided by frequency tuning as described in [21], where a 5 GHz double-sideband radar sensor chip is used. More recently, Phase-Locked-Loop was used to avoid the null point in [22] without requiring the small-angle approximation condition.

11.3.3 Frequency considerations

As previously mentioned, the signal that is received by the radar is given by:

$$R(t) = A_R \cos\left[2\pi f t - \frac{4\pi d_0}{\lambda} - \frac{4\pi x(t)}{\lambda}\right] \tag{11.18}$$

with the phase being:

$$\varphi(\mathrm{t}) = \frac{4\pi x(t)}{\lambda} \tag{11.19}$$

The radar can theoretically operate in any frequency. As can be seen from the phase equation (11.19), the wavelength is inverse proportional to the phase. Therefore, the smaller the wavelength, the bigger the phase and consequently the better the resolution of the system. The wavelength is given by $\lambda = c/f$. This means that a shorter wavelength will come from a higher frequency.

For example, at 2.4 GHz the wavelength is calculated:

$$\lambda = \frac{c}{f} = 12.5 \text{ cm} \tag{11.20}$$

A chest displacement of $x(t) = 1$ cm results in phase change of:

$$\varphi(\text{t}) = \frac{4\pi x(t)}{\lambda} = \frac{4\pi \cdot 1 \text{ cm}}{12.5 \text{ cm}} \cong 1 \text{ rad or } 57.6^\circ \tag{11.21}$$

From theory, peak to peak chest motion due to respiration is 4–12 mm and due to heartbeat 0.5 mm [8]. This results in phase variation from 23° to 69° for respiration and less than 3° for heartbeat. Therefore, a high-performance demodulator is needed to be able to discriminate between respiration and heartbeat. As the frequency increases, the phase change will increase proportionally.

The frequency choice, however, is also based on the internationally reserved bands for industrial, scientific, and medical (ISM) applications, the availability of commercially available RF components and antennas in these frequency bands and the cost thereof. Additionally, the frequency depends on the application, the distance to the target and the environment.

11.3.4 DC offset

One of the main challenges in the receiver is the DC offset in the receiver output. Reflected signal from stationary objects has the same frequency as the LO and mixes with the LO in the mixer, therefore creating a DC component. DC component is also caused by circuit flaws, such as mixers self-mixing [23].

This DC offset in the system is presented in the baseband signals as:

$$B_I(t) = A_I \cos\left(\theta_0 + \frac{4\pi x(t)}{\lambda}\right) + DC_I \tag{11.22}$$

$$B_Q(t) = A_Q \sin\left(\theta_0 + \frac{4\pi x(t)}{\lambda}\right) + DC_Q \tag{11.23}$$

Figure 11.12 depicts a simulated signal with a frequency of 0.2 Hz which is for respiration and 1.5 Hz for heartbeat, denoted as AC. A DC value has been added to the same signal, denoted as AC + DC. White Gaussian noise has been added to the signals to simulate a realistic measurement.

While in other applications this would not be an issue, in respiration and heartbeat detection it plays a major role. Fast Fourier Transform (FFT) can be applied to transform the signals to the frequency domain. From Figure 11.13 it is obvious that the DC component is a large spike around 0 Hz. Respiratory rate in a human subject is between 0.1 and 0.3 Hz which is very close to DC making it difficult to discriminate between the signals. This DC offset may saturate or limit the dynamic range needed for further processing.

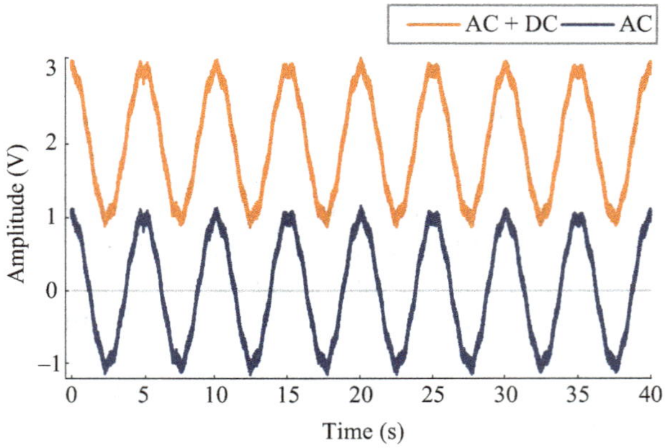

Figure 11.12 Simulated detected signal without and with DC component

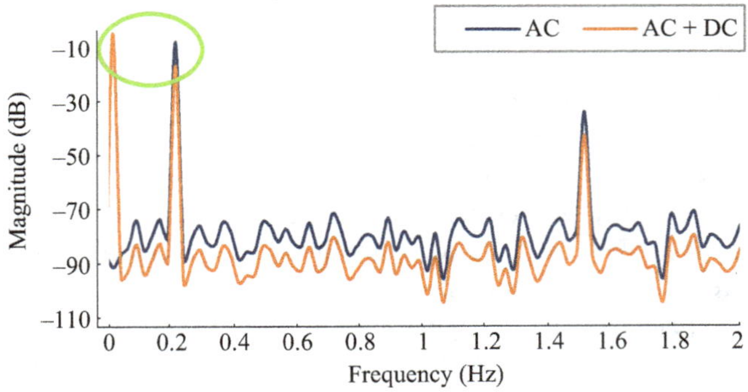

Figure 11.13 FFT plot without DC component and with DC component

DC cancellation techniques include calibrating the DC offset [23], advanced signal processing algorithms [24], complex detection circuits [25] and automated DC offset reduction [26]. In [27,28], Kontou *et al.* implemented detection circuits at baseband level to manually and automatically reduce the DC offset to improve the dynamic range.

11.3.5 Signal processing

11.3.5.1 Phase demodulation

Various signal processing algorithms have been tested. Small angle approximation is used initially in [12,16]. Wavelet transformation was implemented in [29].

In [23], Park *et al.* used a 2.4 GHz quadrature receiver and the data was obtained with arctangent demodulation, where the displacement is determined using the arctangent function of the Q over the I data (Q and I being the outputs).

$$\varphi(t) = \arctan\left(\frac{V_Q + DC_Q}{V_I + DC_I}\right) \tag{11.24}$$

In [20], Li *et al.* tested the complex demodulation for random body movement cancellation. They concluded that it is more suitable for this application, whilst arctangent demodulation is better at eliminating the harmonic and intermodulation interference at high frequencies. More recently, Kalman filtering techniques were implemented in [30].

To extract the phase information, for a quadrature receiver a good solution is to use both I and Q channels, also called linear demodulation. In [8], Boric-Lubecke *et al.* suggest monitoring both signals and choose the one in the optimal point. However, if both signals are varying or if the sign is opposite, I and Q can be added or subtracted, respectively. The steps are to remove the DC from I and Q, calculate the covariance matrix and its eigenvectors and eigenvalues. Multiplying the eigenvectors with the data will give the output with the most variance. Therefore, multidimensional data will be projected in a single dimension.

11.3.5.2 Phase processing

Fourier analysis converts a signal from time to frequency domain to approximate the rate of a periodic event. Fourier method breaks down the original data into windows and calculates the Fourier transform of each one. A series of sinusoidal terms are produced, each with its own amplitude, frequency and phase. A power spectrum is created when plotting the amplitude of each term versus its frequency. This is also the response of the original waveform in the frequency domain.

Another method is peak detection that calculates the time between two consecutive peaks. However, this method is not a good choice for this application due to the noise generated from the background, which can move the location of the peaks [8].

Autocorrelation has also been used, which gives the degree of similarity between observations. This is achieved by calculating the cross correlation of a signal with itself and peak detection is used to determine the heart rate. A disadvantage is that because this processing happens in windows, variability can be lost, and changes faster than 6 bpm cannot be detected in 10 s of data, according to [8].

11.4 Implementation of a CW radar

Figure 11.14 depicts a radar system with quadrature receiver. A microwave signal is generated from a signal generator and is transmitted through the transmit antenna towards a subject. The reflected signal, which is amplitude and phase modulated, is received from the same antenna, and is amplified. It is then going through a splitter and two signals are generated which are down converted to baseband by being mixed with the transmitted signal to benefit from the range correlation effect. This

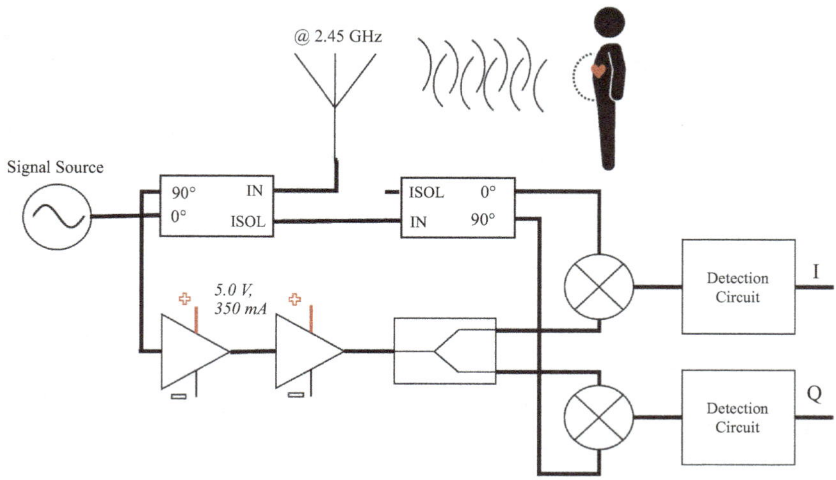

Figure 11.14 *Block diagram of the proposed CW radar-based system for respiration and heartbeat detection implemented in [27,28] by Kontou et al.*

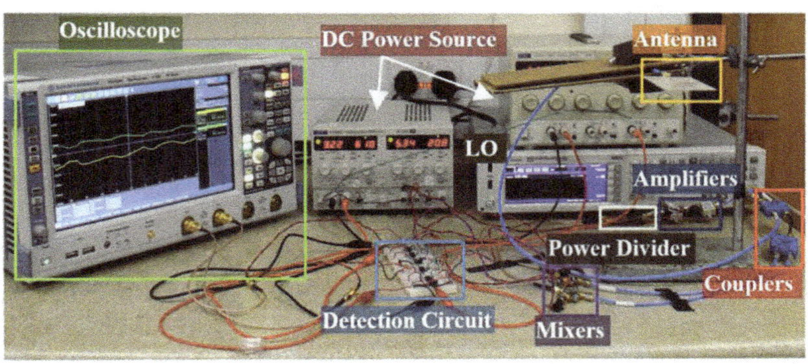

Figure 11.15 *Implementation of the CW radar-based system shown in Figure 11.14 implemented in [27,28] by Kontou et al.*

leads to the two receiver chains one for *I* and one for *Q*. This part is the quadrature receiver.

The system that was just described, was implemented in the microwave lab in Heriot-Watt University using commercially available components which are denoted in Figure 11.15 and shown in Table 11.1.

Only one antenna is used for transmitting and receiving the signals. The antenna is a compact planar end-fire *Yagi-Uda* printed circuit board (PCB) based antenna. The resonant frequency is at 2.45 GHz and can maintain a gain higher than 6 dBi in a free-space environment and the front to back ratio is close to 19 dB.

Table 11.1 Components of the CW radar system

Block	Model	Specifications
Signal generator (LO)	Keysight MXG N5183B	2.45 GHz; Pout: 3 dBm
Mixers	Analog Devices LTC5549	2–14 GHz
Couplers	Anaren Model 10016-3	2–4 GHz
Amplifiers – LNA	Mini-Circuits ZX60-V63+	0.05–6 GHz; Gain: 20 dB

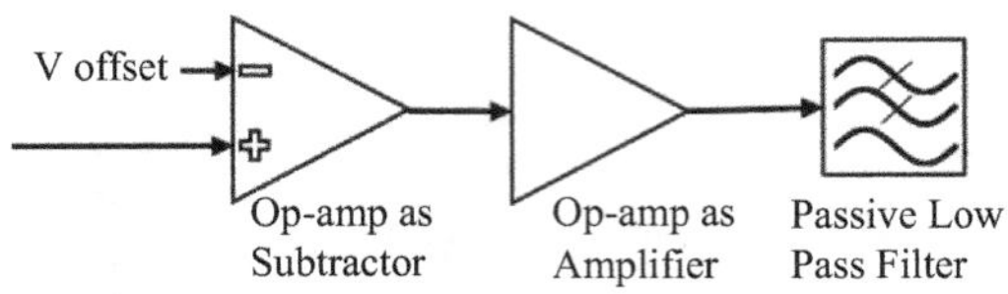

Figure 11.16 Detection circuit to manually tune the signal so that DC offsets can be removed

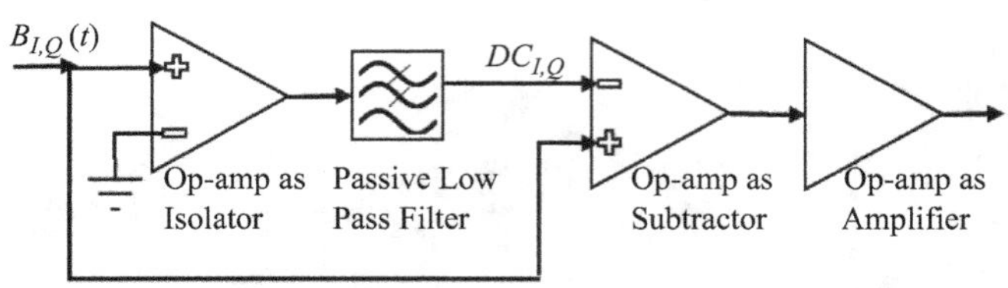

Figure 11.17 Detection circuit that isolates the DC offset from the received signal and subtracts it from the original

A method for DC offset reduction is achieved with the detection circuit depicted in Figure 11.16. Two operational amplifiers (LM741) are used for every circuit and the signal is manually tuned to eliminate the DC offsets. The signal is then amplified to obtain a wider dynamic range and is filtered with a passive low-pass filter to remove the high-frequency components.

A second method for automatic reduction was implemented and is based on operational amplifiers to calibrate the DC offset automatically (Figure 11.17). A detection circuit is used, and the DC components are reduced as follows; the first op-amp followed by a low pass filter allows the isolation of the DC component. That DC component is then subtracted from the original signal with the second op-amp. The result is amplified with the third op-amp to improve the dynamic range.

11.4.1 Experimental results

An illustration of the experimental setup is shown in Figure 11.18. The CW radar is located on a workbench in a lab and its antenna beam is facing a healthy human volunteer. The volunteer was breathing normally. The distance between the radar

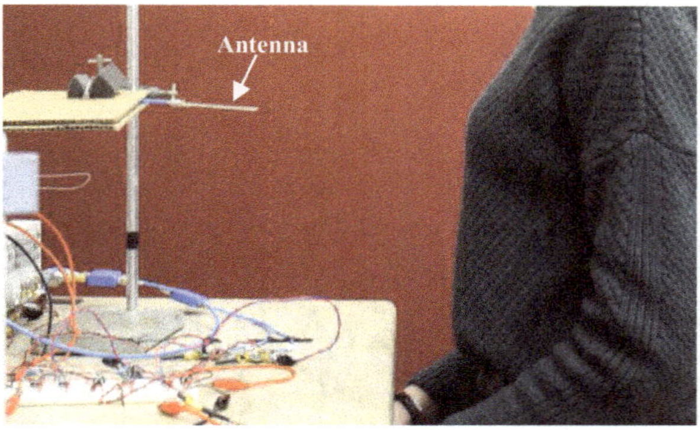

Figure 11.18 Experimental setup for CW radar system: a volunteer is seated at a distance of 25 cm, facing the Planar Yagi-Uda antenna and breathing normally

and the volunteer was varied between 0.25 and 1 m and multiple measurements were carried out. A wearable fitness tracker was used to record the heartbeat as a reference. The volunteer was counting the breaths taken in the measurements. The experiments were approved by Heriot-Watt University and the volunteer had signed the consent form.

Method 1: Tunable DC offset reduction

The following presented measurements were carried out using the detection circuit of Figure 11.16. A Rohde & Schwarz RTO2064 oscilloscope was used to capture and visualize the data. Raw data of a measurement is shown in Figure 11.19. The

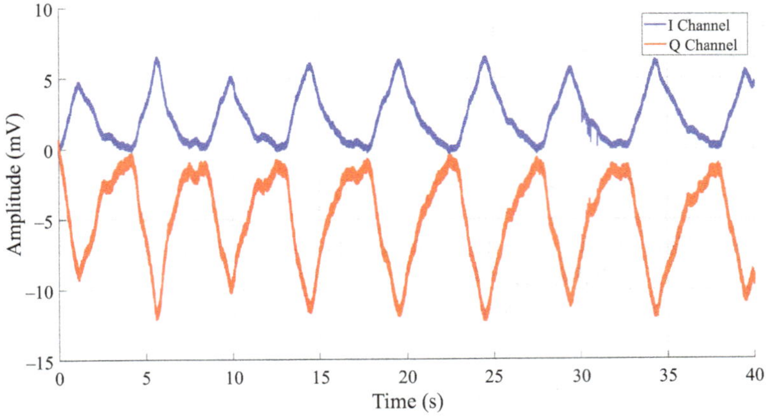

Figure 11.19 Raw I and Q data capture for 0.5 m distance between the antenna and the volunteer

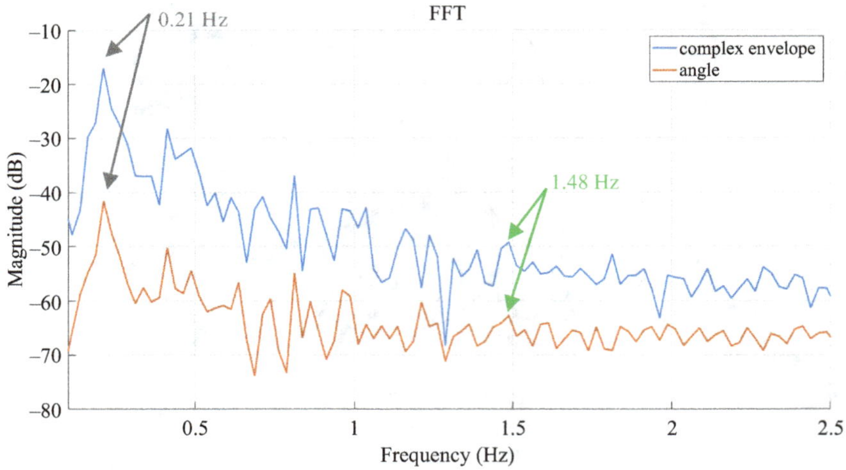

Figure 11.20 FFT of the complex envelope and unwrapped phase of the received signal

volunteer is seated in front of the antenna at a distance of 0.5 cm. The sampling frequency was set to 1 kHz and the measurement lasted 40 s. Each rise in the I channel indicates and inhale and each fall indicates an exhale.

The software Matlab was used to process the recorded data. FFT was applied to the complex envelope and the unwrapped phase of the received signal. Figure 11.20 shows the FFT results. As can be observed, both signals have identical spikes in 0.21 Hz and 1.48 Hz. The volunteer counted approximately 8.5 breaths in 40 s. This is translated in 0.21 Hz which is the highest peak in the plot. The heart rate of the volunteer according to the wearable fitness tracker was 89 beats per minute (bpm). This is 1.48 beats per second (Hz), which is observed in the FFT.

Inverse FFT (IFFT) was applied to recover the heartbeat and respiration pattern. The respiration information is hidden in the range of 0.1−1 Hz. The IFFT signal in these frequencies is depicted in Figure 11.21. The frequency of the signal corresponds to the counted breaths taken.

To recover the heartbeat IFFT was applied to the range 1−30 Hz and is shown in Figure 11.22. The frequency of the heartbeat in the FFT is unclear, however, by filtering out the respiration frequency and applying the IFFT we get a very clear pattern of the heartbeat in the time domain. The measurements are positively correlated with the wearable fitness tracker measurements. Zooming in 0−10 s, 14 peaks can be observed. This means that in 10 s there are 14 heartbeats and, therefore, 89 bpm.

Method 2: Automated DC offset reduction

The following presented measurements were carried out using the detection circuit of Figure 11.17. The measurement setup and signal processing techniques are the same as described in method 1.

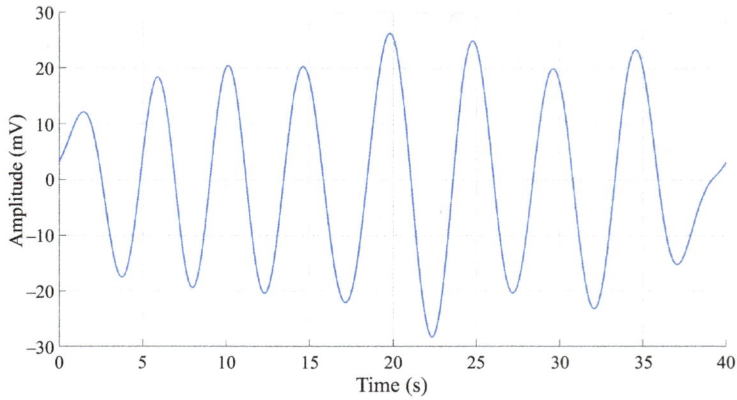

Figure 11.21 IFFT results in the range of 0.1–0.5 Hz where the respiration frequency is, that shows the respiration pattern of this measurement

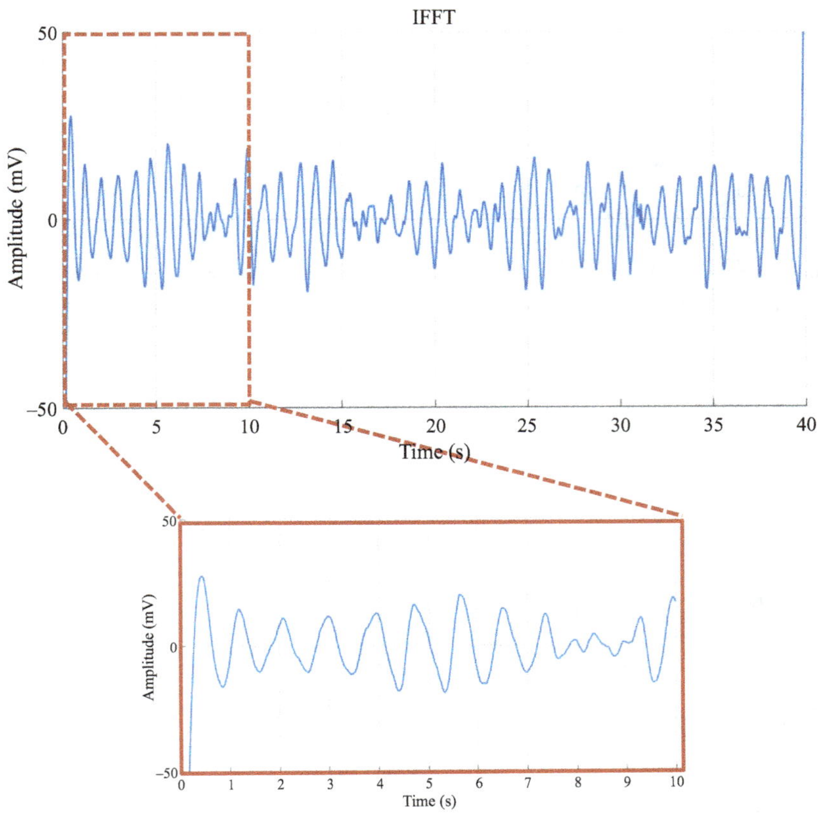

Figure 11.22 IFFT results in the range of 1–30 Hz. Twelve cycles in 10 s which corresponds to the 89 bpm heartbeat frequency

Figure 11.23 depicts raw *I* and *Q* data with reduced DC offset of a measurement for 40 s, sampling frequency 1 KHz and 0.5 m distance between the antenna and the volunteer. In this case, each rise in the signals of the *I* and *Q* indicates an inhale and each drop indicates an exhale.

FFT was applied to the complex envelope of the received signal in Matlab. The FFT plot is depicted in Figure 11.24. The highest peak is at 0.187 Hz which is the breathing frequency. Indeed, the volunteer counted 7.5 breaths in 40 s. A peak at 1.21 Hz is observed as the heartbeat frequency. This is supported by the measurement of the wearable fitness tracker which showed 72 bpm.

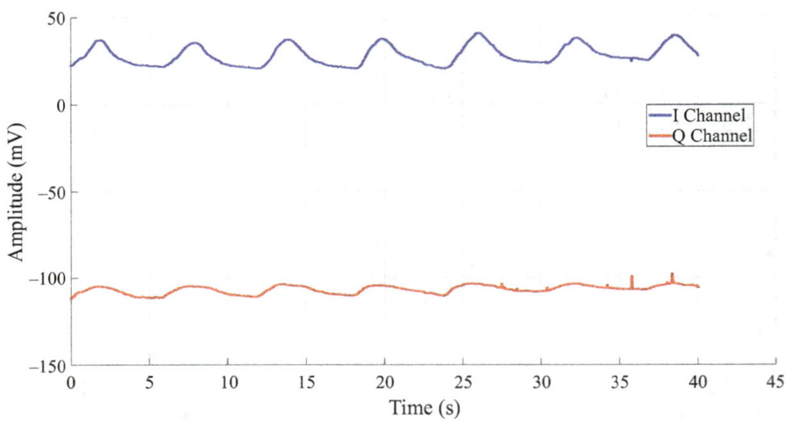

Figure 11.23 Raw I and Q data capture for 0.5 m distance between the antenna and the volunteer

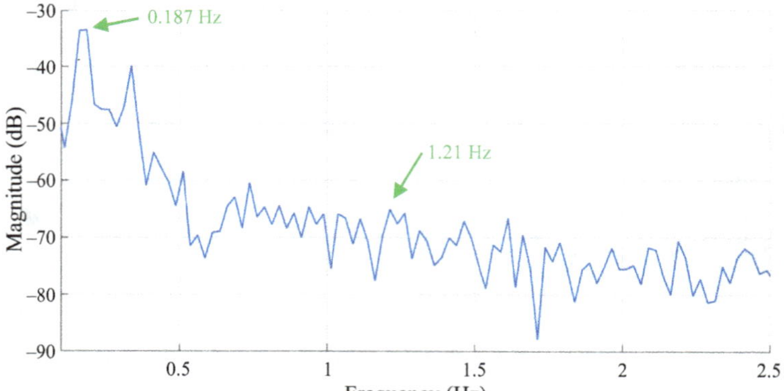

Figure 11.24 FFT of the complex envelope of the received signal showing a peak at 0.187 Hz from respiration and a peak at 1.21 Hz from the heartbeat

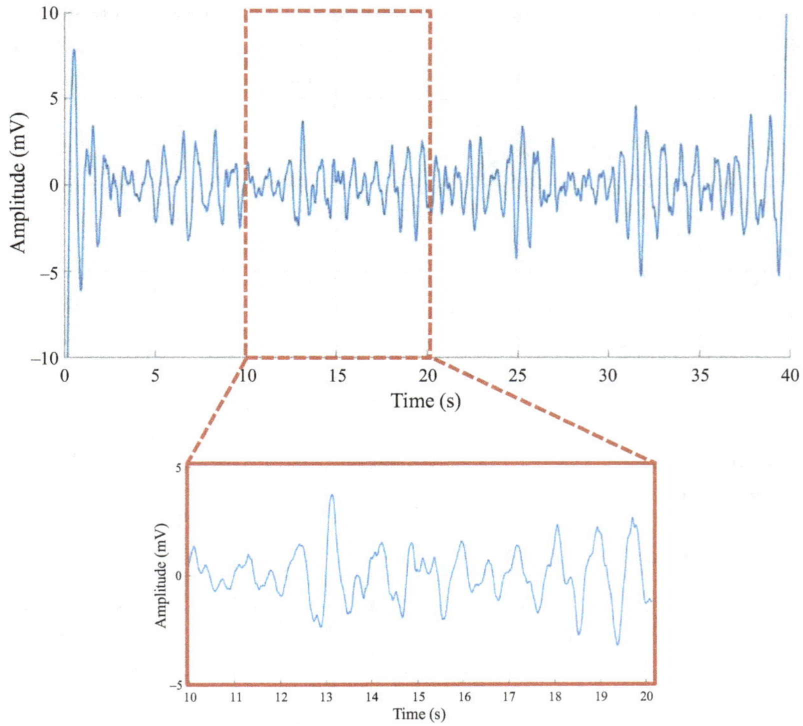

Figure 11.25 IFFT results in the range of 1–30 Hz. Twelve cycles in 10 s which corresponds to the 72 bpm heartbeat frequency

IFFT was applied to the range 1−30 Hz to recover the heartbeat. Zooming in 10−20 s, 12 cycles can be observed in Figure 11.25. This means that in 10 s there are 12 heartbeats and, therefore, 72 bpm.

11.5 Applications

Cardiopulmonary monitoring is achieved mainly with electrocardiogram (ECG) electrodes, which are required to be in contact with the patient. Not only is this method uncomfortable but also it does not allow for continuous monitoring. Many applications such as patients with compromised skin and sleep monitoring, would benefit from the ability to detect heartbeat and respiration remotely. Besides healthcare, non-contact sensing could be applied in other emerging applications such as fatigue monitoring, through-the-wall search and rescue operations from earthquakes, fires, etc. [8].

11.5.1 Patients with compromised skin

Reliable contactless devices could play an important role and be clinically valuable in many circumstances. In the case of burn victims, monitoring their vital signs can be difficult, due to the lack of healthy skin to attach the electrodes. Occasionally, an esophageal ECG has to be used which is very unpleasant [8]. In addition, the continuous monitoring of premature babies may also lead to damage on their skin.

11.5.2 Sleep monitoring

Good sleep is important not only for physical, but also mental health. Specifically, it can affect metabolism, weight, cardiovascular health, immune system function, memory, mood and more [8]. There are many sleep disorders, sleep apnea being one of the most common, during which, airflow is stopped for 10 s or longer affecting the aforementioned physical and mental health parameters. Moreover, it can also affect a person's job or be responsible for vehicle collisions. Another sleep disorder is sudden infant death syndrome (SIDS) which is related to sleep apnea. An extreme episode can lead to choking and instant death. Over the last 15 years SIDS episodes have dropped, however SIDS still remains the third leading cause of infant mortality overall [31]. A wireless device could sense the loss of breathing and provide reliable domestic monitoring.

11.5.3 Elderly monitoring

With recent technology and constantly improving healthcare, the elderly population has increased. Thus, the need for monitoring the elderly has also increased. They can receive faster, optimized, and personalized care in the long term. Moreover, such a system eliminates the use of frequent visits to the doctor, such as the case of patients bearing chronic diseases, which has a negative effect on the patient's mind-set and prevents them from leading a normal life. At the same time, an alert could be issued in emergencies for the medical personnel to attend to the patient, providing assisted living opportunities.

11.6 Summary

Radar systems can detect vital signs from a distance by sensing the chest displacement. The later comes from the heartbeat and respiration. This chapter presents the theoretical background of this application along with the main characteristics of such a system. Limitations and challenges are presented. An implementation of a CW radar-based system in order to detect the heartbeat of a person is presented and demonstrated. Radar technology has the potential to replace traditional methods to monitor vital signs.

References

[1] C.-H. Hsu and J. C. Chow, "Design and clinic monitoring of a newly developed non-attached infant apnea monitor", *Biomedical Engineering: Applications, Basis and Communications*, vol. 17, no. 03, pp. 126–134, 2005.

[2] Y. Mitsunari, Y. Hasegawa, M. Matsushima, T. Kawabe, and M. Shikida, "Development of small-footprint thermal sensor detecting airflow at mouth in baby", *Multidisciplinary Digital Publishing Institute Proceedings*, vol. 1, no. 4, p. 359, 2017.

[3] M. I. Skolnik, *Radar Handbook,* New York, NY: McGraw-Hill Professional, 1970, ISBN: 0070579083.

[4] J. C. Lin, "Noninvasive microwave measurement of respiration", *Proceedings of the IEEE*, vol. 63, no. 10, pp. 1530–1530, 1975.

[5] S. M. Sharpe, J. Seals, A. H. MacDonald, and S. R. Crowgey, "Non-Contact Vital Signs Monitor", *ed: Google Patents*, 1990.

[6] E. M. Staderini, "UWB radars in medicine", *IEEE Aerospace and Electronic Systems Magazine*, vol. 17, no. 1, pp. 13–18, 2002, doi: 10.1109/62.978359.

[7] C. Starr, C. Evers, and L. Starr, *Biology Today and Tomorrow with Physiology*, USA: Cengage Learning, 2020.

[8] O. Boric-Lubecke, V. M. Lubecke, A. D. Droitcour, B.-K. Park, and A. Singh, *Doppler Radar Physiological Sensing*, New York, NY: Wiley Online Library, 2016.

[9] R. N. Fogoros, "What is an Electrocardiogram (ECG or EKG), *Verywell Health*, Aug 5, 2021. Available from https://www.verywellhealth.com/the-electrocardiogram-ecg-1745304

[10] C. Li, V. M. Lubecke, O. Boric-Lubecke, and J. Lin, "A review on recent advances in Doppler radar sensors for noncontact healthcare monitoring", *IEEE Transactions on Microwave Theory and Techniques*, vol. 61, no. 5, pp. 2046–2060, 2013.

[11] B. Razavi, *Design of Analog CMOS Integrated Circuits*, London: McGraw-Hill, 2001.

[12] A. D. Droitcour, O. Boric-Lubecke, V. M. Lubecke, J. Lin, and G. T. Kovacs, "Range correlation and I/Q performance benefits in single-chip silicon Doppler radars for noncontact cardiopulmonary monitoring", *IEEE Transactions on Microwave Theory and Techniques*, vol. 52, no. 3, pp. 838–848, 2004.

[13] J. Seals, "Electromagnetic vital signs monitor", in *Engineering Experiment Station*, Atlanta, GA: Georgia Institute of Technology, 1984.

[14] A. D. Droitcour, O. Boric-Lubecke, V. M. Lubecke, and J. Lin, "0.25/spl mu/m CMOS and BiCMOS single-chip direct-conversion Doppler radars for remote sensing of vital signs", in *2002 IEEE International Solid-State Circuits Conference. Digest of Technical Papers (Cat. No. 02CH37315)*, 2002, vol. 1, pp. 348–349, New York, NY: IEEE.

[15] A. Vergara, N. Petrochilos, O. Boric-Lubecke, A. Host-Madsen, and V. Lubecke, "Blind source separation of human body motion using direct conversion Doppler radar", in *2008 IEEE MTT-S International Microwave Symposium Digest*, 2008, pp. 1321–1324, New York, NY: IEEE.

[16] W. Massagram, V. M. Lubecke, A. Høst-Madsen, and O. Boric-Lubecke, "Assessment of heart rate variability and respiratory sinus arrhythmia via Doppler radar", *IEEE Transactions on Microwave Theory and Techniques*, vol. 57, no. 10, pp. 2542–2549, 2009.

[17] C. Gu, C. Li, J. Lin, J. Long, J. Huangfu, and L. Ran, "Instrument-based noncontact Doppler radar vital sign detection system using heterodyne digital quadrature demodulation architecture", *IEEE Transactions on Instrumentation and Measurement*, vol. 59, no. 6, pp. 1580–1588, 2009.

[18] Y. Wang, Q. Liu, and A. E. Fathy, "CW and pulse–Doppler radar processing based on FPGA for human sensing applications", *IEEE Transactions on Geoscience and Remote Sensing*, vol. 51, no. 5, pp. 3097–3107, 2012.

[19] Y. Xiao, J. Lin, O. Boric-Lubecke, and V. M. Lubecke, "A Ka-band low power Doppler radar system for remote detection of cardiopulmonary motion", in *2005 IEEE Engineering in Medicine and Biology 27th Annual Conference*, 2006, pp. 7151–7154, New York, NY: IEEE.

[20] C. Li and J. Lin, "Random body movement cancellation in Doppler radar vital sign detection", *IEEE Transactions on Microwave Theory and Techniques*, vol. 56, no. 12, pp. 3143–3152, 2008.

[21] C. Li, Y. Xiao, and J. Lin, "A 5GHz double-sideband radar sensor chip in 0.18 μ CMOS for non-contact vital sign detection", *IEEE Microwave and Wireless Components Letters*, vol. 18, no. 7, pp. 494–496, 2008.

[22] M. Mercuri, Y.-H. Liu, I. Lorato, T. Torfs, A. Bourdoux, and C. Van Hoof, "Frequency-tracking CW Doppler radar solving small-angle approximation and null point issues in non-contact vital signs monitoring", *IEEE Transactions on Biomedical Circuits and Systems*, vol. 11, no. 3, pp. 671–680, 2017.

[23] B.-K. Park, O. Boric-Lubecke, and V. M. Lubecke, "Arctangent demodulation with DC offset compensation in quadrature Doppler radar receiver systems", *IEEE Transactions on Microwave Theory and Techniques*, vol. 55, no. 5, pp.1073–1079, 2007.

[24] M. C. Huang, J. J. Liu, W. Xu, C. Gu, C. Li, and M. Sarrafzadeh, "A self-calibrating radar sensor system for measuring vital signs", *IEEE Transactions on Biomedical Circuits and Systems*, vol. 10, no. 2, 2015, pp. 352–363.

[25] C. Y. Hsu, C. Y. Chuang, F. K. Wang, T. S. Horng and L. T. Hwang, "Detection of vital signs for multiple subjects by using self-injection-locked radar and mutually injection-locked beam scanning array", in *IEEE 2017 MTT-S Intl' Microwave Symposium*, pp. 991–994.

[26] C. Gu and C. Li, "DC coupled CW radar sensor using fine-tuning adaptive feedback loop", *Electronics Letters*, vol. 48, no. 6, 2012, pp. 344–345.

[27] P. Kontou, S. Ben Smida, M. Dragone, S. Nikolaou, and D.E. Anagnostou, "CW radar based system with automated DC offset reduction for heartbeat detection", in *2020 IEEE USNC-CNC-URSI North American Radio Science Meeting (Joint with AP-S Symposium)*, July 2020, pp. 73–74.

[28] P. Kontou, S. Ben Smida, S. N. Daskalakis, M. Dragone, S. Nikolaou, and D. E. Anagnostou, "Heartbeat and respiration detection using a low complexity CW radar system", in *50th IEEE European Microwave Week (EuMW), Utrecht Netherlands*, January 2021, pp. 929–932.

[29] G. Ossberger, T. Buchegger, E. Schimback, A. Stelzer, and R. Weigel, "Non-invasive respiratory movement detection and monitoring of hidden humans using ultra wideband pulse radar", in *2004 International Workshop on Ultra Wideband Systems Joint with Conference on Ultra Wideband Systems and Technologies. Joint UWBST & IWUWBS 2004 (IEEE Cat. No. 04EX812)*, 2004, pp. 395–399, New York, NY: IEEE.

[30] K. Ramachandra, *Kalman Filtering Techniques for Radar Tracking*, London: CRC Press, 2018.

[31] B. E. Hamilton, D. L. Hoyert, J. A. Martin, D. M. Strobino, and B. Guyer, "Annual summary of vital statistics: 2010–2011", *Pediatrics*, vol. 131, no. 3, pp. 548–558, 2013.

Chapter 12

Therapeutic applications of electromagnetic waves

Koichi Ito[1]

12.1 Introduction

Recently, various types of medical and healthcare applications of electromagnetic waves have widely been investigated and reported.

Typical recent applications include:

1. Information/wireless power transmission
   - Wearable or implantable vital data sensor
   - Wireless telemedicine/patient monitoring system
   - Advanced wireless capsule endoscopy
2. Diagnosis
   - High intensity magnetic resonance imaging (MRI)
   - Microwave computed tomography (CT) for cancer detection
   - Wireless sleep monitor/electrocardiogram (ECG) monitor
3. Treatment
   - Thermal therapy (hyperthermia, ablation, etc.)
   - Surgical device (coagulation device, microwave knife, etc.)
   - Electric/magnetic brain or nerve stimulator
   - Cancer treatment based on non-thermal effects

This chapter focuses on therapeutic applications of electromagnetic waves. Among the examples listed above in (3), thermal therapy and surgical devices utilize temperature elevation in the human tissue caused by electromagnetic energy. On the contrary, brain/nerve stimulators and a few types of cancer treatment are based on non-thermal effects. These topics have been actively studied and discussed within some scientific organizations such as International Union of Radio Science (URSI) Commission K [1] and COST (European Cooperation in Science and Technology) MyWAVE [2].

[1]Center for Frontier Medical Engineering, Chiba University, Chiba, Japan

Section 12.2 introduces some of the typical therapeutic applications. Section 12.3 describes new aspects to two major ways of evaluation and validation of medical devices. Finally, Section 12.4 exemplifies a couple of further challenges.

12.2 Typical applications

12.2.1 Hyperthermia

Hyperthermia is one of the practical modalities for cancer treatment, utilizing the difference of thermal sensitivity between tumor and normal tissue [3]. A target tumor should be heated up to the therapeutic temperature between 42 °C and 45 °C without overheating the surrounding normal tissues. Moreover, therapeutic effect of other cancer treatments such as radiation therapy and chemotherapy can be enhanced by combining with hyperthermia.

This subsection further focuses on specific antennas for interstitial microwave hyperthermia. However, antennas and the technique for interstitial heating can also be applied to microwave ablation.

Antennas for interstitial heating must be thin enough. Practically, the diameter of an antenna should be less than 3 mm. Interstitial microwave hyperthermia is usually applied for treatment of large-volumed and deep-seated tumors. During the treatment, an array applicator composed of multiple thin microwave antennas are inserted directly into a target tumor. The number of antennas is decided with the tumor size and shape.

Interstitial microwave hyperthermia can be utilized as an adjuvant therapy to conventional interstitial radiation therapy. In that case, catheters are commonly used for both therapies as shown in Figure 12.1. First, thin microwave antennas are inserted into the catheters. After heating, only the antennas are pulled out from the catheters. Then, radiation sources such as the iridium 192 are automatically inserted into the catheters with a high-dose rate after loading system.

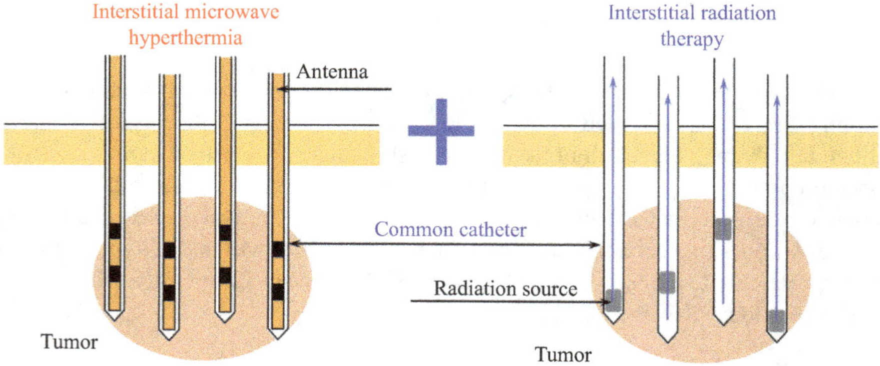

Figure 12.1 Combination of interstitial microwave hyperthermia and interstitial radiation therapy

It is essential that a therapeutic heating pattern must cover the whole target tumor. Regarding the perpendicular direction to the antenna, it is easy to control a heating pattern by adjusting the number of antennas and antenna positions. Concerning the control of a heating pattern along the antenna, it is realized to some extent by modifying the structure of antenna elements.

A schematic diagram of interstitial heating system is shown in Figure 12.2. The temperature in the tumor can be monitored by means of inserted temperature sensors, and the output power of a microwave generator can be automatically adjusted by feedback control.

Different types of interstitial microwave antennas have been developed and reported [4,5]. Figure 12.3 shows some examples of various antennas for interstitial heating. Most antennas are made of thin coaxial cables and designed with specific mechanism to suppress the electric current flowing back on the outer conductor of the coaxial cables. Such outer-surface electric currents may cause unwanted heating of healthy tissue along the antennas. For example, a quarter-wavelength sleeve is attached to the outer conductor as shown in Figure 12.3(b).

The author has been studying a coaxial-slot antenna [6] shown in Figure 12.3 (f) and (g). Figure 12.4 shows the basic structure of the coaxial-slot antenna. The antenna is composed of a thin semi-rigid coaxial cable. Some ring slots are cut in the outer conductor of a thin coaxial cable and the tip of the cable is short-circuited. By adjusting the number of slots or slot positions, its heating pattern along the antenna can be controlled to some degree. For clinical use, the antenna is inserted into a catheter made of polytetrafluoroethylene (PTFE). The operating frequency is 2.45 GHz, which is one of the Industrial, Scientific, and Medical (ISM) bands, for this particular case.

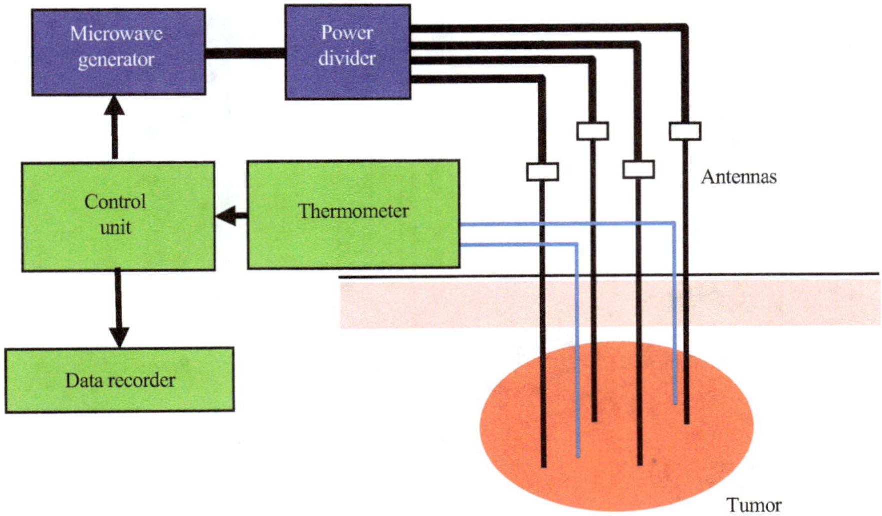

Figure 12.2 Schematic diagram of interstitial heating system

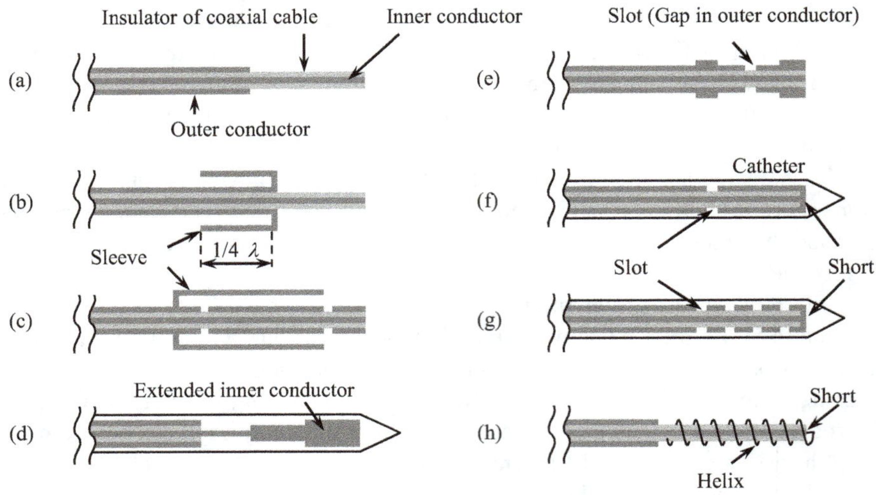

Figure 12.3 Various antennas for interstitial heating

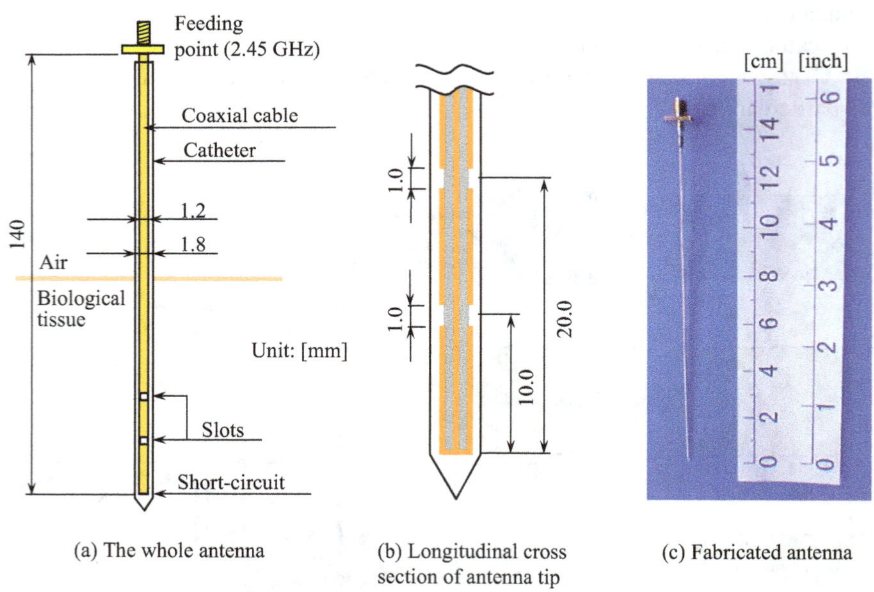

Figure 12.4 Basic structure of coaxial-slot antenna. Reprinted with permission from [6]

Regarding antennas used for telecommunication and broadcasting, radiation pattern, radiation efficiency and antenna gain are important factors. Meanwhile, specific absorption rate (SAR [W/kg]) and temperature distributions in a human body are essential parameters of the antennas for thermal treatment.

Figure 12.5 shows the flowchart of computer simulation for calculating the temperature distribution around the coaxial-slot antenna inserted into the tissue. After calculating an electric field around the antenna, an SAR distribution around the antenna can be derived, where, σ: conductivity of the tissue [S/m], ρ: density of the tissue [kg/m^3] and E: electric field (rms) [V/m]. Finally, by solving the bioheat transfer equation, a temperature distribution in the tissue can be obtained, where, T: temperature [°C], t: time [s], c: specific heat [J/kg·K], κ: thermal conductivity [W/m·K], ρ_b: density of the blood [kg/m^3], c_b: specific heat of the blood [J/kg·K], T_b: temperature of the blood [°C] and F: blood flow rate [m^3/kg·s].

Figure 12.6 shows two examples of calculated SAR distributions around antennas with a single slot and two slots. The coaxial-slot antenna is inserted into a liver tissue ($\varepsilon_r = 43.03$, $\sigma = 1.69$ S/m at 2.45 GHz) for instance. The observation plane is a longitudinal plane (x–z plane) that includes the antenna. Calculated results of the coaxial-slot antenna with a single slot and with two slots are shown for comparison. In both cases, the antenna insertion depth was set as 70 mm which seems sufficient for most clinical cases. From Figure 12.6(a), the high SAR region tailed off along the antenna. On the contrary, in Figure 12.6(b), the high SAR region was confined to the tip of the antenna. Comparing both results, a more localized SAR profile around the tip of the antenna is achieved by use of a coaxial-slot antenna with two slots.

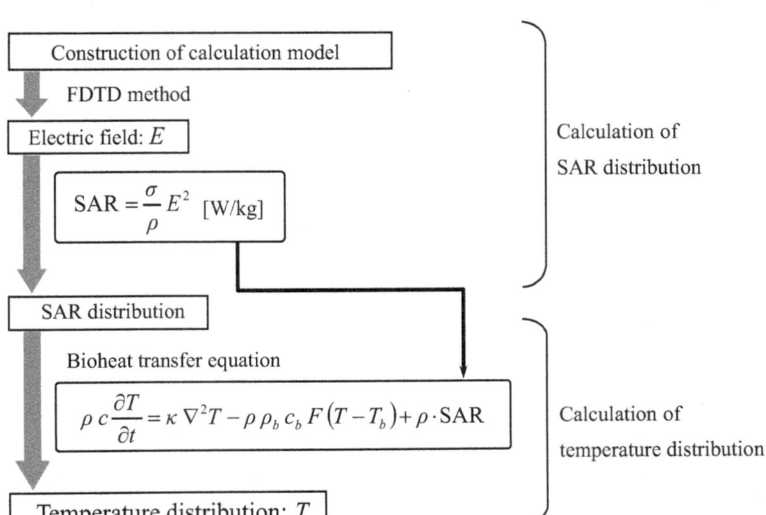

Figure 12.5 Procedure of temperature calculation. Reprinted with permission from [6]

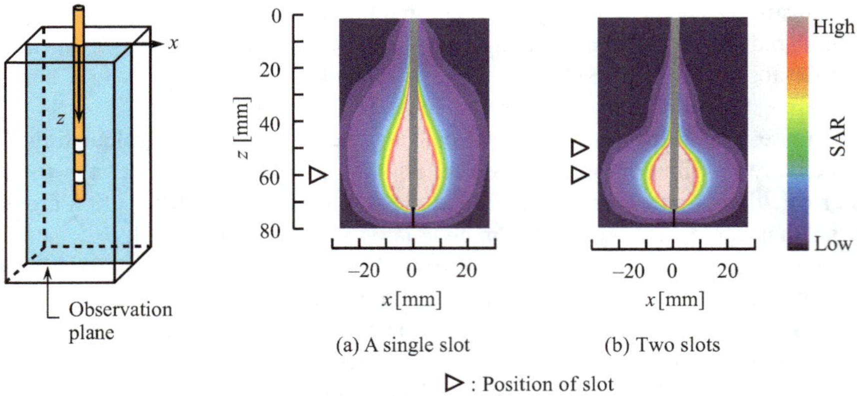

Figure 12.6 Calculated SAR distributions of coaxial-slot antennas

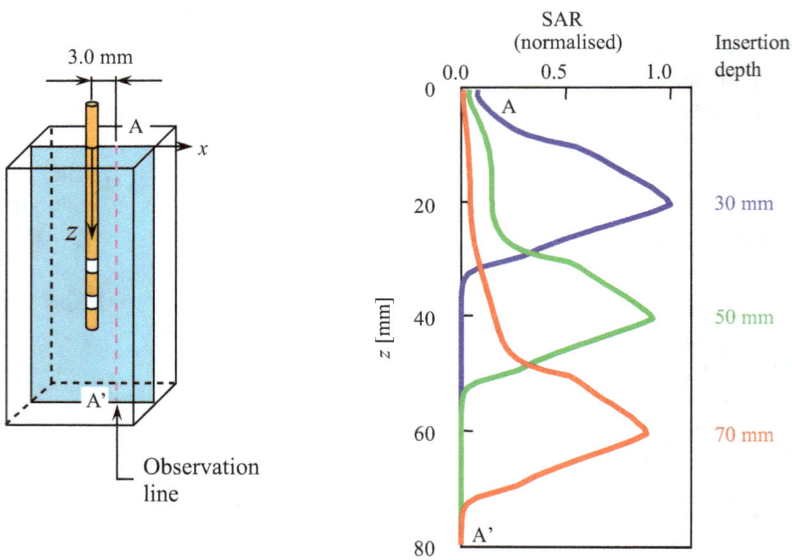

Figure 12.7 SAR distributions with different insertion depths (coaxial-slot antenna with two slots)

In actual treatments, it is very important not only to generate a localized heating region but also to have an independence of the antenna insertion depth on the heating pattern. An influence of the insertion depth on SAR distribution was simulated for the coaxial-slot antenna with two slots. Figure 12.7 shows calculated SAR profiles along the SAR observation line, which is located at a distance of 3 mm away from the center of the antenna in the x–z plane. The insertion depth was changed from 30 mm to 70 mm every 20 mm. From Figure 12.7, it is

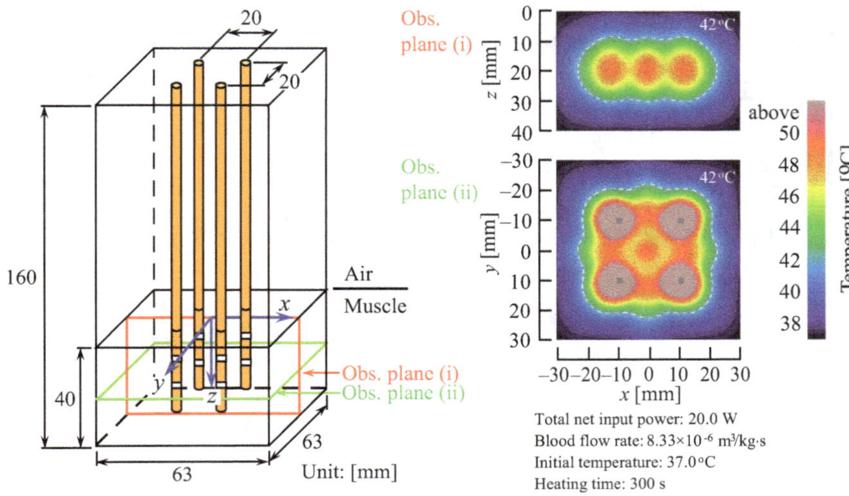

Figure 12.8 A four-element array applicator and its calculated temperature distributions

confirmed that SAR profiles are independent of the antenna insertion depth. Consequently, the SAR value around $z = 0$ (the surface) for the insertion depth of 30 mm was less than 10% of the peak value. It means that a hotspot at the surface caused by the antenna can be neglected. This result is essential for such interstitial heating because a stable heating pattern can be realized regardless of the insertion depth.

Figure 12.8 shows a calculation model of an array applicator consisting of four antennas with two observation planes and its calculated temperature distributions. An antenna spacing and an antenna insertion depth are 20 mm and 30 mm, respectively. In the calculation, the net input power to the antenna is 5.0 W for each element and the heating time is 600 s. In addition, the initial temperature of the biological tissue is set to 37 °C. A relatively large heating region was generated in and around the array applicator. The size of the effective heating region was about 20 mm in a vertical plane and about 40 mm in a horizontal plane. Furthermore, a hotspot at the surface of the tissue was not observed. Hence, it is considered that skin injuries will be well suppressed by use of the proposed antennas.

12.2.2 Microwave ablation

Microwave ablation (MWA) has been applied mainly for treatment of small-sized tumors such as hepatocellular carcinoma [7]. During the treatment, a thin antenna is directly inserted into the tumor to produce heated region which includes the whole cancer cells as shown in Figure 12.9. Unlike hyperthermia, the therapeutic temperature for ablation is basically above 60 °C and its treatment time is around a few minutes. Typical operating frequencies are 915 MHz and 2.45 GHz.

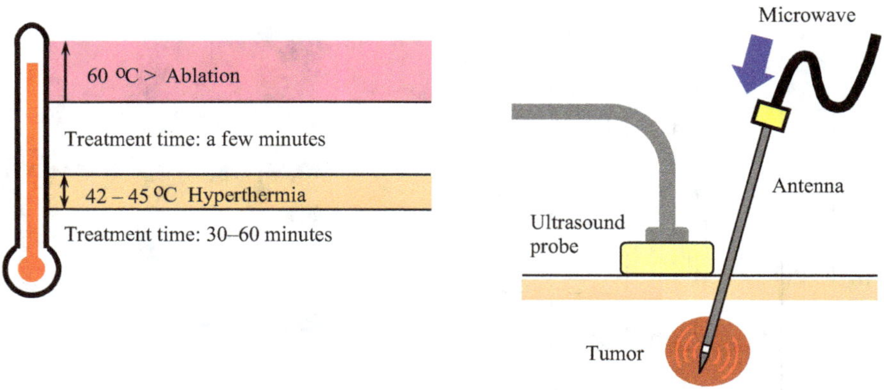

Figure 12.9 Schematic diagram of microwave ablation

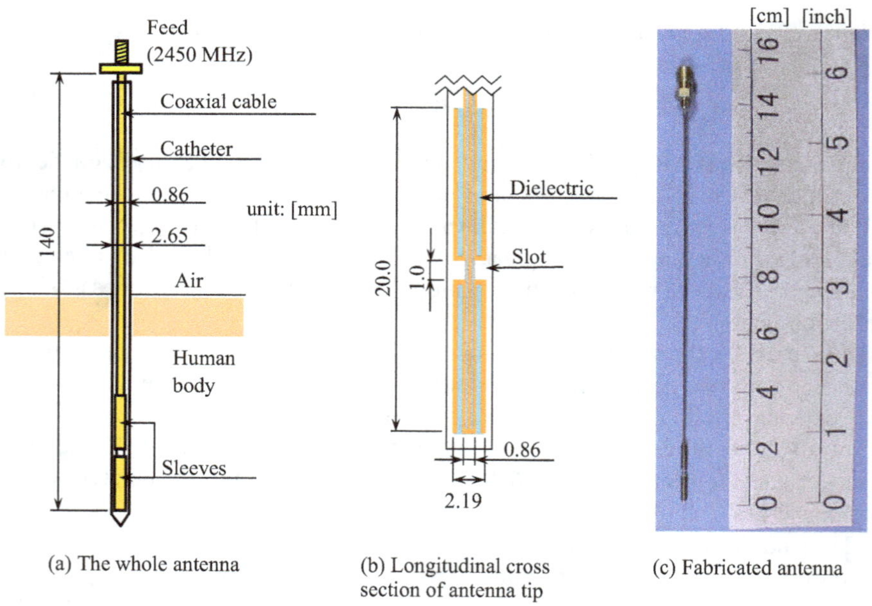

Figure 12.10 Basic structure of coaxial-dipole antenna

In order to generate a controllable heating region along the antenna axis, a coaxial-dipole antenna [8] was developed and tested. Figure 12.10 shows the basic structure of a coaxial-dipole antenna. The size of a heating region along the antenna axis can be controlled by changing the length of the sleeves.

Conventionally, heating performance of antennas for MWA has usually been evaluated experimentally by using a piece of meat as shown in Figure 12.11. It is

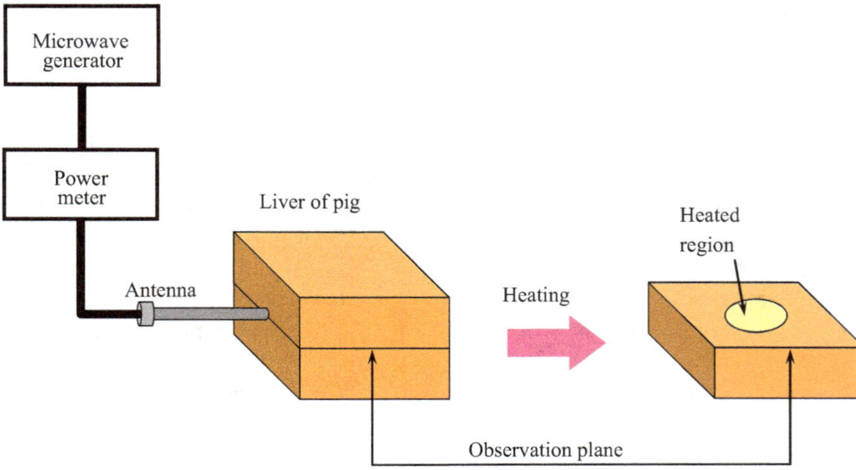

Figure 12.11 Set-up of ablation experiment with meat

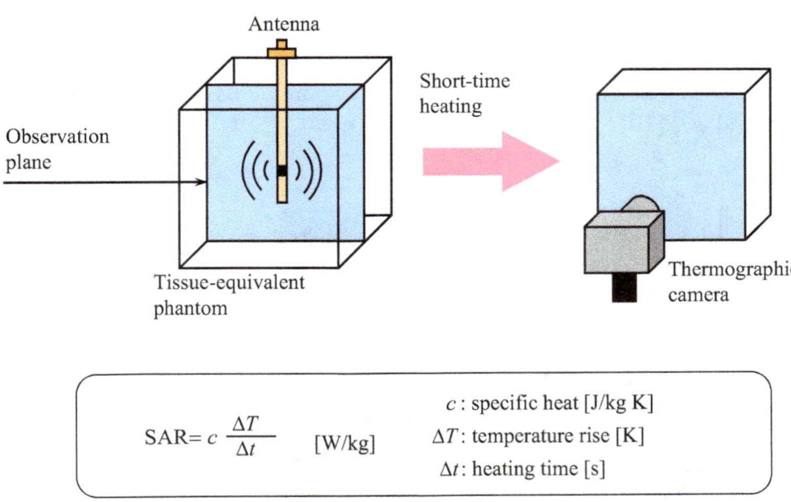

Figure 12.12 SAR measurement system using the thermographic method

rather easy to do such experiments, but difficult to obtain quantitative data and to secure repeatability.

Figure 12.12 illustrates an SAR measurement system using the thermographic method [9]. First, an antenna is placed between the pre-cut phantoms and a high-power microwave energy (20–30 W for example) is supplied to the antenna. After a short time period of heating, the inner surface of the phantom is observed by an infrared camera to capture a thermal image. When the heating time Δt is short enough, typically one minute or less, this technique gives good estimation of SAR distribution.

Figure 12.13 shows three different evaluation results of a coaxial-dipole antenna shown in Figure 12.10. Figure 12.13(b) illustrates the measured SAR distribution by use of the thermographic method which indicated a good agreement with the calculated result shown in Figure 12.13(a). On the contrary, Figure 12.13 (c) shows a heating pattern observed with a set-up shown in Figure 12.11. It should be noted that such a heating pattern largely depends on the experimental conditions.

Most antennas for MWA generate axisymmetric heating patterns around the antennas because of their axisymmetric structure. However, a directional heating pattern may be required for some specific scenarios where, for example, a target tumor is close to an organ which should not be ablated. Different types of MWA antennas with such directional heating patterns have been developed and reported [10–12].

Figure 12.14 illustrates an example of an MWA antenna with directional heating pattern [10]. The applicator consists of a coaxial antenna with a hemi-cylindrical reflector to control the deposition of the electromagnetic energy toward the direction of the target with minimal heating of tissue in the reverse direction. The coaxial antenna and reflector were inserted in a 3.5-mm (outer diameter) catheter filled with circulating water. The antenna is optimized to be well matched at 2.45 GHz. Heating patterns can be controlled by changing its reflector aperture angle (2θ). Figure 12.14 bottom left shows that 2θ is nearly 180°. Figure 12.15 demonstrates simulated temperature patterns in porcine tissue for the two cases of $\theta = 90°$ and 70° [10].

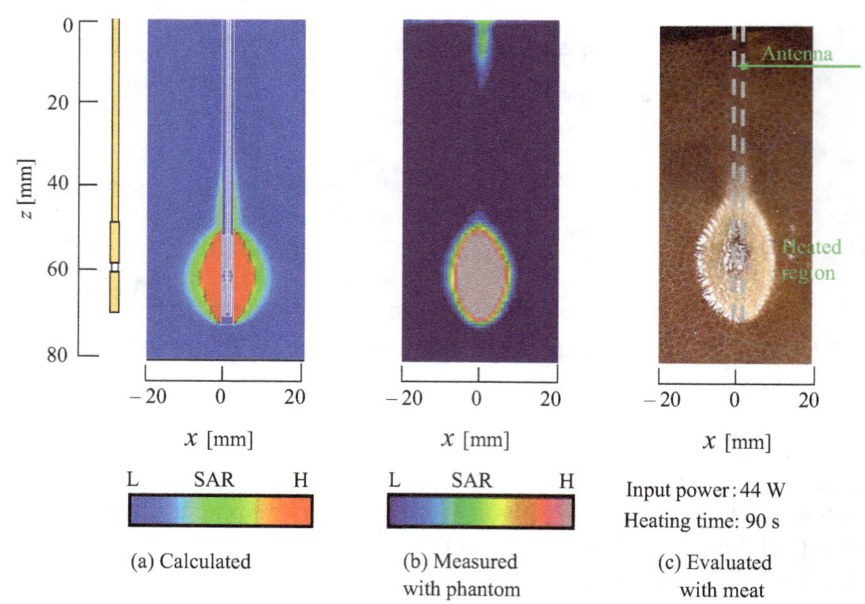

Figure 12.13 Different evaluation results of coaxial-dipole antenna

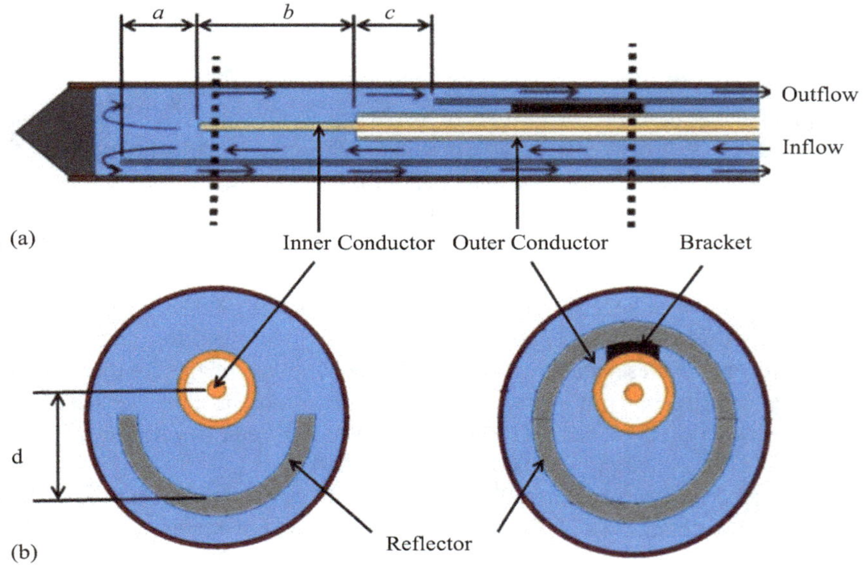

Figure 12.14 Directional antenna with (a) longitudinal side view and (b) transverse view taken through two cross sections. Reprinted with permission from [10]

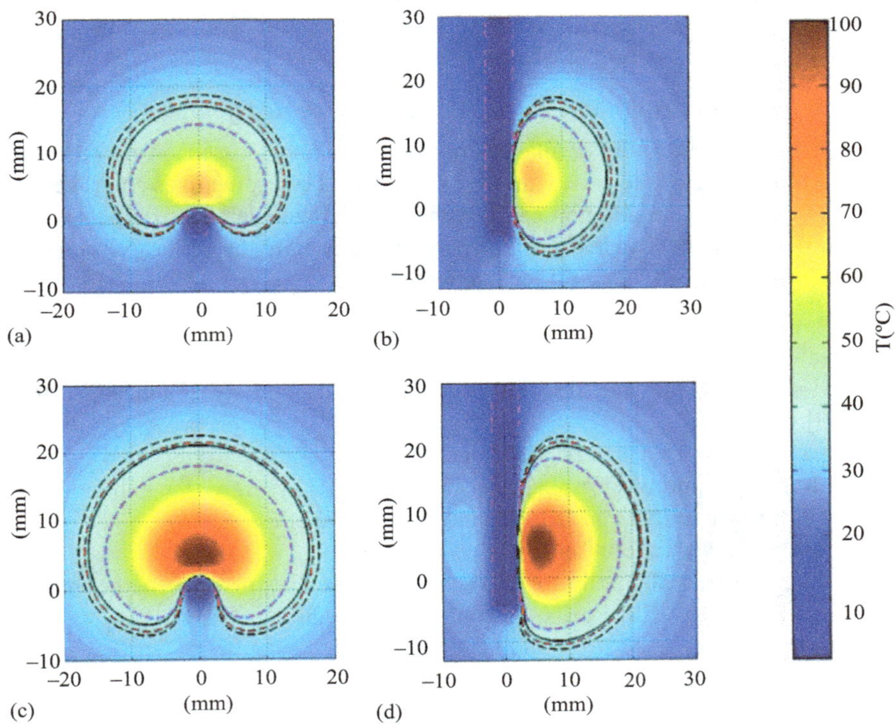

Figure 12.15 Simulated temperature patterns for directional antenna in porcine tissue. (a) $\theta = 90^o$ and (c) $\theta = 70^o$ in transverse plane, and (b) $\theta = 90^o$ and (d) $\theta = 70^o$ in coronal plane (2θ: reflector aperture angle). Reprinted with permission from [10]

In recent years, with the advent of sophisticated medical imaging technologies and an increasing demand for further improving quality of life (QOL) of patients, image-guided MWA has emerged as a minimally invasive therapeutic modality for the treatment of not only unresectable tumors but also cardiac arrhythmias, neuromodulation, and other medical applications.

During the treatment, one of key techniques to MWA is how to guide an antenna to the right position. Computed tomography (CT) has been widely employed as the guidance. Alternatively, ultrasound (US) has been used as the guidance to avoid additional exposure to ionizing radiation and to reduce the cost [13]. However, for these conventional imaging techniques, image contrast may be sometimes insufficient and therefore treatment may not become perfect. More recently, magnetic resonance (MR) guided MWA has been developed as a promising modality, although the cost is rather high [14].

In terms of electromagnetic ablation, besides MWA, other modalities such as radio frequency ablation (RFA) and laser ablation (LA) have been widely applied to clinical use. RFA has two different systems, namely, monopolar and bipolar systems. A monopolar RFA system requires an external electrode to form an electric current path. Both MWA and RFA have advantages and disadvantages in clinical treatment. Comparison between MWA and RFA has been discussed and summarized in, for example [15,16] from clinical viewpoint. LA is another type of minimally invasive modality utilizing the laser light and an optical fiber [17]. MR-guided LA is an option for the treatment of brain tumors in particular.

12.2.3 Cancer treatment based on non-thermal effects

There are lots of literature available on cancer treatment based on non-thermal effects of electromagnetic waves, although they have been sometimes controversial for many years. Electroporation-based techniques utilize short electric pulses that can induce important changes in cell physiology, especially the permeabilization of the cell membrane [18]. Typical techniques include:

- Electrochemotherapy (ECT)
  A safe, easy, and efficient technique for the treatment of solid tumors that uses cell-permeabilizing electrical pulses to enhance the activity of anticancer drug [19].
- Nanosecond pulsed electric fields (nsPEF)
  A drug free, purely electrical cancer therapy. They allow the inhibition of tumor growth, and it can target intracellular organelles.
- Irreversible electroporation (IRE)
  IRE is a new ablation procedure using pulses that provoke the permanent permeabilization of the cells resulting in their death.

Figure 12.16 depicts different ECT electrodes which are suitable for different clinical challenges. Adjustable electrodes allow better support for the needles. After drug administration, the electric pulses are administered through the electrodes that are inserted into the tumor tissue. The electrodes are systematically moved, and pulses applied again, in order to cover the entire tumor area [19].

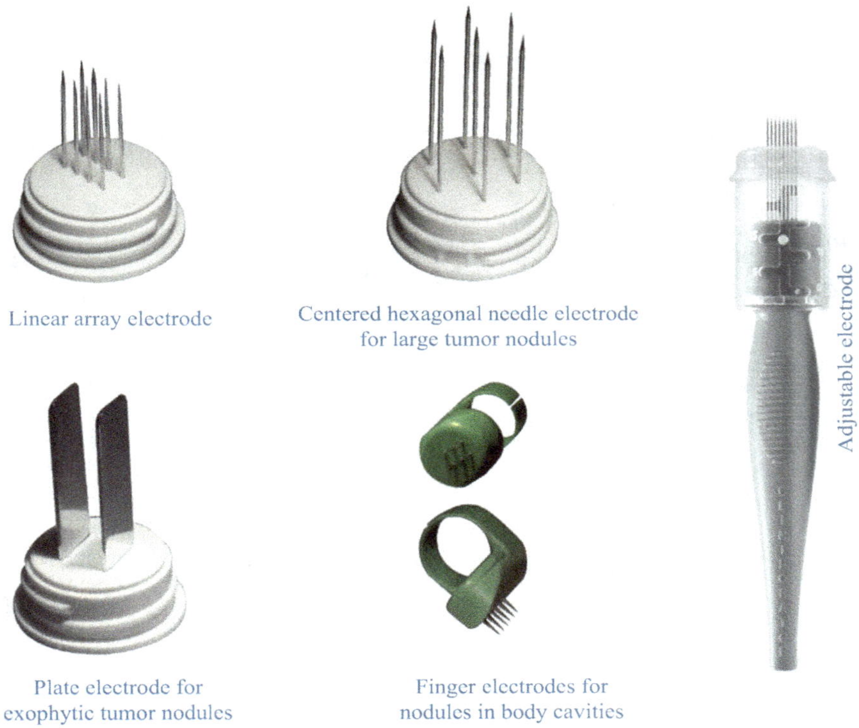

Figure 12.16 Different ECT electrodes suitable for different clinical challenges. Adjustable electrodes allow better support for the needles. Reprinted from [19]

ECT is now in routine clinical use to treat cutaneous metastases. This treatment utilizes electric pulses to permeabilize cell membranes in tumors, thus allowing a dramatic increase of the cytotoxicity of anti-cancer agents. Response rates, often after only one treatment, are very high across all tumor types. The most frequent indications are cutaneous metastases from malignant melanoma and breast cancer [19].

Only recently, Wust *et al.* [20] presented a literature review of preclinical and clinical data on non-thermal antiproliferative effects of various electromagnetic field applications. Further, they suggested and evaluated plausible biophysical and electrophysiological models to decipher non-thermal antiproliferative membrane effects.

Figure 12.17 illustrates a model of a calcium (Ca^{2+}) channel placed in the membrane. The ion channels are assumed to act like diodes because ions can only be directed along the electrochemical gradient (for calcium from extra- to intracellular) [20].

Figure 12.18 left shows a typical tumor environment characterized by isolated tumor cells surrounded by extracellular water. Figure 12.18 right illustrates an

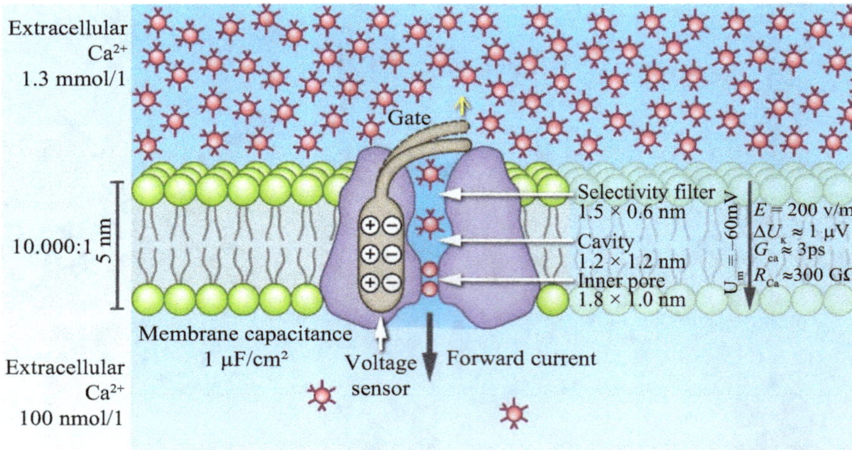

Figure 12.17 Model of a calcium channel placed in the membrane. The ion channels act like diodes with forward direction along the electrochemical gradient (for calcium from extra- to intracellular). Reprinted from [20]

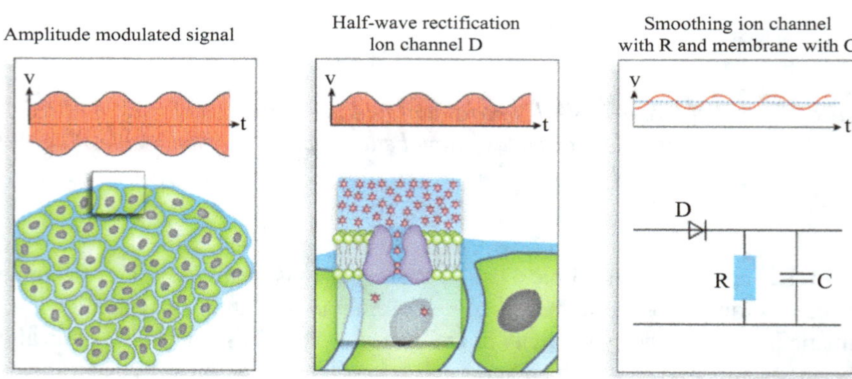

Figure 12.18 Left: Typical tumor environment characterized by isolated tumor cells surrounded by extracellular water. Middle: Simplified model of ion channels in the membrane (Figure 12.17). Right: Equivalent rectifier circuit model of an ion channel. Reprinted from [20]

equivalent rectifier circuit model of an ion channel [20]. The voltage associated with the RF E-field along the channel (i.e., perpendicular to the membrane) is rectified and smoothed. In the case of a pure sinusoidal RF wave ($E = 200$ V/m assumed), a direct voltage of around 1 μV is emerging (see dotted line). The voltage may induce ion fluxes that can potentially cause relevant disequilibrium of most ions. This may result in potential contribution to non-thermal effects. When an RF wave is amplitude-modulated, the rectifier circuit generates a mean DC voltage with a superposed sinus oscillating with the modulation frequency in the kHz range (red solid line).

The above-mentioned ideas and models regarding non-thermal effects could improve future treatments in oncology, although they need further investigation and validation.

12.3 Evaluation and validation

Basically, antennas or wireless devices for medical applications are used close to or in a human body. Therefore, effects of the human body should be included in the development and evaluation of such antennas or devices [21].

Section 12.3 briefly describes new aspects to two major ways of evaluation and validation of antennas or wireless devices, namely, numerical simulation with digital phantoms and experiment with physical phantoms.

12.3.1 Numerical simulation with digital phantoms

Unlike antennas for communications or broadcasting, in general, there are five different steps for the development and evaluation of antennas or wireless devices for medical applications. Namely,

(a) Numerical simulation
(b) Phantom measurement
(c) Meat experiment
(d) Animal experiment
(e) Clinical trial

As for the first step, numerical simulation is generally conducted with sophisticated and precise digital animal or human-body models [22]. Many different types of software as well as digital human-body models are commercially available these days. For instance, Figure 12.19 shows Japanese male and female

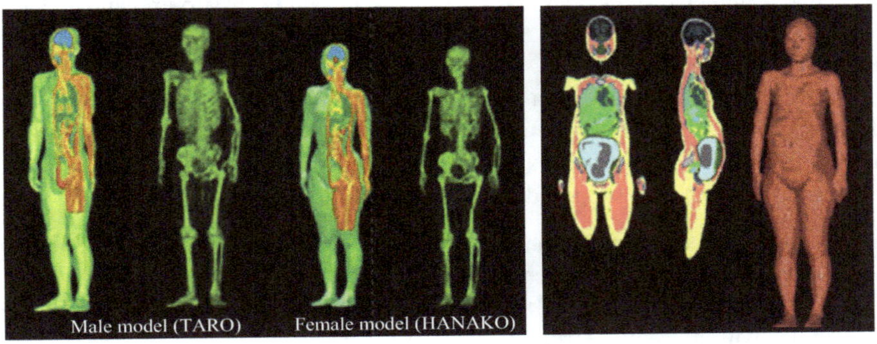

Figure 12.19 Left: Adult male and female voxel models. Right: Pregnant woman voxel model. Copied from [23]

digital models developed by a Japanese research group led by National Institute of Information and Communications Technology (NICT) [23].

The Numerical Breast Phantom Repository [24] is another useful example. This online repository has been developed and provided by the University of Wisconsin Cross-Disciplinary Electromagnetics Laboratory. It contains a number of anatomically realistic MRI-derived numerical breast phantoms for breast cancer detection and treatment applications. The breast tissues in these digital phantoms have the realistic ultrawideband dielectric properties. These phantoms will be of use to the bioelectromagnetics research community free of charge.

Furthermore, multiphysics simulation is a powerful tool for the development and evaluation of medical devices under certain circumstances. Simulation of temperature distribution generated by an antenna for hyperthermia shown in Figure 12.5 is a good example.

Recently, the FDA (Food and Drug Administration in US) has announced the Medical Device Development Tool (MDDT) program [25]. According to this program, once the FDA qualifies tools, sponsors of medical devices can utilize the tools in the development and evaluation of the devices. Sophisticated numerical simulation with a qualified MDDT is expected to eliminate or mitigate much of the risk as well as uncertainty in product development including animal experiments. Therefore, it may accelerate and promote the R&D of medical devices. It also may help bridge the gap between research of medical devices and the delivery of devices to patients.

12.3.2 Experiment with physical phantoms

Following numerical simulation with digital phantoms, experiments using physical human-body phantoms are indispensable for validation of the numerical simulation results or to minimize animal experiments particularly for the R&D of implantable antennas and devices.

Various kinds of physical phantoms have widely been developed to meet different requirements (see Chapter 2 for details). Physical phantoms are typically categorized into four groups, namely, liquid, gel, semi-solid and solid phantoms [22]. In particular, semi-solid phantoms are considered suitable to experiments for implantable antennas or devices. This is because it is quite easy to implant antennas or devices into the right place in the semi-solid phantoms and to hold them without any additional fixtures.

Before moving forward to a clinical trial, it is recommended that the medical devices or systems be tested on reference physical phantoms in order to evaluate and compare their performances under controlled realistic configurations. These reference phantoms are supposed to meet a few following requirements:

1. the size and shape of the phantoms should be similar to those of the human tissues or body parts to be considered,
2. the dielectric properties of the phantoms should mimic those of the human tissues within specific errors, and
3. the structure and dielectric properties of the phantoms should be stable over certain time or easily reproducible for a single-use case.

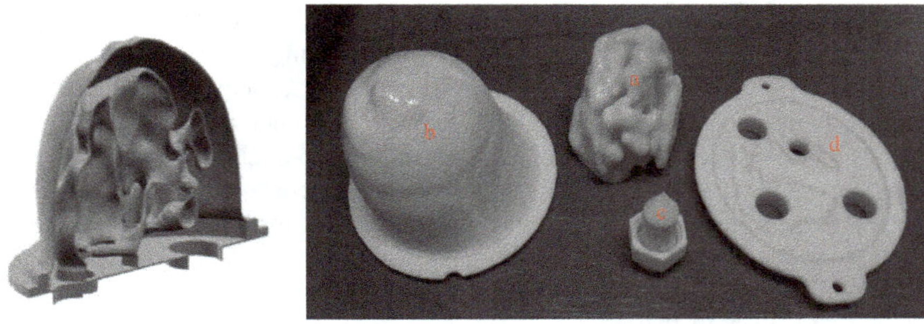

Figure 12.20 *Left: Sagittal section of the breast reference phantom. Right: The different parts of the breast phantom: (a) The inner part contains the fibroglandular or heterogeneous mix tissue mimicking material (TMM), (b) the outer shell contains the fatty TMM, (c) the removable inclusion contains the tumor-like TMM, and (d) the support plate holds the different parts in place. Reprinted from [26]*

Figure 12.20 shows an example of a breast reference phantom [26]. The structure is made of acrylonitrile butadiene styrene (ABS) and built up by using a 3D printer. The data file was obtained by modifying original file which comes from the University of Wisconsin–Madison [24]. It has already been used as a reference phantom by several laboratories to test their measurement systems in the framework of COST Action TD1301 MiMed [27,28].

Physical phantoms sometimes need to mimic not only dielectric properties but also mechanical properties of the human tissues simultaneously to meet specific requirements. For this purpose, semi-solid phantoms are usually employed. Gelatin-based tissue-mimicking materials have been proposed to make more realistic breast phantoms including healthy and cancerous tissue models [29].

12.4 Further challenges

12.4.1 Design of MWA antennas with change of tissue parameters

As described in Section 12.2.2, it is essential for development of MWA antennas and clinical treatment planning to evaluate numerically distributions of SAR as well as temperature in human tissues generated by the antennas. For such numerical evaluation, biological tissues are modeled with their physical properties including dielectric and thermal parameters. Unlike conventional hyperthermia, a significant change of dielectric and thermal parameters must be considered during ablation treatment because the targeted tissue gets desiccated or charred with temperature well exceeding 60 °C. It is obvious that the water content ratio over the targeted tissue or charring the tissue significantly affects physical properties of the tissue.

Several groups have studied and reported numerical modeling and experimental validation of microwave ablation incorporating water vaporization or dehydration [30,31]. Figure 12.21 demonstrates a relationship between measured dielectric constants of swine liver and its water content ratio which strongly depends on the tissue temperature [31]. The relationship obtained was employed in numerical simulation of temperature distributions generated by a coaxial-slot antenna shown in Figure 12.4 (with a single slot for this particular case). Specifically, the relationship was implemented in the calculation procedure shown in Figure 12.5 which was repeated until the end of heating at a time interval of 3 s. Figure 12.22 illustrates simulated temperature and water content ratio under two conditions (condition 1: proposed model, and condition 2: conventional model with fixed dielectric and thermal constants of the tissue) [31]. These results imply that the temperature distribution obtained under the condition 1 was closer to the measured values than those under the condition 2 which ignored a change of the water content ratio of the tissue.

For further improvement, other factors such as a change of blood flow rate and tissue contraction with temperature should be incorporated as well.

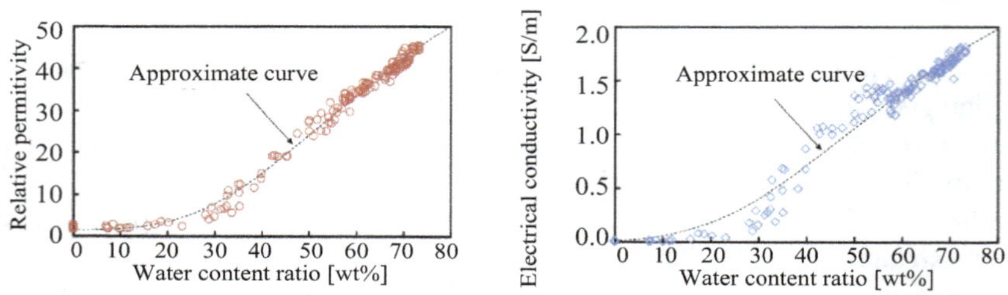

(a) Relative permittivity (b) Electrical conductivity

Figure 12.21 Measured dielectric constants of swine liver versus its water content ratio [31]. Copyright © 2016 IEICE.

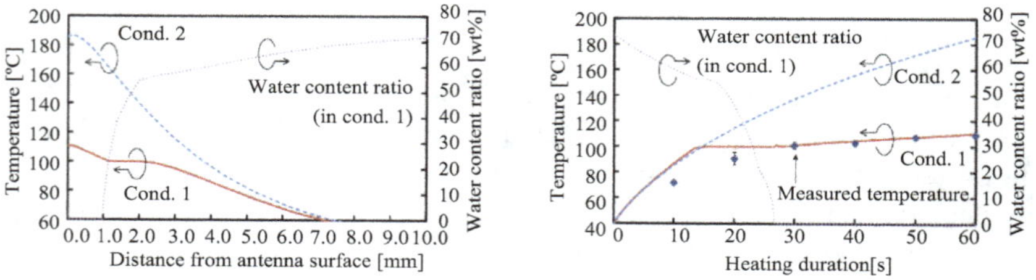

(a) Distributions of temperature and water content ratio along perpendicular line started from the slot after 60 - second heating

(b) Time variation of temperature and water content ratio at the slot point. Measured temperatures are also plotted

Figure 12.22 Simulated temperature and water content ratio under two conditions [31]. Copyright © 2016 IEICE.

12.4.2 Microwave theranostics for cancer treatment

The term "theranostics" is used as a combination of "therapeutics" and "diagnostics." In a clinical situation, drugs, radiation, electromagnetic waves or other techniques are sometimes combined to simultaneously or sequentially diagnose and treat medical conditions.

For example, a compact and thin microwave theranostic device was developed and reported in [32] which can act as an implantable sensor to detect malignant tissue and as an applicator for thermal ablation.

As a proof of concept, a prototype of a microwave theranostic device (length: 12 mm, width: 2 mm) was fabricated as shown in Figure 12.23 [32]. A coplanar waveguide (CPW) was structured on the frontside and two oval-shaped Split Ring Resonators (SRRs) were made on the backside. The SRRs are working in a frequency range between 8 GHz and 12 GHz. For the detection mode, resonance frequency changes due to tissue abnormalities are evaluated. In the treatment mode, SRRs perform microwave ablation. Figure 12.24 shows a piece of liver tissue after thermal ablation. The ablation zone can be extracted by evaluating the color change of the tissue [32].

Such theranostic devices seem promising for cancer treatment in the future. They would be quite beneficial not only to hospitals but also to patients, although much further work should be done to realize these devices in clinical use.

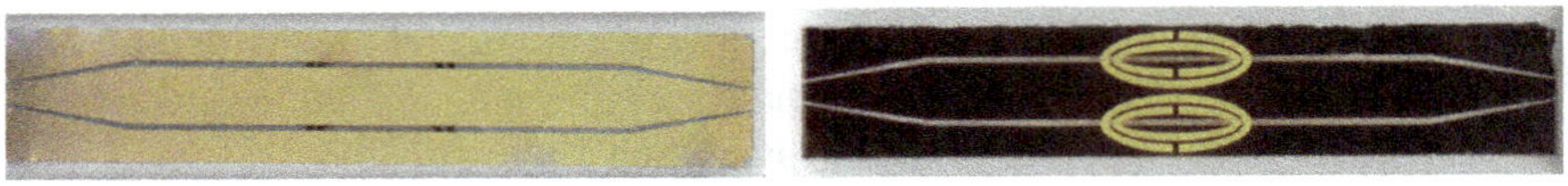

(a) (b)

Figure 12.23 Prototype of microwave theranostic device (length: 12 mm, width: 2 mm). (a) Frontside is structured with CPW line and (b) backside is structured with two oval SRRs. Reprinted from [32]

(a) (b) (c)

Figure 12.24 Liver tissue after thermal ablation. The ablation zone can be extracted by evaluating the color of the tissue. (a) Ablated liver tissue in the container of the holding device, (b) front side of ablated liver tissue, and (c) cross section of liver tissue to extract penetration depth. Reprinted from [32]

References

[1] URSI (International Union of Radio Science) Commission K [online]. Available from https://www.ursi.org/commission.php?id=K [accessed 30 Sep 2021].

[2] COST (European Cooperation in Science and Technology) MyWAVE [online]. Available from https://www.um.edu.mt/projects/mywave/ [accessed 30 Sep 2021].

[3] Seegenschmiedt M.H., Fessenden P., and Vernon C.C. (eds.). *Thermoradiotherapy and Thermochemotherapy*. Berlin: Springer-Verlag; 1995.

[4] Lin J.C., and Wang Y.J. 'Interstitial microwave antennas for thermal therapy'. *International Journal of Hyperthermia*. 1987; 3(1): 37–47

[5] Hamada L., Saito K., Yoshimura H., and Ito K. 'Dielectric-loaded coaxial-slot antenna for interstitial microwave hyperthermia: longitudinal control of heating patterns'. *International Journal of Hyperthermia*. 2000; 16(3): 219–29.

[6] Ito K., and Saito K. 'Antenna issues in microwave thermal therapies' in Chen Z.N. (ed.). *Antennas for Portable Devices*. West Sussex, England: John Wiley & Sons; 2007. pp. 169–96.

[7] Saito K., and Ito K. 'Preliminary investigation of numerical estimation of coagulated region generated by interstitial microwave antenna'. *International Journal of Hyperthermia*. 2017; 33(1): 69–73.

[8] Iwata K., Udagawa K., Wu M.S., Ito K., and Kasai H. 'A basic study of coaxial-dipole applicator for microwave interstitial hyperthermia'. In *Proceedings of 12th Annual meeting of the Japanese Society of Hyperthermic Oncology*; Tokyo, Japan, Sep 1995. pp. 230–31.

[9] Okano Y., Ito K., Ida I., and Takahashi M. 'The SAR evaluation method by a combination of thermographic experiments and biological tissue-equivalent phantoms'. *IEEE Transactions on Microwave Theory and Techniques*. 2000; 48(11): 2094–103.

[10] McWilliams B.T., Schnell E.E., Curto S., Fahrbach T.M., and Prakash P. 'A directional interstitial antenna for microwave tissue ablation: theoretical and experimental investigation'. *IEEE Transactions on Biomedical Engineering*. 2015; 62(9): 2144–50.

[11] Mohtashami Y., Hagness S.C., and Behdad N. 'A hybrid slot/monopole antenna with directional heating patterns for microwave ablation'. *IEEE Transactions on Antennas and Propagation*. 2017; 65(8): 3889–96.

[12] Sebek J., Curto S., Eaton-Evans J., *et al.* 'Feasibility assessment of microwave ablation for treating esophageal varices'. *Journal of Medical Devices*. 2017; 11(3): 031013 (8 pages).

[13] Yu J., Wu H., Meng X.W., *et al.* 'Ultrasound-guided percutaneous microwave ablation of central intraductal papilloma: a prospective pilot study'. *International Journal of Hyperthermia*. 2019; 36(1): 605–11.

[14] Li Z., Wang C., Si G., *et al.* 'Image-guided microwave ablation of hepatocellular carcinoma (≤ 5.0 cm): is MR guidance more effective than CT guidance?' *BMC Cancer*. 2021; 21: 366 (9 pages).

[15] Vorländer C., Kohlhase K.D., Korkusuz Y., *et al.* 'Comparison between microwave ablation and bipolar radiofrequency ablation in benign thyroid nodules: differences in energy transmission, duration of application and applied shots'. *International Journal of Hyperthermia*. 2018; 35(1): 216–25.
[16] Aufranca V., Farouil G., Abdel-Rehimb M., *et al.* 'Percutaneous thermal ablation of primary and secondary lung tumours: comparison between microwave and radiofrequency ablation'. *Diagnostic and Interventional Imaging*. 2019; 100(12): 781–91.
[17] Schena E., Saccomandi P., and Fong Y. 'Laser ablation for cancer: past, present and future'. *Journal of Functional Biomaterials*. 2017; 8(2): 19 (15 pages).
[18] Al-Sakere B., André F., Bernat C., *et al.* 'Tumour ablation with irreversible electroporation'. *PLoS One*. 2007; 2(11): e1135 (8 pages).
[19] Gehl J., Sersa G., Matthiessen L.W., *et al.* 'Updated standard operating procedures for electrochemotherapy of cutaneous tumours and skin metastases'. *Acta Oncologica*. 2018; 57(7): 874–82.
[20] Wust P., Stein U., and Ghadjar P. 'Non-thermal membrane effects of electromagnetic fields and therapeutic applications in oncology'. *International Journal of Hyperthermia*. 2021; 38(1): 715–31.
[21] Ito K., Takahashi M., and Saito K. 'Small antennas used in the vicinity of human body'. *IEICE Transactions on Communications*. 2016; E99-B (1): 9–18.
[22] Hao Y., Ito K., and Hall P.S. 'Electromagnetic properties and modeling of the human body'. In Hall P.S., Hao Y. (eds.). *Antennas and Propagation for Body-Centric Wireless Communications*. Norwood, MA: Artech House; 2012. pp. 17–61.
[23] NICT. Information on the provision of whole-body voxel human models [online]. Available from https://emc.nict.go.jp/bio/data/index_e.html [accessed 30 Sep 2021].
[24] Numerical Breast Phantom Repository, University of Wisconsin–Madison [online]. Available from https://uwcem.ece.wisc.edu/phantomRepository.html [accessed 30 Sep 2021].
[25] Medical Device Development Tools (MDDT), Food and Drug Administration, USA [online]. Available from https://www.fda.gov/medical-devices/science-and-research-medical-devices/medical-device-development-tools-mddt [accessed 30 Sep 2021].
[26] Joachimowicz N., Duchêne B., Conessa C., and Meyer O. 'Anthropomorphic breast and head phantoms for microwave imaging'. *Diagnostics*. 2018; 8(4): 85 (12 pages).
[27] COST MiMed [online]. Available from https://www.cost.eu/actions/TD1301/ [accessed 30 Sep 2021].
[28] Conceição R.C., Rodrigues D.B., Oliveira B.L., Koutsoupidou M., and Ruvio G. 'Overview of microwave medical applications in Europe since the beginning of the COST Action TD1301 – MiMed'. In *Proceedings of the 11th European Conference on Antennas and Propagation*; Paris, France, Mar 2017. pp. 2714–718.

[29] Cannatà A., Di Meo S., Morganti S., Matrone G., and Pasian M. 'Gelatin-based tissue-mimicking materials for breast phantoms: Dielectric and mechanical characterization'. In *Proceedings of the 34th General Assembly and Scientific Symposium of the International Union of Radio Science*; Rome, Italy, Sep 2021. We-K04-PM3-2

[30] Chiang J., Birla S., Bedoya M., Jones D., Subbiah J., and Brace C.L. 'Modeling and validation of microwave ablations with internal vaporization'. *IEEE Transactions on Biomedical Engineering*. 2015; 62(2): 657–63.

[31] Endo Y., Saito K., and Ito K. 'Temperature analysis of liver tissue in microwave coagulation therapy considering tissue dehydration by heating'. *IEICE Transactions on Electronics*. 2016; E99-C(2): 257–65.

[32] Reimann C., Puentes M., Maasch M., *et al.* 'Planar microwave sensor for theranostic therapy of organic tissue based on oval split ring resonators'. *Sensors*. 2016; 16(9): 1450–62.

Chapter 13

Optoelectronic sensing of physiological monitoring from contact to noncontact and point to imaging

Sijung Hu[1] and Yasmin Elsahar[1]

13.1 Summary

Opto-physiological monitoring (OPM) considers biological tissue as a set of optical media, and studies how light interacts within biological tissue, where the optical properties of the latter reflect the mechanical, physical, and biochemical functions of the living organism. This kind of physiological monitoring includes optoelectronic sensor-based contact monitoring and image sensor-based remote monitoring.

OPM has a clear scientific approach, with interest in the photonics and dynamics governing the interaction of light and biological tissues, as well as the practical engineering implementation required to translate laboratory results into commercial products. OPM research, including contact and imaging sensing work, is mainly specialized in photonics based biomedical engineering, i.e., photoplethysmography (PPG), tissue optics engineering simulation, light scattering, etc., and looks in depth at tissue optics, microelectronics, software design and systems integration at component and systems levels, combined with a problem solving approach allowing for a clear and rapid progression from fundamental research ideas to industrial prototypes [1].

The Beer–Lambert (B–L) model underpins the working principle of the current generation of PPG devices as the measurement site is treated as a blood-filled cuvette with no scattering effects and the light sources are assumed to be monochromatic. Accounting for motion artefacts is one of the key challenges in PPG-based biomedical monitoring techniques. The reduction of this phenomenon has been investigated through different approaches to improve the signal to noise ratio (SNR) and to obtain a robust data readout for reliable health assessment.

Wearable reflection-mode optoelectronic sensors (in contact with the skin, in the form of bands or watches e.g., Apple Watch, Fitbit, Samsung Galaxy, Huawei Watch GT, Fossil Gen, Mobvoi) have been recently gaining widespread attention.

[1]Wolfson School of Mechanical, Electrical, and Manufacturing Engineering, Loughborough University, Loughborough, UK

This is due to their potential to offer continuous measurement and monitoring of valuable physiological status information, including heart rate (HR), oxygen saturation (SpO_2) and blood pressure. Such devices could be used to monitor and to protect the health of the users.

An imaging photoplethysmography (iPPG) as an emerging and non-contact imaging sensing technology has rapidly developing and spreading out with its a great potential of vial sign monitoring and assessment associated with a massive demand of imaging technologies, i.e., smartphone, automotive, etc. [2]. However, the performance of these PPG devices has not yet reached the required level of health monitoring and assessment to comply with the relevant clinical monitoring standards, due to their unreliable measurements.

OPM has provided a new paradigm of the B–L model driven current generation of PPG devices. This chapter details optoelectronic sensing of OPM including contact and non-contact imaging, and introduces the principles of OPM, a new approach that overcomes the limitations of PPG-based sensors. The chapter covers the following aspects:

(I) Review of the measurement capabilities of present PPG sensors.
(II) Navigation of OPM principles for improved performance of the optoelectronic sensor and its system.
(III) Outline of the emerging OPM technologies of contact and non-contact monitoring.
(IV) Application case studies

13.2 Opto-physiological interaction

13.2.1 Radiative transport theorem

Light propagation in tissue from the wave perspective can be described by using multiple scattering electromagnetic theory [3]. Tissue can be considered a random medium whose permittivity, $\varepsilon(r)$, fluctuates with respect to position in 3D space. However, the lack of readily applied models means that this formulation finds little applicability in tissue optics [4]. The remaining models rely on the photon perspective of light, where the rigorous mathematical description provided by the Radiative Transport Theorem (RTT) contrasts with the simplistic B–L model. The following formulation of the radiative transport equation considers monochromatic light and disregards polarization [5]:

$$\frac{\partial I(r,s)}{\partial s} = -\mu_t I(r,s) + \frac{\mu_s}{4\pi}\int_{4\pi} I(r,s')f(s,s')d\Omega' \quad (13.1)$$

The first term represents the net change in energy due to energy flow, where is the radiance at a position r in direction s. The second term represents the radiance lost due to absorption and scattering, where $\mu_t = \mu_s + \mu_a$, μ_s is the scattering coefficient and μ_a is the absorption coefficient. The last term represents the gain in radiance due to scattered light incident from all directions, where is the phase function that describes anisotropic scattering and $d\Omega$ is the solid angle. An anisotropy factor g represents the mean cosine of the scattering angle $(s.s')$ and is used to describe the scattering anisotropy of a medium, where $g = 0$ represents isotropic scattering and $g \rightarrow 1$ as scattering becomes more forward peaked. The diffusion approximation, also known as the P1 approximation [6], assumes isotropic scattering through all media by using the transport scattering coefficient $\mu_{s'} = \mu_s(1 - g)$.

13.2.2 Multi-layered tissue structure with capillary and peripheral blood vessels

Referring to light propagation of RTT, the OPM can be formulated along with the B–L law based PPG to alternate approaches to the modeling of absorption, scattering, and optical path length in living tissue. The models presented in [7] and below are a progression from the standard B–L Model used for PPG in pulse oximetry, allowing the inclusion of the fast-growing tissue optics data. Assuming that the absorption is proportionally increased according to the effective path length increase due to scattering, it is possible to model the attenuation as a factor of absorption and mean optical path length:

$$I(\lambda) = I_0 e^{(\mu_a(\lambda) \times L)} \tag{13.2}$$

where μ_a is the absorption coefficient, and the mean path length L is a function of the reduced scattering coefficient μ'_s. Treating the measuring site as a geometrically static structure and the absorption coefficient as a dynamic optical property $\mu_a(t)$ with a static component $\mu_a(static)$ yields:

$$\mu_a(\lambda, t) = \mu_{a(static)}(\lambda) \times (1 + \sigma\rho(wt)) \tag{13.3}$$

where $\rho(wt)$ is the function of a normalized pulsatile waveform and σ is a coefficient of pulsatility. The dynamic component of the absorption can be attributed to the fractional change in blood volume arising from arterial blood pulsations, and the static component can be attributed to the combination of the non-pulsatile fraction of arterial blood volume and the venous blood. In this case, venous blood is treated as a static component of absorption and assuming the case where measurements are taken from a subject at rest. The inclusion of time variance in the model can be achieved in a manner more representative of the physical system (real tissue) through the use of a time-varying mean path length $L\rho(wt)$:

$$I(\lambda) = I \times \exp(-\mu_a(\lambda)L\rho(wt)) \tag{13.4}$$

The concepts underlying OPM modeling are considered in the context of transmission and reflection-mode PPG. These are summarized in Sections 13.1.2.1 and 13.1.2.2.

13.2.2.1 Transmission mode

An effective OPM model for transmission-mode PPG on a human finger was proposed and verified through a designated protocol [3], based on the use of mean path

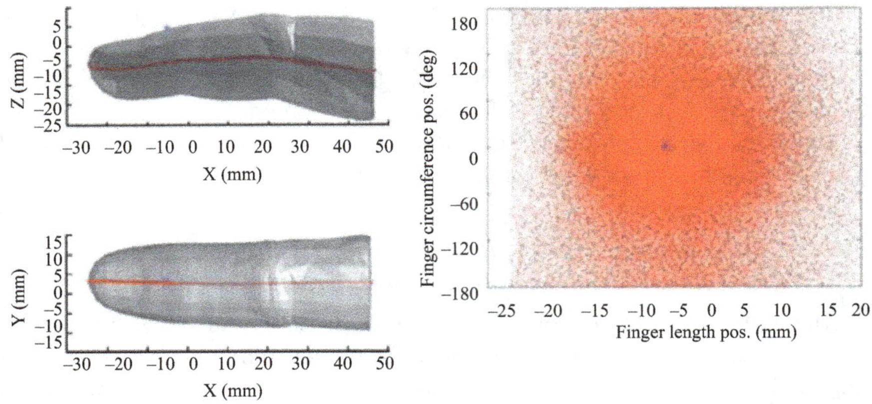

Figure 13.1 Outer skin model with central finger axis highlighted and resultant plot after conversion of transmitted ray coordinates, where the light source position is marked by a star, from [7]

length to account for scattering effects [8,9]. The model considers a set of physiologically significant layers on the measuring site and their corresponding absorption (μ_a), scattering (represented as detector-dependent mean path lengths *L(det))* and pulsatility coefficients (σ), as a function of wavelength (λ), layer (i), longitudinal (x) and circumferential (θ) positions:

$$\frac{I_{(AC)(peak)}}{I_{DC}}(\lambda, x\theta) = \exp\left(\sum_{i=1}^{n}((1+\sigma(\lambda,i))\mu_a(\lambda,i)L_{det}(\lambda,i))\right) \tag{13.5}$$

The data conversion algorithms map the distribution of ray exit points in 3D space (x, y, z) to a cylindrical coordinate system, as would be seen on a flattened outer skin surface (x, y). This is achieved by generating a central axis in 3D coordinates along the center of the finger model and converting the cross-sectional coordinates into equivalent angles, resulting in a surface plot of position along length in mm versus position along circumference in degrees, as shown in Figure 13.1.

13.2.2.2 Reflection mode

In the refection model of tissue, the individual layers are assumed to be homogeneous, and each layer is characterized by a bulk absorption coefficient μ_a, scattering coefficient μ_s, anisotropy factor g and refractive index n. Contrary to this assumption, the layers that make up tissue are neither homogeneous nor well-defined and planar. Nonetheless, results of a previous study [10] indicate that the macroscopic optical properties of whole tissue samples and tissue homogenates are similar. The radiative transfer equation (RTE) provides a highly accurate mathematical description of the interaction of light with a physical system.

To model reflection-mode PPG, the layered absorption μ_a as a function of wavelength λ and layer k, and the detector-dependent pulsatile mean path length L_{AC} as a function of layer k and time t are considered for an arbitrary detector position, where the dynamic component $I_{(AC)(k)}$, representing specific layer blood perfusion can be derived as:

$$I_{(AC)}(k,t) = I_{(AC)} \times \frac{\mu_{a(AC)}(k,\lambda) \times L_{(AC)}(k,t)}{\sum_{i=1}^{n}(\mu_{a(AC)}(i,\lambda) \times L_{(AC)}(i,t)} \tag{13.6}$$

where $I_{(AC)}$ is the pulsatile component of the reflection signal acquired from the arbitrary detector position. By considering all detector positions over the designated tissue, 3D mapping of segment blood perfusion can be achieved. Consequently, the static blood perfusion component of a specific layer k can be derived with its weight $w_{(DC)}(k)$ and the total static component as:

$$\begin{aligned} I_{(DC)}(K) &= I_{(DC)} \times w_{(DC)}(k) = I_{(DC)} \times \frac{\exp[-\mu_{a(DC)}(k,\lambda) \times L_{DC}(k)]}{\Pi_i \exp[-\mu_{a(DC)}(i,\lambda) \times L_{DC}(i)]} \\ &= \frac{I_{(DC)}}{\Pi_{i' \neq i}^{k} \exp[-\mu_{a(DC)}(i',\lambda) \times L_{DC}(i')]} \end{aligned} \tag{13.7}$$

where $\mu_a(DC)$ is the wavelength-dependent absorption coefficient of static component of tissue in mm^{-1}, and $L_{(DC)}$ is the static component of mean path length (MPL) in mm. Similarly, the perfusion $I_{(AC)}$ of a specific layer k can be expressed as the product of the total perfusion and $w_{(AC)}(k)$:

$$\begin{aligned} I_{(AC)}(k) &= I_{(AC)} \times w_{(AC)}(k) = I_{(AC)} \times \frac{\mu_{a(AC)}(k,\lambda) \times L_{(AC)}(k)}{\sum_i \mu_{a(AC)}(i,\lambda) \times L_{(AC)}(i)} \\ &= \frac{I_{(AC)} \times L_{(AC)}(k)}{\sum_i L_{(AC)}(i)} \end{aligned} \tag{13.8}$$

where $\mu_{a(AC)}$ is the wavelength-dependent absorption coefficient of arterial blood in mm^{-1}, and, $L_{(AC)}$ is the dynamic component of mean path length in mm [7].

13.2.3 Tissue optical properties (μ_a, μ_s, g) along with multi-layered tissue

Hemoglobin is the dominant absorber of light in the dermis. Normal adult hemoglobin (Hb A) is a protein consisting four polypeptide chains, each of which is bound to a heme. Within the visible region, Hb A contains three distinctive peaks. The dominant peak is in the blue region of the spectrum and is referred to as the Soret peak or Soret band. Two further peaks can be distinguished in the green-yellow region, between 500 and 600 nm, that in combination with the Soret band cause Hb A to appear red. Melanins are ordinarily contained within the epidermis and produce an absorption spectrum that gradually decreases from the ultraviolet (UV) to the infrared (IR) regions [11]. The optical window of human tissue in which the illumination is greatly absorbed by melanin is within the ultraviolet band (10–400 nm), whereas the water substance is strongly absorbed in the range of longer wavelength. Generally, the optical window (405–1,064 nm) is often selected for opto-physiological measurements, such as the typical application of pulse oximeter [12].

On the other hand, scattering describes a change in the direction, polarization or phase of light and is commonly portrayed as either a surface effect (such as reflection or refraction) or as an interaction with a small region whose optical properties differ from its surroundings (particulate scatter). It has been estimated that 4–7% of visible light is reflected from the surface of the skin, independent of wavelength and skin color. The primary sources of particulate scatter within the skin are filamentous proteins. Keratins are the filamentous proteins of the epidermis and form this layer's major constituent, whereas collagen is the principal filamentous protein of the dermis and occupies approximately 18–30% of its volume. Further scatter is attributed to melanosomes in the epidermis, cell nuclei, cell walls and many other structures in the skin that occur in smaller numbers [11]. The absorption and scattering coefficients for the 633 nm and 850 nm wavelengths are presented in Table 13.1.

Table 13.1 Optical properties, absorption and scattering coefficients, at selected wavelengths, from [13]

	633 nm		**850 nm**		**References**
	$\mu_s(mm^{-1})$	**$\mu_{s'}(mm^{-1})$**	**$\mu_s(mm^{-1})$**	**$\mu_{s'}(mm^{-1})$**	
Epidermis	0.43	2.25	0.22	1.69	
Dermis	0.27	3.37	0.14	2.52	Tuchin [14]
Dermal with plexus superficialis	0.33	3.46	0.12	2.59	$\mu_{s'(850)} = 0.75\mu_{s(633)'}$ [15]
Dermis	0.27	3.37	0.14	2.53	$\mu_{a(850)} = 0.50\mu_{a(633)}$ [15]
Dermis with plexus profundus	0.34	3.49	0.12	2.62	
Subcutaneous fat	0	1.2	0	0.98	Tuchin [14] $\mu_{s'(850)} = 0.82\mu_{s'(633)}$ [15]
Nerve fiber	0.06	1.8	0.08	1.7	Tuchin [14]
Muscle	0.1	0.53	0.05	0.35	Tuchin [14]
Whole blood $SaO_2 = 97\%$, hct = 0.41	0.66	1.7	1.3	1.58	Tuchin [14] $\mu_{s'(850)} = 0.75\mu_{s(633)'}$ [16] $\mu_{a(850)} = 1.97\mu_{a(633)}$ [17]
	Reflection (%)	Absorption (%)	Reflection (%)	Absorption (%)	
Bone	0.54	0.22	0.54	0.18	Ugnell [18]

13.2.4 Modeling of opto-physiological interaction in contact and non-contact

An accurate 3D anatomical model (Zygote Media Group, USA) of a male adult finger composed of bones, tendons, nerves, arteries, veins, and outer skin was hired in this investigation. The platform can be sectioned into three phases, each of which rely on an available software package, as shown in Figure 13.2. 3D Studio MAX v.9.0 (Autodesk, USA) was employed to make modifications to the base model prior to simulation. These included the creation of five-layer skin, the creation of subcutaneous fat as a filler between the skin layers and all other internals, as well as the generation of separate files for each of the objects in the model. MC ray tracing was performed by OptiCAD 10.0 (Opticad Corporation, USA). Post-processing was performed in MATLAB® 7.1 (Mathworks Co., USA) algorithms customized for the output data of OptiCAD [13].

The simulation results presented in [13] are representative of the in vivo measurement of a transmission-mode pulse oximeter on a measuring site with an arterial oxygen saturation of 97%. Sensor responses were generated by scanning a square bucket of arbitrary size in 0.1 mm steps from -1 to 1 mm longitudinally and from 0 to 360° circumferentially. All results are averaged over the longitudinal axis. The SNR in a photoplethysmographic signal is proportional to the peak-to-peak AC intensity. Figure 13.3 shows that the strongest pulsatile signals can be found at $\pm 60°$ to 90° of the light source position, being approximately 3 times those found at the standard sensor position of transmission pulse oximetry ($\pm 180°$).

Figure 13.4 shows the ratio of ratios R generated from both a 2 mm sq. and a 4 mm sq. sensor and scaled to 97% of SpO_2 ($R = 0.5$). A scaling factor $S = \alpha/\beta \approx 0.46$ was required to achieve the required saturation. It can be seen that R drops by up to 0.2, representing a 5% increase in SpO_2. This reflects the expected unreliability of oxygen saturation readings between $\pm 45°$ of the light source position due to the simulation only taking into account the pulsatility of the main blood vessels. Conversely, saturation readings show stability at 97% with an error of $\pm 2.5\%$ for the 2 mm sensor and of $\pm 1\%$ for the 4 mm sensor in the remaining circumference.

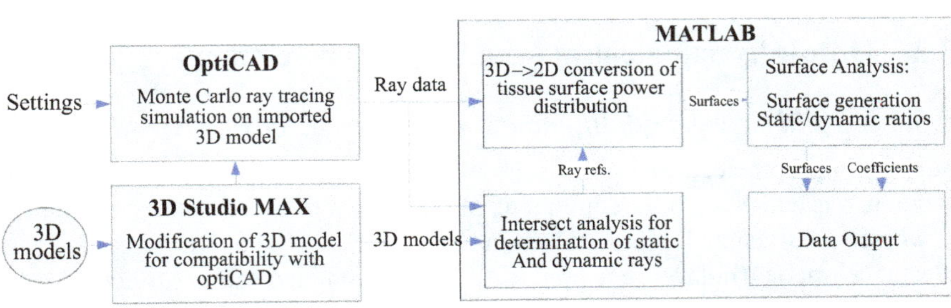

Figure 13.2 Block diagram for the MC ray tracing platform, from [13]

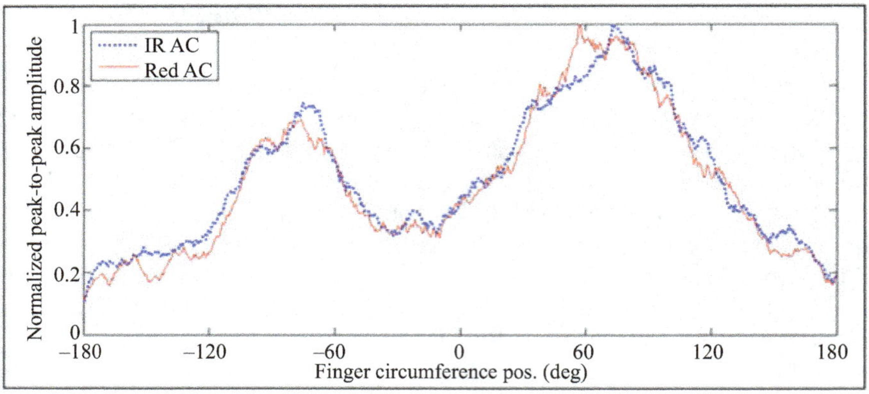

Figure 13.3 Normalized peak-to-peak AC signals with red illumination (633 nm) and IR illumination (850 nm), from [13]

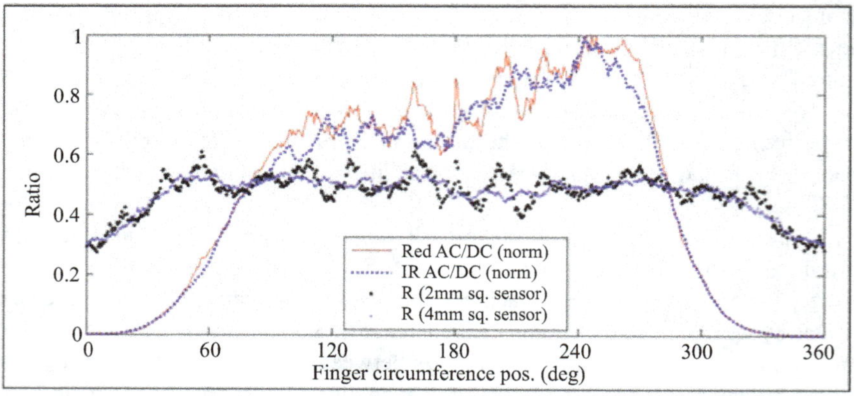

Figure 13.4 Normalized AC/DC ratios and ratio of ratios for data scaled to 97% SpO_2 ($R = 0.5$) *Red AC/DC with 633 nm illumination, IR AC/DC with 850 nm illumination, and the active areas of photo sensor: 2 mm*^{2} *(in black dot) or 4 mm*^{2} *(in blue dot), from [13]*

13.3 Opto-physiological monitoring and assessment

13.3.1 Multi-wavelength illuminations for optoelectronic sensor

The Carelight optoelectronic patch sensor (OEPS) is a miniaturized and wearable device. It was constructed for continuous measurement of blood volume changes in the vessels, corresponding to pulsations of the heart. The OEPS operates in reflectance mode (light source and detector are on the same surface) and is suitable for being attached or adhered to different locations on the human body, for instance, forehead, palm, earlobe and wrist as shown in Figure 13.5 [19].

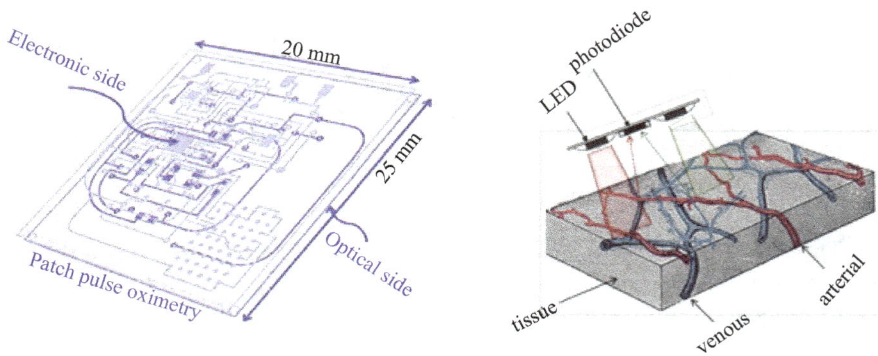

Figure 13.5 Overview of the wearable OEPS system design along with initial prototype, from [19]

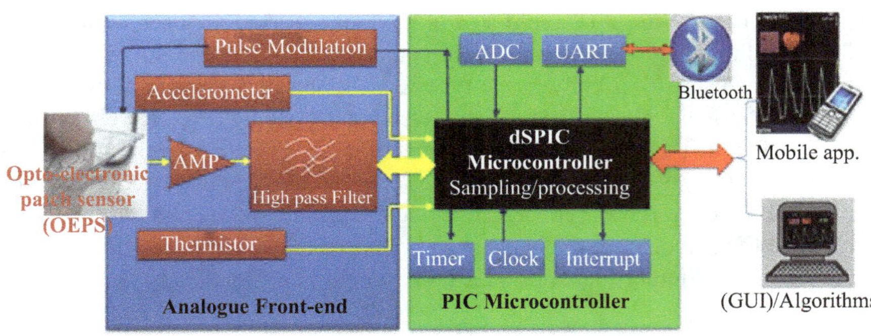

Figure 13.6 Overview of OEPS electronic system for continuous physiological monitoring, from [19]

The schematic of the OEPS consisting of an opto-electronic sensor, a 3MA and a body temperature sensor, as presented in Figure 13.6, shows all integrated within one piece of electronic printed circuit board (PCB). The opto-electronic sensor uses multi-wavelength illumination sources, including, green 525 nm, red 660 nm and infrared (IR) 990 nm (JMSienna Co., Ltd), and a low-profile photodiode (BPW34SR18R, Osram, GmbH) [19].

The complete system, as shown in Figure 13.6, consists of:

(I) A customized wearable sensor comprising opto-electronics to capture physiological signals, an analogue three-axis MEMS accelerometer (3MA, ADXL337, Analog Devices Inc., Norwood, MA, USA) to characterize motion, a body temperature sensor (SMD3 0402, Measurement Specialties), and a suitable analogue front-end with pre-amplification and high-pass filtering.

(II) A PIC Microcontroller (Microchip Technology Inc., Chandler, AZ, USA) to implement low pass filtering, sampling of signals (ADC) and to process the captured signals for preparation of data for wireless transfer to a smartphone via a Bluetooth module (UART protocol).

(III) A mobile device or Bluetooth-enabled computer to receive the data, to optionally process the desired signals, and finally to display the available data in a suitable graphical form.

The OEPS was constructed as a reflection opto-electronic sensor, utilizing OPM principles, together with a specific optoelectronic system to reduce the impact of motion disturbance. The optimal optical design for the OEPS is shown in Figure 13.7.

The optimum separation of each of a plurality of light sources from a photodetector, based on modeled optical path lengths for light travelling from each light source, through the body tissue type to be monitored, to the photodetector; and locating light sources of different wavelengths at different distances from the photodetector. A reasonable distance between illumination sources and photosensor is required in order to avoid a saturation signal from a closer source and a reduced amplitude signal from a far light source. A non-invasive OEPS is ideally configured for use with a body tissue type to be monitored so as to allow optical data, relating to the physiological parameters being studied, to be reliably and accurately obtained. High gain, low noise, and ultra-low power consumption in the OEPS electronic system was determined for continuous and real time monitoring.

The OEPS operates in reflectance mode (i.e., multi-wavelength illumination sources and the detector are on the same side) to make it suitable for attachment or adherence onto different locations of a human body, i.e., forehead, palm, earlobe

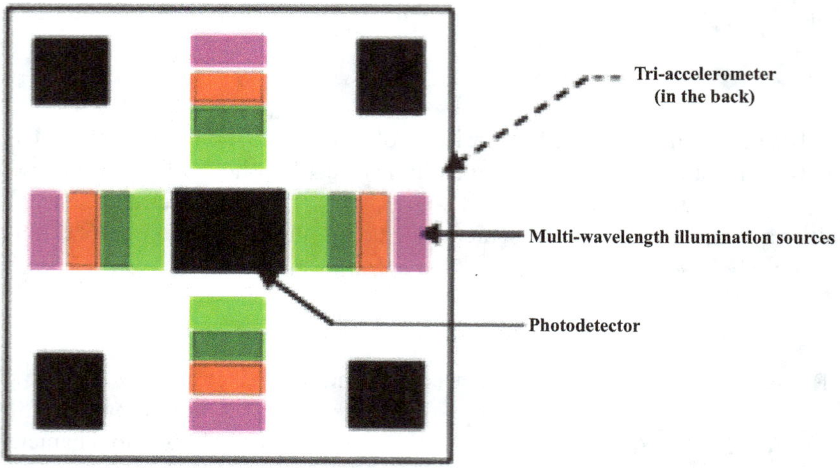

Figure 13.7 Schematic diagram of opto-electronic patch sensor (OEPS) with accelerometer, from [20]

and wrist. However, the amplitude of pulsatile signal in a reflectance mode is smaller than a conventional transmission mode; thus, a trans-impedance is required in order to convert photocurrent intensity into a proportional suitable voltage to enhance the signals generated from the photodiode [20,21]. Multi-wavelength illumination sources, i.e., LEDs of green 525 nm, red 650 nm and IR 870 nm (JMSienna Co., Ltd., Palo Alto, CA, USA) were used for the OEPS including, a low-profile PiN photodiode (BPW34SR18R, Osram, GmbH). The analogue front-end consists of a pre-amplifier, a multiplexer and a low pass filter to performer-processing of the analogue signals gained from the OEPS to be prepared for analogue-to-digital conversion (ADC). A3MA (ADXL337, Analog Devices Co., Palo Alto, MA, USA) was employed to detect any motion from the body and the signal generated from the 3MA was used as a reference against MA. A dsPIC Microcontroller (dsPIC33FJ64GS610, Microchip Co., Chandler, AZ, USA) was hired to implement digital signal processing (DSP) and communication via a Universal Asynchronous Receiver/Transmitter (UART) protocol.

Since there is a variation in depth of penetration and absorption over the range of wavelengths, the intensity of the light penetrating through the skin and emerging from the tissue can be represented as the banana shape effect. Figure 13.8 shows the light path distribution of a typical reflection model. It is the light incident on the skin surface; I_0 is the intensity of light penetrating into the skin, I_r is the reflected light intensity and I is the intensity of light emerging from the tissue through the banana effect. Referring to OPM principle, the light intensity passing through different layers of skin decays exponentially [20].

The light back to the photo-detector can be separated into a non-dynamic (DC) component that results from non-pulsatile tissues and a dynamic (AC) component due to the presence of pulsatile blood volume changes. OEPS typically captures back-scattered light with movement because the OEPS can capture from other mirrored wavelength elements when one of the elements is being misaligned with the vessels. In contrast, Figure 13.9 shows a standard commercial probe movement inducing and causing changes in the sensor position, changing the detected back-scattered light [20].

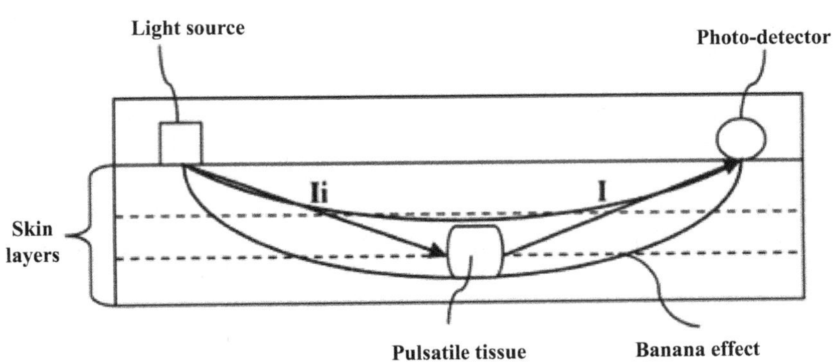

Figure 13.8 Banana shape effect on reflection mode OEPS, from [20]

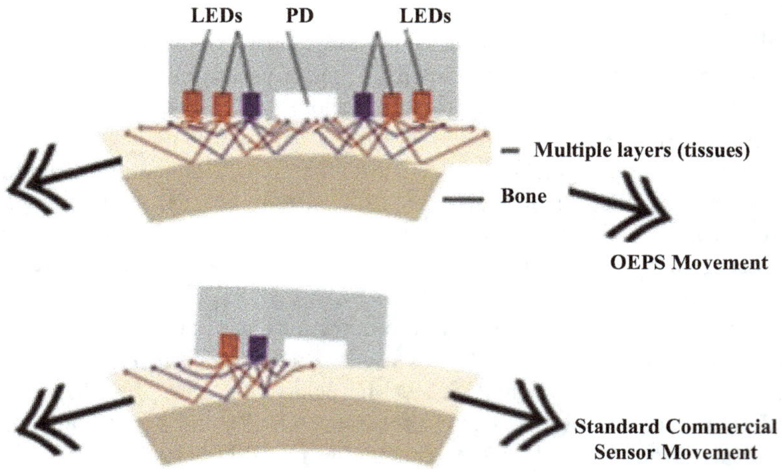

(a)

(b)

Figure 13.9 Sensor displacement altering backscattered light, from [20]

Referring to the AC amplitude, the LED intensities are auto-adjusted accordingly to the strength of the signal received so that high quality signals are acquired. Since the alternative component (AC) is usually less than 10% of the static component (DC), the raw signal, acquired from the PD, needs to be amplified by the Pre-Amp [15]. The multiplexed and the demultiplexed signals should be separated. MUX is also used as a demultiplexer to extract pulse waveform from each channel of illuminations, as the output of the PD is a pulse train of multi-wavelength illuminations.

The common mode rejection of the differential amplifier (DA) is driven to a LPF to eliminate high frequency (HF), such as power supply frequency and electromagnetic. In addition, the LPF is playing an important role in preventing the aliasing components from being sampled. The signal from the LPF is then transferred to the DsPIC processor, to be later converted to a digital form through an analogue-to-digital convert (ADC). Wireless transmission is achieved by a Bluetooth module that receives the digital PPG signals from the central processor, through the USART protocol, to send them eventually to a PC [12].

A time multiplexing algorithm is employed in the DsPIC to asynchronously switch the LEDs to "ON" and "OFF." The four channel illuminations (green, orange, red and infrared LEDs) are multiplexed at the frequency of 1 kHz. Figure 13.10 is a schematic diagram of time switching for multiplexed illumination of four LED channels and demultiplexing of the signals from the photodiode. The time sequence is implementing a timer interrupt for accurate performance. DsPIC is used to switch the LEDs, alternating them between two states ("OFF," "ON") providing only a single wavelength LED illumination at a time.

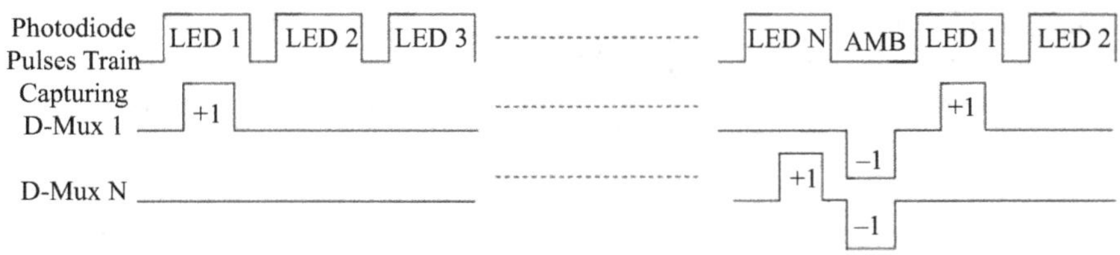

Figure 13.10 Schematic diagram of time switching in multiplexed illumination of LEDs and demultiplexed signals from a photodiode, from [12]

On the other hand, the switching time between these illumination sources is crucial to distinguish between different LED wavelengths, otherwise the PD produces only interference pulse trains with a combination of multi-wavelength illuminations. The output of the PD requires demultiplexing where a specific wavelength illumination is extracted from these combined signals at a certain time. After cancelling the effect of ambient light and common mode rejection, the signals from these individual LED illuminations, are acquired, respectively, and passed on to a LPF (f_c = 15 Hz) where they are converted to digital signals inside the DsPIC processor [12].

13.3.2 Influence of motion artifacts and ambient light during the measurement

The automatic differential motion artefact cancellation technique used with the system is shown in Figure 13.11. The technique behind this method uses the

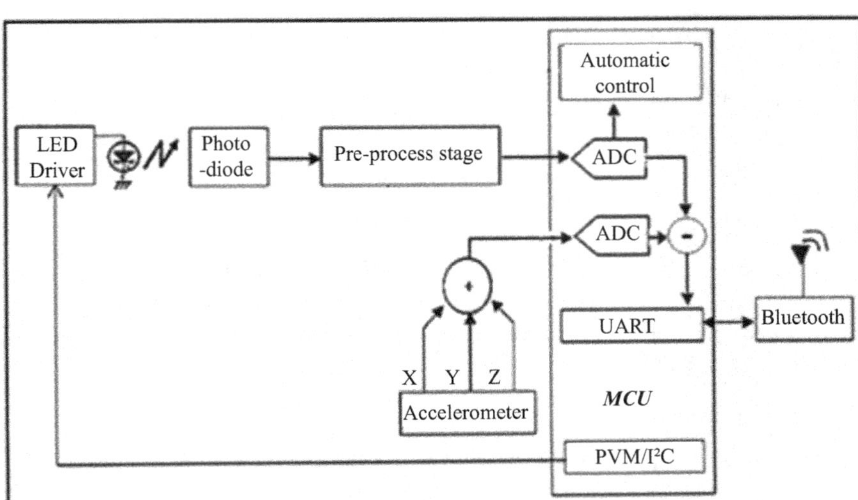

Figure 13.11 System design of automatic differential measurement technique, from [22]

corrupted pulsatile signal as a primary input and the motion sensor signal as a reference input, which is correlated with the noise in the primary input. The two correlated signals cancel each other out and an uncorrelated signal thus is a precise pulsatile signal, remains. Both signals resulting from the accelerometer and the sensor are digitized. In the absence of movement, no signal comes out from the accelerometer, and only one channel of ADC of the detection sensor is executed. For example, when movement takes place, the accelerometer produces a signal proportional to the body movement and regardless of the direction of the movement, since the system design uses a summation circuit [22].

The accelerometer can detect forces movement on all three axes in respect of gravitational acceleration. This force results from any movement that could occur. Since the accelerometer which was used in the present design is analogue chip, the output is converted in the appropriate voltage level. The pic16F1783 microcontroller build-in 12-bit ADC modules; the output values is in the range of 0–4,095, where 212 = 4,095 steps. Moreover, each ADC module has a reference voltage that the conversion depends on and which can be set by software. The reference voltage in this design is 3.3 V.

The ADC of the accelerometer channel results in the following values for the three axes of accelerometer x, y and z: ADCx = A, ADCy = B, and ADCz = C, respectively. The value of A = 2,024, B = 3,467, and C = 4,051. To convert these ADC values to proper voltage, the following formula is used:

$$V = \frac{(ADC, axis)}{4{,}095} \times Vref \tag{13.9}$$

By applying (13.9) to the values of A, B, and C,

$$Vx = \frac{2{,}024}{4{,}095} \times 3.3 = 1.63 \text{ V}$$

$$Vy = \frac{3{,}467}{4{,}095} \times 3.3 = 2.79 \text{ V}$$

$$Vz = \frac{2{,}751}{4{,}095} \times 3.3 = 2.21 \text{ V}$$

Each accelerometer includes a bias voltage level or offset voltage. This voltage is equal to 1.6 V and corresponds to 0 g, as stated in the specifications of the accelerometer. Therefore, the above calculation changes to the following in respect of voltage shifts from zero-g voltage:

$$Vx, 0g = 1.63 \text{ V} - 1.6 \text{ V} = 0.03 \text{ V}$$

$$Vy, 0g = 2.76 \text{ V} - 1.6 \text{ V} = 1.16 \text{ V}$$

$$Vz, 0g = 2.21 \text{ V} - 1.6 \text{ V} = 0.61 \text{ V}$$

The above values are the accelerometer reading in volts and could be converted to the force values that correspond to the movement by using accelerometer sensitivity, as specified in the datasheet of the accelerometer. The sensitivity of the accelerometer for all axes is 7.2 mV/g. The conversions of forces in gravitational acceleration are shown as the following:

$$Gx = \frac{0.03}{0.0072} = 4.2 \text{ V}$$

$$Gy = \frac{1.16}{0.0072} = 161.1 \text{ V}$$

$$Gz = \frac{0.61}{0.0072} = 84.7 \text{ V}$$

All processes can be combined as shown in (13.8):

$$g, axis = \frac{\frac{(ADC,axis)}{4095} \times Vref - V, 0g}{Sensitivity\frac{mv}{g}} \tag{13.10}$$

[22].

13.3.3 *Multi-signals generated from multi-wavelength illumination optoelectronic sensor*

This section describes the approaches taken for the measurement of HR, SpO_2 as well as providing a description of the sensing and processing systems. HR measurement uses the signal obtained at a single wavelength of OPM illumination to determine the systolic and the diastolic phase of the heart. In particular, a measurement of the separation in time between successive peaks or successive troughs can be used. This is shown in Figure 13.12.

Assuming that times of the first and second peaks are t_1 and t_2, respectively, then the heart rate is given by

$$HR \text{ (bpm)} = \frac{60}{t_2 - t_1} \tag{13.11}$$

On the other hand, the oxygen saturation of the blood is the degree to which oxygen is chemically combined with hemoglobin. Absorption of hemoglobin under red and IR illumination shows that Hb absorbs relatively more red light than HbO_2 and Hb relatively less IR light than HbO_2. SpO_2 can be determined as the fractional ratio between the concentration of oxygenated hemoglobin and the total hemoglobin present in the blood, as shown in the following equation:

$$SpO_2\% = \frac{HbO_2}{[HbO_2] + [Hb]} \times 100\% \tag{13.12}$$

where [Hb] is the concentration of the deoxyhemoglobin form and $[HbO_2]$ is the concentration of the oxy-hemoglobin form. In general, SpO_2 can be worked out using a standard PPG signal from measurements obtained from illuminating the vascular bed with light of two separate wavelengths [23]. The transmission of

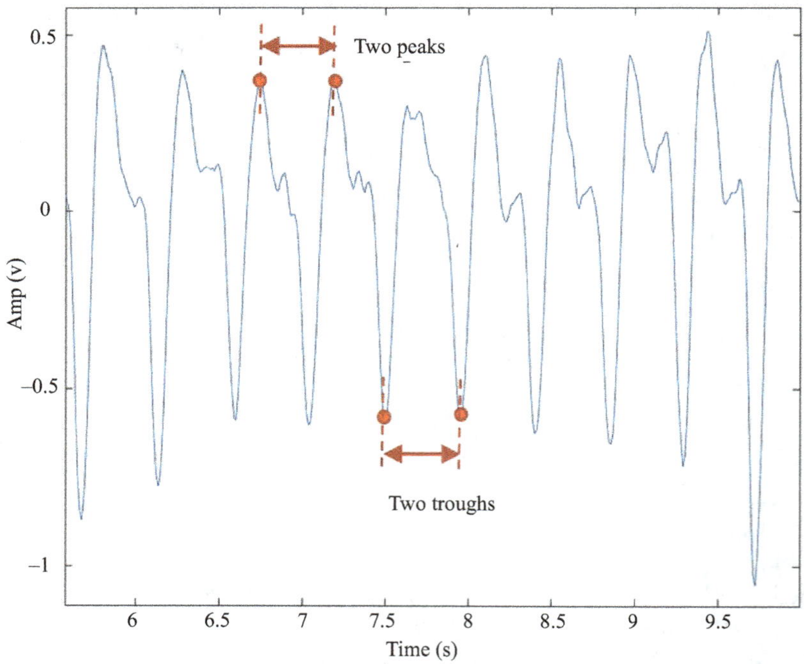

Figure 13.12 Pulsatile changes showing measurements obtained from successive peaks or troughs [23]

illumination decays exponentially with the absorption coefficient and the optical path length. The tissues absorb light differently during the diastolic and systolic phases. In [24], green and orange wavelengths were used to extract the SpO_2 and their absorption is expressed as per (13.13) and (13.14) as:

$$I_{h,g} = I_{in,G}\exp^{-\mu(\lambda_G)d_{G\min}} \tag{13.13}$$

$$I_{l,g} = I_{in,G}\exp^{-\mu(\lambda_G)d_{G\max}} \tag{13.14}$$

where $I_{h,G}$ and $I_{l,G}$ are the intensity of the green light transmission during the diastolic phase and systolic phases, respectively, $I_{in,G}$ is the intensity of incident green light, μ is the absorption coefficient, λ is the wavelength of the green light (525 nm) and $d_{G\min}$ and $d_{G\max}$ are the minimum and maximum optical path lengths, respectively.

By taking the logarithm of the ratio of transmissions in the diastolic and systolic phases, and, assuming the absorption coefficient is known, the difference between the optical path lengths for the green illumination can be determined by:

$$ln\frac{I_{h,G}}{I_{l,G}} = -\mu(\lambda_G)(d_{G\max} - d_{G\min}) = -\mu(\lambda_G)\Delta d_G \tag{13.15}$$

Similarly, for the orange illumination, the difference between the optical path lengths is given by (13.14) as:

$$ln\frac{I_{h,O}}{I_{l,O}} = -\mu(\lambda_O)(d_{O\max} - d_{O\min}) = -\mu(\lambda_O)\Delta d_O \tag{13.16}$$

The absorption ratio of green and orange wavelengths (*R*) can be calculated as follows in (13.15):

$$R = \frac{ln\frac{I_{h,G}}{I_{l,G}}}{ln\frac{I_{h,O}}{I_{l,O}}} = \frac{\mu(\lambda_G)\Delta d_G}{\mu(\lambda_O)\Delta d_O} \tag{13.17}$$

From the previous equation, $ln(Ih, G/Il, G)$ is the arterial transmittance variation under incident green light $\mu(\lambda G)$ and the arterial transmittance variation under incident orange light $\mu(\lambda O)$. If it is assumed that both sources of illumination (green and orange) are placed at the same position on the palm, then the optical path lengths (light travel to the skin) are approximately the same $(\Delta d_G \approx \Delta d_O)$. The lengths of the green and orange return optical paths from the arterial blood flow to the detector will also be similar. As a result, the absorption ratio depends principally on the relative absorption coefficients obtained at the two wavelengths and is largely independent of the optical path lengths:

$$R = \frac{\mu(\lambda_G)}{\mu(\lambda_O)} \tag{13.18}$$

The absorption ratio can be estimated from a "ratio of ratios" measurement as follows:

$$R = \frac{\mu(\lambda_G)}{\mu(\lambda_O)} = \frac{\frac{AC_{green}}{DC_{green}}}{\frac{AC_{orange}}{DC_{orange}}} \tag{13.19}$$

The oxygen saturation can then be calculated from (13.18) as:

$$SpO_2 = A - BR \tag{13.20}$$

where A and B are coefficients that are obtained from a best-fit linear regression equation relating to the standard pulse oximeter series reading against the ratio R of OEPS.

13.4 Contact and non-contact measurement

13.4.1 Configuration of the OEPS system

An overview of the contact OEPS, including the OPM model and the sensor design is presented in [25]. The OEPS sensor can capture radiation originating from each of the plurality of light sources, after light from each light source has traveled through the body tissue being monitored. In this embodiment, the light sources are distributed

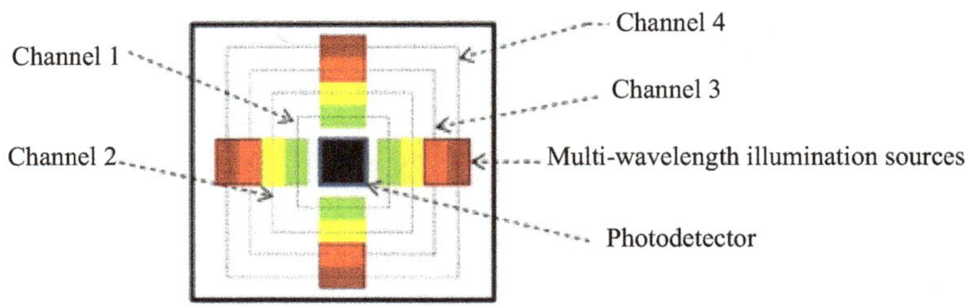

Figure 13.13 Multi-wavelength opto-electronic patch sensor (OEPS). Each channel has four LEDs: Channel 1, green LEDs; Channel 2, orange LED; Channel 3 red LEDs; and Channel 4, IR LEDs, from [26]

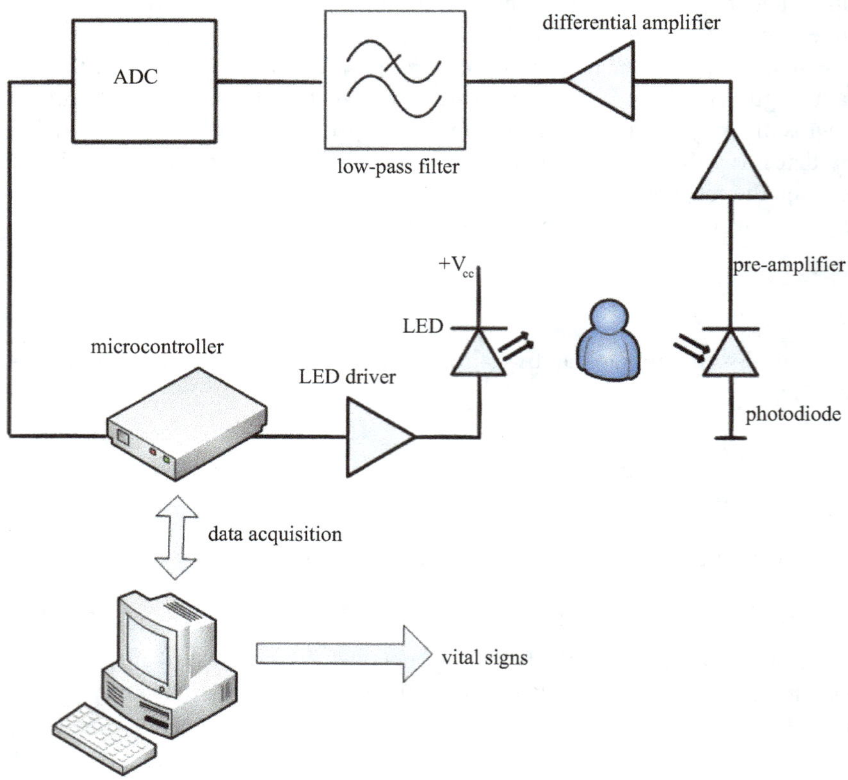

Figure 13.14 Block diagram for the OEPS circuit, from [26]

around the photodetector, extending in four rows which are aligned with each side of the photodetector. It can be understood that each row of light sources will be configured to generate light at one of four different wavelengths. Each row is therefore positioned on the circumference of a circle 34 that is centered on the photodetector, such that the radius of each circle is based on the optimum separation which has been determined for light at the wavelength of the respective light sources in that row.

Figure 13.13 shows the OEPS has a measurement of 18 mm × 18 mm × 0.1 mm and includes 16 LEDs as illumination sources. The LEDs are configured in four channels, each channel containing four LEDs. The peak wavelengths of the channels are 525 nm (green), 590 nm (orange), 650 nm (red), and 870 nm (infra-red), and the returned light is captured using a silicon photo diode (PD) with an active area of 1.69 mm^2 (Hamamatsu photonics K. K., Japan) [26].

The required LED intensity is provided from voltage output from a PIC-based microcontroller (MCU). This voltage is converted by the LED driver to a suitable current to supply to the individual LEDs of the OEPS, as shown in Figure 13.14. The signal reflected from the subject is passed through a pre-amplifier, a differential amplifier and a low pass filter to reduce high-frequency noise content such as the power supply frequency. An analogue-to-digital convertor (ADC, USB-6009, National Instruments Inc., Austin, TX, USA) is then used to supply a suitable signal for analysis by the MCU. The acquired data are sent to the PC where appropriate processing is carried out using LabVIEW (National Instrument Inc., Austin, TX, USA) to extract the required vital signs [26].

13.4.2 Configuration of non-contact image sensing system

When the PPG measurements move from contact to non-contact sensing placement, the conditions of non-contact sensing configuration need to be considered in a completely situation with its requirement for the end users. A non-contact imaging PPG (iPPG) referring the OPM driven B–L model [27] was taken into account to describe the relationship between the light trans-illumination on a designated biological tissue and the back light received after transmission through it. The non-contact iPPG measurement depends upon the varying light absorption due to arterial blood volume change during a cardiac cycle [28]. Figure 13.15 shows a

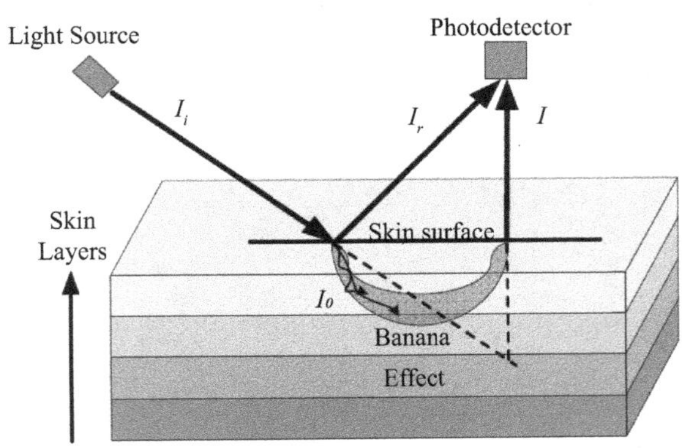

Figure 13.15 Simplified representation of the light distribution, from [28]

simplified diagram of the light distribution of the non-contact PPG sensing model. I_i is the light incident on the skin surface, I_0 is the intensity of light penetrating the skin, I_r is the reflected light intensity, and I is the intensity of light emerging from the tissue through the banana effect [29]. The light reaching the photodetector can be separated into static (DC) and dynamic (AC) terms. DC consists of the reflected component (I_r) and the static component (from non-variable absorbers, i.e., tissue, bone, static blood, etc.), while AC represents the variation of light absorption due to the change of arterial blood volume. I, thus, can be expressed as:

$$I = I_0 \exp(-\mu_{eff} r) = I_0 \exp[-\mu_{dynamic} d(t) - \mu_{static} m] \tag{13.21}$$

where $\mu_{dynamic}$ is the extinction coefficient of the dynamic component, and μ_{static} is the extinction coefficient of static components. The light path length through the dynamic and static components are $d(t)$ and m, respectively.

The AC term becomes independent of the light source intensity by normalizing the AC with respect to the DC:

$$\frac{AC}{DC} = \frac{I_0 \exp(-\mu_{static} m)\mu_{dynamic} d(t)}{I_0 \exp(-\mu_{static} m)} = \mu_{dynamic} d(t) \tag{13.22}$$

The amplitude of the normalized AC is proportional to the time-variant arterial light path length, which in turn depends on the characteristics of the banana shape. In practice, the reflected light is significant, resulting in an additive term in the denominator of equation (13.22), the minimization of which results in a maximized dynamic term [28].

To reach a non-contact physiological monitoring, imaging photoplethysmography (iPPG) has been emerging to achieve remote blood perfusion mapping in one or multi-region-of-interests (ROIs) with providing a 3D segment blood perfusion mapping (x, y, z) based on the extended multi-layered opto-physiological modeling rather than from a point measurement. In a functional iPPG model, the layered absorption (μ_a), the detector-dependent pulsatile mean path lengths (r), as a function of wavelength (λ), layer (k), and the detector position are considered. For a specific detector position, the dynamic component $AC(k)$ representing specific layer blood perfusion can be derived as:

$$AC(k) = AC \times \frac{\mu_{a,blood}(k, \lambda) \times r(k, t)}{\sum_{i=1}^{n} \mu_{a,blood}(i, \lambda) \times r(i, t)} \tag{13.23}$$

with an amplitude

$$ac(i) = ac \times \frac{\mu_{a,blood}(k, \lambda) \times r(k)}{\sum_{i=1}^{n} \mu_{a,blood}(i, \lambda) \times r(i)} \tag{13.24}$$

where AC represents the blood perfusion in a certain time period (t), and ac states the pulsatile component corresponding amplitude from the imaging PPG signals in

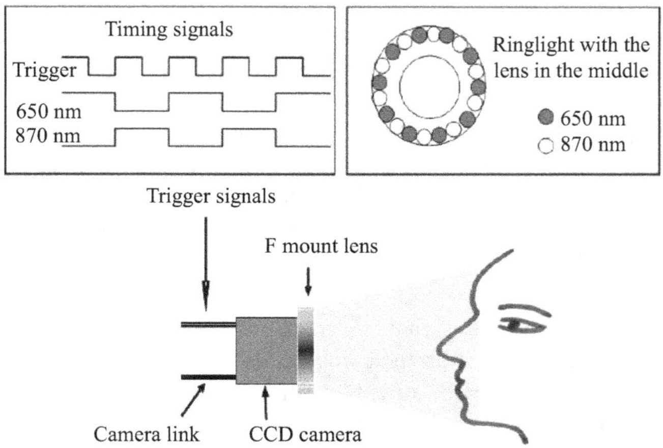

Figure 13.16 Illustration of the iPPG system. The depicting illumination and the imaging geometry (bottom); the relative timing signals (upper right); and the RCLED arrangement of the dual-wavelength ring light (upper left), from [28]

a certain time period (t). When moving the detector all over the specific tissue or ROIs, 3D mapping of blood perfusion can be segmented in the designated layered tissues.

The iPPG system setup as shown in Figure 13.16, a bespoke dual-wavelength LED ring light source (λ_1: 650 nm, λ_2:870 nm) with a parabolic reflector (DIA: 18 cm, O.L.: 6.5 cm, B&Q, UK) to provide a relevant uniform light, was mounted around the camera lens. The ring light sources comprised of 20 LEDs: a peak wavelength of 650 nm and another peak wavelength of 870 nm (TRC650SMD0603, WelTek Co. Ltd., Changhua, Taiwan), both LEDs having 10 nm of bandwidths, to provide 1.0–1.3 mW illumination power. A control circuit with a microcontroller (PIC16F876A, MicroChip Inc.) was developed to alternately power each wavelength group of these LEDs and to ensure the light output power constant duration for each wavelength. The relative illumination timings are illustrated in the upper inset, where the camera was synchronized every time to make sure there was a switch of wavelength. The frame rate of the camera was set at 30 fps (15 fps for each wavelength). The arbitrary areas of the skin surface were observed from a few square millimeters to several square centimeters. The camera was mounted and fixed on an optical bench with the lens positioned approximately 13 cm away from the human face (Figure 13.16) [28].

13.4.3 Importance of multiplexing multi-wavelength illumination and demultiplexing multi-signals

To illuminate the specific four LED channels in sequence, a technique is required to provide time-division multiplexing of the individual LED driver signals as well as ensuring that the signal being received at the PD can be correctly identified with the specific LED being driven. A time-division multiplexing algorithm was implemented on the MCU to illuminate the LEDs individually in a strict sequence using a frequency of 1 kHz. Note that, at every fifth clock pulse, none of the LEDs are driven, thereby allowing capture of information resulting from only ambient lighting conditions and providing immunity from optical interference. Following the sequence of LED illumination, the MCU controls the four-channel sample-hold circuitry to separate the original signals of the LEDs into separate channels. The MCU generates all the pulse signals, co-ordinates the acquisition of the channels captured by the ADC, and identifies these as originating from the PD signals captured for specific LEDs.

Demultiplexing received signals from the PD extracts each specific wavelength illumination response from the combined signal by isolating one time period of the PD. After cancelling the effect of ambient light, the signals from these individual LED illuminations are acquired, respectively, and passed through a low-pass filter (LPF) ($f_c = 15$ Hz), after which they are converted to four channels of digital signals [26].

The digitally controlled LED drivers are an essential part of ensuring the normal operation of the system. Imprecise control of the electrical current would increase switching noise caused by the circuitry. Thus, precise control of the LED drivers and the independent driving of the devices is necessary.

Optical power control can be achieved by biasing the LED with a resistor, but this method can only provide a constant current if the LED voltage drop and the supply voltage are constant, which limits the LED optical power. A more stable solution is to provide an active circuit that delivers a constant and predetermined current to the LED, thus ensuring that the current through the LED is maintained regardless of the LED drop and this is also an important factor for driver stability. Such an active current control methodology includes a group of four LEDs whose load current is independent of the voltage drop. This not only ensures constant LED power under switching conditions but also makes it possible to drive a wider range of devices without modification [26].

13.5 Case studies

13.5.1 Case 1: wrist model for smart watch

This case study presents an effective engineering approach for human vital signs monitoring as increasingly demanded by personal healthcare. The aim of this study was to research how to capture critical physiological parameters efficiently through a well-constructed electronic system and a robust multi-wavelength illumination optoelectronic patch sensor (OPES, named Carelight) [30]. Sixteen healthy active males (aged 24 $\pm$ 3 years (mean $\pm$ standard deviation)) volunteered to take part in the present investigation which was approved by the Loughborough University Ethical Advisory Committee. All participants had the procedures and risks associated with involvement in the study explained to them and their written consent to participant in the study was obtained.

The data presented and analyzed in this work correspond to the HR measurements. Measurements were taken while participants were seated, while they walked and ran on a treadmill and cycled on a cycle ergometer. Height (179.8 ± 4.1 cm), body mass (74.9 ± 7.9 kg), and body mass index (23.1 ± 2.0 kg/m^2) and blood pressure (systolic 130 ± 10 mmHg, diastolic 67 ± 9 mmHg) for individual participants as well as the room temperature (24.4 ± 0.5 °C) and humidity (33 ± 2%), were also measured during the experiment. The heart rate was monitored using a chest Strap and Acti-Graph's Bluetooth® Smart wGT3X-BT wireless (ActiGraph, LLC) as well as Mio-Alpha (Physical Enterprises Inc.) devices as a reference. The Carelight sensor was placed on the palm of the participant's left hand while the Acti-Graph was worn on the left-hand wrist. The Carelight sensor and the Acti-Graph commercial devices captured data simultaneously to facilitate the processing signals of reference indications, respectively.

The participants were asked to rest in a seated, upright position for 3 min and the heart rate was then recorded over a 1-min period. Once a resting heart rate had been obtained, the participants completed the treadmill (Technogym Excite Med 700, Gambettola, Italy) exercise protocol. This protocol consisted of three, 6 min stages completed at 4 km/h, 7 km/h, and 8.5 km/h. In the fourth min of each stage, the heart rate was measured while participants moved their arms freely. The participants were then asked to rest their left arm on the treadmill railing in order to reduce movement of the sensors and heart rate was measured during the sixth minute of each stage. Participants then competed to the cycling protocol which consisted of four, 4 min stages cycling (Monark Ergomedic 874E, Vansbro, Sweden) at 60 rpm against an applied mass of 1, 1.5, 2.0, and 2.5 kg. During the final minute of each stage, the heart rate was recorded.

Figure 13.17 shows the Carelight sensor signals comprising of multi-wavelength illumination sources (green, red, infrared red, and even more), i.e., pulsatile waveforms. Figure 13.18 shows pulsatile variations that were captured from the palm by using Carelight for different wavelengths during the stationary state where the participant was in the sitting position.

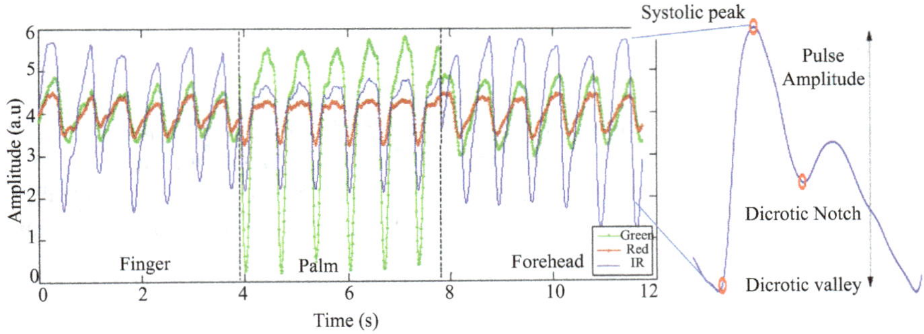

Figure 13.17 Carelight sensor comprising of multi wavelength illumination sources (green, red, infrared red and even more) generating quality pulsatile waveforms

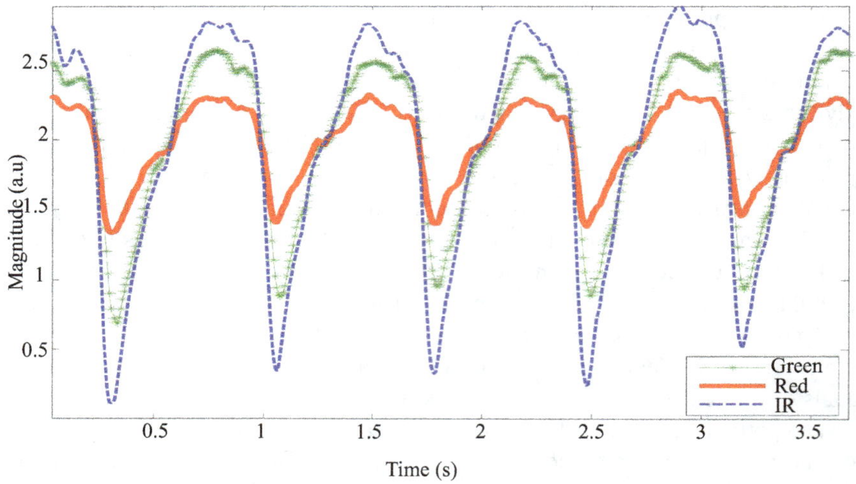

Figure 13.18 Detected pulsatile signals of different wavelengths by Carelight, from [30]

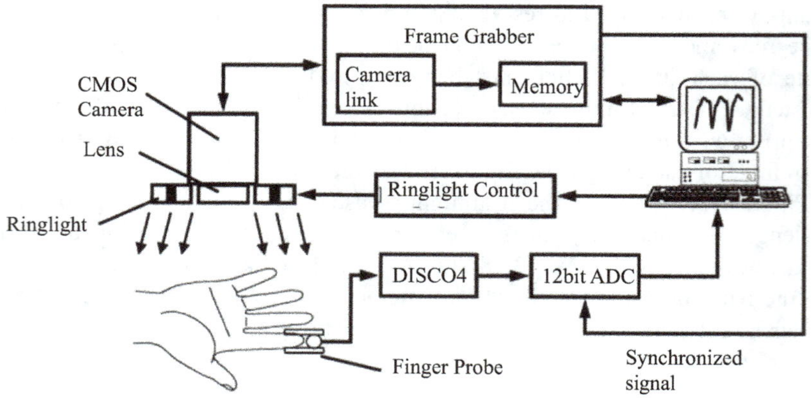

Figure 13.19 Experimental setup of the non-contact iPPG system, from [31]

Comparisons of HR between measurement methods of (Acti-Graph and OEPS) and (Mio-Alpha and OEPS) were performed using an independent *t*-test. The result showed no significant difference between methods ($p < 0.05$). The correlation analysis demonstrated a strong relationship in HR measured using the Acti-Graph and OEPS as well as Mio-Alpha and OEPS (both $r = 0.96$ and $p < 0.0001$).

13.5.2 Case 2: imaging PPG

The iPPG system for the case study in [31] is schematically presented in Figure 13.19. A monochrome CMOS camera (Model: EoSens MC1360-63, Mikrotron, Germany) with a maximum resolution of 1280 × 1024 pixels was focused on the palm of the participant's right hand using an industry-standard F-mount zoom lens (Model: Nikkor 20 mm f/2.8D, Nikon, Tokyo, Japan). The pixels were encoded in a 10 bit gray scale, allowing the camera to detect the weak pulsations of the microvascular tissue bed. A custom built infrared (λ = 870 nm) ringlight illumination source (RIS) comprising 100-RCLEDs (Model: 0603SMD, JMSienna) fitted into a ring reflector, was mounted around the lens to collimate and homogenize the distribution of light. The RIS was current-controlled and synchronized with the camera capture cycle. The distance between camera lens/RIS and skin was approximately 350 mm. In all experiments, a commercial pulse oximeter contact sensor (Model: P871RA, Viamed, UK) was placed on the index finger of the right hand to measure the participant's pulse signal for validation. The analog signal outputs from the contact sensor were digitized using a DISCO4 (Dialog Devices, UK) data acquisition system incorporating a 12-bit A/D converter running at a sample rate of 128 Hz. The images captured from the camera were synchronized to the physiological signals acquired from the contact sensor via an additional signal from the camera to the A/D converter.

All measurements were made in a temperature-controlled darkroom (27±1 °C) by a trained operator. Upon arrival, the subjects were required to sit on an adjustable chair and rest for at least 10 min before a blood pressure measurement were taken from the left arm using a clinically validated digital blood pressure monitor (Model: M6, Omron, Japan). Afterwards, participants were asked to sit at ease and rest their right hand as still as possible during the examination. The right palm of each participant was exposed to the infrared RIS and recorded with the CMOS camera. For each subject, images were recorded for 34 s at a rate of 50 fps and the exposure time of 15 ms. The size of the raw image was 640 × 480 pixels. Contact PPG was recorded from the finger probe for about 3 min to verify that the heart rate was stable. Approximately 60 s later, the camera was started, at which time, a synchronization signal was activated for simultaneous recording with the finger probe measurements. Then, the subject was asked to ride a cycling machine (Model: XR-580, PowerTrek, UK) at a speed of 15 km/h (exercise 1) for 5 min. Immediately after the exercise, participants' blood pressure and the second 34 s video was acquired. After 10 min rest, another 5 min exercise was performed at 25 km/h (exercise 2) and the post-exercise data acquisition procedures were repeated. A final measurement was made after a further 10 minute rest period. Figure 13.20 shows the measured variables: heart rate (HR), systolic blood pressure (SBP), and diastolic blood pressure (DBP).

To demonstrate the influence of exercise on the cardiovascular system, ANOVA was conducted on each of the measured variables. Post hoc analysis with Duncan's test was also employed. The results are shown in Figure 13.20. ANOVA SBP and HR showed a significant influence of exercise ($F = 7.608, p = 0.001$ and $F = 12.666, p = 0.000$). Post hoc tests revealed that compared to the rest condition, the HR and SBP were significantly higher after both exercise levels (ex1 vs. rest, $p < 0.05$, ex2 vs. rest, $p < 0.01$). In addition, a significant difference between different exercise levels was also observed in HR and SBP. Specifically, higher HR ($p = 0.004$) and SBP ($p = 0.039$) were revealed after exercise 2 (25 km/h)

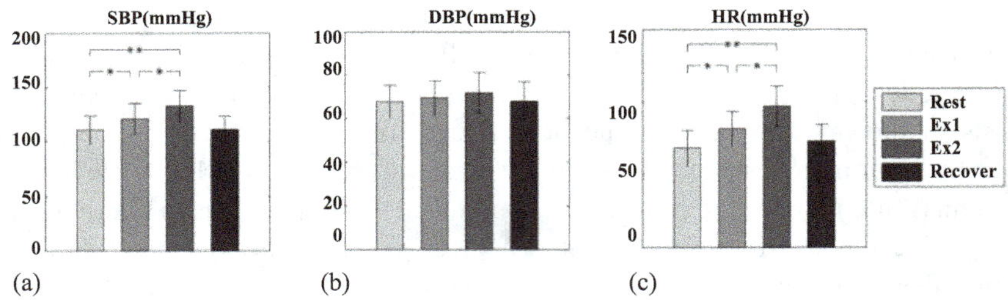

*Figure 13.20 Effects of exercise on blood pressure and heart rate. Each bar represents the average SBP (a), DBP (b), and HR (c) for each state (rest, exercise 1, exercise 2, and recovery). Error bars show standard deviations. Significant statistical differences are indicated with * for p < 0.05 and ** for p < 0.01, from [31]*

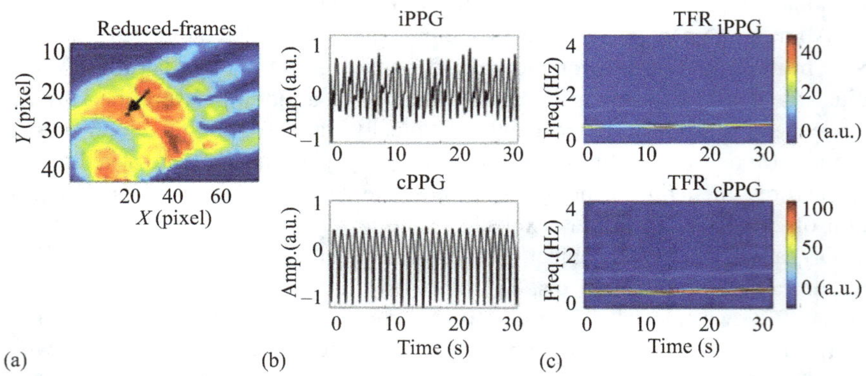

Figure 13.21 A representative figure showing a reduced frame with greyscale sidebar showing the signal amplitude (a), PPG signals (b), and the corresponding TFR results (c). The upper TFR trace is from the non-contact imaging PPG and lower is from contact PPG with the greyscale indicating the power intensity. The signal is for a subject (male, age = 55 years) at rest condition. The position from which the iPPG signal was obtained is highlighted with a black box ($x = 20, y = 25$) and an arrow, from [31]

compared to the moderate exercise level (15 km/h). After 10 min of rest, the hemodynamic parameters all returned to the rest level. No significant effect of exercise on DBP was observed.

Figure 13.21 is an example of the PPG signals obtained from a single subject and the TFR, in which the HR frequency and second harmonic components are shown. Additionally, the HR derived from the TFR plot is found to be in excellent agreement with the HR obtained from the commercial pulse oximeter sensor readings. A slowly varying HR is also detected from the TFR trace which indicates the potential usage of TFR in revealing the time-varying HR during the exercise.

To evaluate the reliability and stability of the OPI system, Bland–Altman plots were employed to compare the agreement between the imaging-PPG (iPPG) and contact-PPG (cPPG) measurements. The HRiPPG is obtained through averaging the HR within all the sub-windows (64 $\times$ 48 pixels). Historical estimations of the HR were used to define a threshold for maximum deviation in HR among different regions. The HR calculation was started in the middle of the processed frames ($x = 32$, $y = 24$), using the HR within this region as a reference. Successive calculations were then performed where, if the difference between the current HR estimation and the reference value exceeded the threshold (9 bpm in this study), the algorithm isolated it as a non-tissue or corrupt area and rejected the invalid HR.

A layered tissue model was employed in a Monte Carlo simulation to obtain the contribution of each layer to the output signals by accounting for the absorption (i.e., photon packet intensity) and scattering (i.e., optical path length) properties of tissue. The minima and maxima of the PPG signals were identified using a custom algorithm in MATLAB. The amplitude of the PPG signal was then determined for each pulse and the mean peak-to-peak amplitude served as the ac part. The slowly varying baseline was obtained by low pass filtering the PPG signal (<0.5 Hz), and the mean value of the baseline was taken as the dc component. Accordingly, the ac/dc ratio for each ROI could be calculated.

Figure 13.22 shows a layered blood perfusion map for a subject at rest. Within the six-layer tissue model, the blood fraction was set to zero within the outermost layer, i.e., epidermis and dermis, which served as a wavelength-dependent attenuator. The contribution of the blood fraction to the three inner layers, i.e., papillary plexus, dermis, and cutaneous plexus is shown. Specifically, the dermal layer contributes the most of the output signals. A stronger perfusion in the fingers is also revealed.

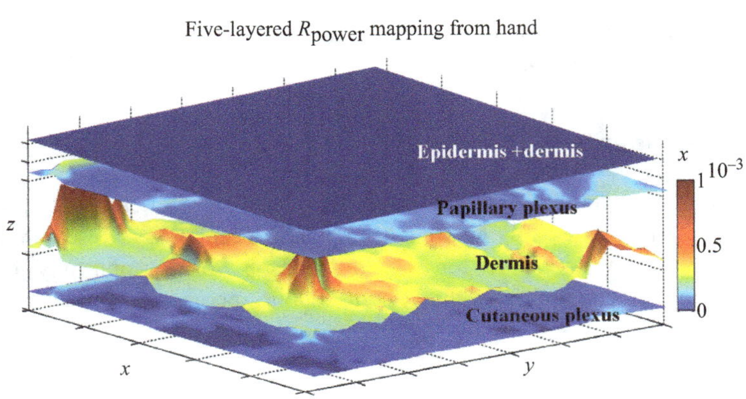

Figure 13.22 The layered power mapping of ROIip-dependent R_{power} of the multi-layer skin tissue under 870 nm illumination, from [32]

13.6 Conclusion

Opto-physiological monitoring (OPM) relies on a process of fine-tuning the accuracy of characterization for both simulation and empirical validation based upon light propagation of RTT to exclusively interpret opto-physiological interaction. The OPM engineering approach along with the realization of contact optoelectronic sensing and non-contact imaging demonstrates its functionality at a qualitative level and provides a path toward increasingly comprehensive descriptions of the mechanisms involved in PPG. In the OEPS engineering setup, the light source and the photodetector are placed on the same side of skin surface. The geometrical configuration in a such reflection Carelight sensor has well considered how to consolidate the findings of light trans-illumination into multi-wavelength optoelectronic sensing realization. The technique used in non-contact measurement not only considered the above principles but also the combination of the incident angle and the illumination distance, because these two geometrical parameters are to decide the optimal range which determines the quality of PPG signal captured in the setup. The iPPG technique performs comparably with a conventional contact device for HR measurements to acquire PPG waveforms. The potential of a non-contact iPPG system to measure vital signs, i.e., heart rate, respiration rate, oxygen saturation, and display PPG waveforms by non-contact means is not limited by the configuration of the present system, as machine learning (ML) and deep learning (DL) algorithms along with could increase practicability of non-contact applications.

The highlights between established PPG and the OPM sensors lie in their underpinning principles of light trans-illuminating tissue, such as:

(I) The B–L law-based PPG sensors take the measurement site as a blood-filled cuvette with no scattering effects, and treat the light sources assumed to be monochromatic.

(II) OPM-based sensors take the measurement site as dynamic variations of tissue optical properties, e.g., absorption μ_a, scattering μ_s and isotopic factor with time being.

(III) More specifically, the optical pathlength $l(t)$ is also a function of time along with the blood pulsatile variation.

Acknowledgments

The authors are grateful to all PhD graduates, research associates, and visiting scholars from Photonics Engineering and Health Technology Research Group, Loughborough University, UK, for their enthusiastic work on opto-physiological monitoring and assessment from the year of 2006 to the present time.

References

[1] Hu S. Opto-physiological monitoring. In: Scholarly Community Encyclopedia. Available from: https://encyclopedia.pub/3338

[2] Rasche S, Huhle R, Junghans E, *et al.* Association of remote imaging photoplethysmography and cutaneous perfusion in volunteers. *Sci. Rep. [Internet].* 2020;10(1):1–9. Available from: https://doi.org/10.1038/s41598-020-73531-0

[3] Peris VA. Opto-physiological modelling of pulse oximetry. *PhD Thesis.* Loughborough University, Loughborough; 2008.

[4] Patterson MS, Wilson BC, Wyman DR, The propagation of optical radiation in tissue I. Models of radiation transport and their application. *Lasers Med. Sci.* 1991;6:155–68 .

[5] Chandrasekhar S, *Radiative Transfer*, Dover, New York, NY; 1960.

[6] Chen B, Stamnes K, Stamnes JJ. Validity of the diffusion approximation in bio-optical imaging. *Appl. Opt.* 2001;40(34):6356–66.

[7] Hu S, Azorin-Peris V, Zheng J. Opto-physiological modeling applied to photoplethysmographic cardiovascular assessment. *J. Healthc. Eng.* 2013;4(4):505–28.

[8] Delpy DT, Cope M, van der Zee P, Arridge S, Wray S, Wyatt J. Estimation of optical pathlength through tissue from direct time of flight measurement. *Phys. Med. Biol.* 1988;33(12):1433–42.

[9] Hiraoka M, Firbank M, Essenpreis M, *et al.* A Monte Carlo investigation of optical pathlength in inhomogeneous tissue and its application to near-infrared spectroscopy. *Phys. Med. Biol.* 1993, 38:1859–76.

[10] Takatani S, Davies C, Sakakibara N, *et al.* Experimental and clinical evaluation of a noninvasive reflectance pulse oximeter sensor. *J. Clin. Monitor.* 1992, 8(4):257–66.

[11] Lister T, Wright PA, Chappell PH. Optical properties of human skin. *J Biomed Opt.* 2012;17(9):909011.

[12] Yan L, Hu S, Alharbi S, Blanos P. A multiplexed electronic architecture for opto-electronic patch sensor to effectively monitor heart rate and oxygen saturation. *Proc. SPIE BiOS* 2018;1050115(February):39.

[13] Peris VA, Hu S. A Monte Carlo platform for the optical modelling of pulse oximetry. Proceedings of SPIE (6446). *J. Phys. Conf. Ser.* 2007;85(1).

[14] Tuchin V. *Tissue Optics: Light Scattering Methods and Instruments for Medical Diagnosis*, SPIE Press, Washington, DC, 2000.

[15] Webster JG, *Design of Pulse Oximeters*, Bristol, UK: Institute of Physics, 1997.

[16] Roggan A, Friebel M, Döschel K, Hahn A, Müller G. Optical properties of circulating human blood in the wavelength range 400–2500 nm. *J. Biomed. Opt.* 1999;4:36–46.

[17] Prahl S. [Internet]. 1999. Available from: http://omlc.ogi.edu/spectra/hemoglobin/index.html

[18] Ugnell A, Oberg P. The optical properties of the cochlear bone. *Med. Eng. Phys.* 1997;19:630–36.

[19] Alzahrani A, Hu S, Azorin-Peris V. A comparative study of physiological monitoring with a wearable opto-electronic patch sensor (OEPS) for motion reduction. *Biosensors.* 2015;5(2):288–307.

[20] Alzahrani A, Hu S, Azorin-Peris V, *et al.* A multi-channel opto-electronic sensor to accurately monitor heart rate against motion artefact during exercise. *Sensors (Switzerland).* 2015;15(10):25681–702.

[21] Venema B, Blanik N, Blazek V, Schiefer J, Leonhardt S. Evaluating innovative in-ear pulse oximetry for unobtrusive cardiovascular and pulmonary monitoring during sleep. *IEEE J. Trans. Eng. Heal.* 2013(1):2700208.

[22] Alzahrani A, Hu S. An effective solution to reduce motion artefact in new generation reflectance pulse oximeter. In: 2013 Saudi International Electronics, Communications and Photonics Conference SIECPC 2013. 2013, pp. 1–5.

[23] Alharbi S, Hu S, Mulvaney D, Blanos P. An applicable approach for extracting human heart rate and oxygen saturation during physical movements using a multi-wavelength illumination optoelectronic sensor system. In: *Proc SPIE BiOS, 2018*, San Francisco, CA, USA; 2018, p. 27.

[24] Alharbi S, Hu S, Mulvaney D, *et al.* Oxygen saturation measurements from green and orange illuminations of multi-wavelength optoelectronic patch sensors. *Sensors (Switzerland).* 2019;19(1):1–17.

[25] Peris VA, Hu S. Opto-physiological sensor and method of assembly (Granted Patent, GB2522298, 17 Nov., 2015, https://www.ipo.gov.uk/p-ipsum/Case/ApplicationNumber/GB1418267.9)

[26] Yan L, Yu Y, Hu S, *et al.* Illumination adaptation in a multi-wavelength opto-electronic patch sensor. *Sensors (Switzerland).* 2020;20(17):1–12.

[27] Wang Y, Gao M, Guo D. Design of 16-bits constant-current LED driver. In: Proceedings – 2010 International Conference on Anti-Counterfeiting, Secur Identification, Chengdu, China, 18–20 July 2010 ASID. 2010;53–6.

[28] Hu S, Peris VA, Echiadis A, Zheng J, Shi P. Development of effective photoplethysmographic measurement techniques: from contact to non-contact and from point to imaging. In: Proceedings of the 31st Annual International Conference of the IEEE Engineering in Medicine and Biology Society: Engineering the Future of Biomedicine EMBC 2009. 2009;6550–3.

[29] Zhao C, Xie X, Liu S. Multioutput LED drivers with precise passive current. *IEEE Trans. Power Electron.* 2013;28(3):1438–48.

[30] Alzahrani A, Hu S, Azorin-Peris V, *et al.* Enhancement of absorption and resistance of motion utilizing a multi-channel opto-electronic sensor to effectively monitor physiological signs during sport exercise. In: Design and Quality for Biomedical Technologies VIII, Proceedings of SPIE Vol 9315, 93150O. 2015.

[31] Sun Y, Hu S, Azorin-Peris V, *et al.* Detection of physiological changes after exercise via a remote optophysiological imaging system. In: SPIE 7891, Design and Quality for Biomedical Technologies IV, 2011; p. 78910E.

[32] Zheng J. Opto-physiological modelling of imaging photoplethysmography. PhD Thesis. Loughborough University, Loughborough; 2010.

Index

www.ingramcontent.com/pod-product-compliance
Lightning Source LLC
LaVergne TN
LVHW011845200926
848484LV00002B/403

* 9 7 8 1 8 3 9 5 3 3 4 9 5 *